AF817767

*Routledge Revivals*

# The Social Framework of Agriculture

# The Social Framework of Agriculture

Harold H. Mann

Edited by
Daniel Thorner

Routledge
Taylor & Francis Group

First published in 1968 by Frank Cass and Company Limited

This edition first published in 2018 by Routledge
2 Park Square, Milton Park, Abingdon, Oxon, OX14 4RN
and by Routledge
52 Vanderbilt Avenue, New York, NY 10017, USA

*Routledge is an imprint of the Taylor & Francis Group, an informa business*

© 1967 by Robin H. Mann

All rights reserved. No part of this book may be reprinted or reproduced or utilised in any form or by any electronic, mechanical, or other means, now known or hereafter invented, including photocopying and recording, or in any information storage or retrieval system, without permission in writing from the publishers.

**Publisher's Note**
The publisher has gone to great lengths to ensure the quality of this reprint but points out that some imperfections in the original copies may be apparent.

**Disclaimer**
The publisher has made every effort to trace copyright holders and welcomes correspondence from those they have been unable to contact.

A Library of Congress record exists under ISBN:

ISBN 13: 978-0-367-15045-7 (hbk)
ISBN 13: 978-0-429-05469-3 (ebk)

THE SOCIAL FRAMEWORK OF AGRICULTURE

HAROLD H. MANN, ABOUT 1915.

# THE SOCIAL FRAMEWORK OF AGRICULTURE

India · Middle East · England

HAROLD H. MANN

*Edited by Daniel Thorner*

FRANK CASS & CO. LTD.

1968

First published in India in 1967 by
Vora & Co. Publishers Pvt. Ltd.

First published in Great Britain in 1968 by
FRANK CASS AND COMPANY LIMITED
67 Great Russell Street, London WC1

© 1967 Robin H. Mann

*Printed in India*

## PREFACE

Dr. Harold H. Mann was during his lifetime an acknow ledged authority on applied science and agriculture in England, the Middle East and India, but it is less widely known that he was equally distinguished by his work in the social sciences. He not only pioneered modern-style village surveys in both England and India, but also modern-style urban surveys and studies in India. There he broke new ground in his remarkable first-hand researches on agricultural labour, village economics, depressed or "Untouchable" classes in town and country, and human and industrial relations in India's first steeltown, Jamshedpur. In the text of this book we reproduce thirty-five of Dr. Mann's papers—in whole, in part, or in summary.

Dr. Mann, then, was of those rare men who are as much at home in the social as in the natural sciences, whether applied west or east of Suez. In the essays at the beginning of the book, Mr. F. C. Bawden, F.R.S., Director of the Rothamsted Experimental Station, outlines Dr. Mann's contribution to agriculture and the natural sciences, and I sketch the significance of his work in what he liked to call his hobby: the social sciences.

It is a measure of Dr. Mann's stature that this division of subject matter between agricultural science on the one hand and the social studies on the other was alien to him. His own social investigations in town and country, he felt, fell within the limits of what he considered natural science The method in both subjects, he contended, should be the same. The principal difference was that the social studies were so much more difficult. He held that the most complex phenomena on earth were the phenomena of society. Half a century of study convinced him that the fundamental difficulty in applying science to agriculture was social.

It is only natural that after Dr. Mann's death the suggestion came from several quarters that a selection of his studies be republished. I am indeed honoured by the fact that the task, really the privilege, of editing this volume fell to me. But I would never have dared to undertake it had I not been assured of the guidance and collaboration of Dr. Mann's old friends and colleagues, Mr. Bawden of Rothamsted and Mr. C. A. Thorold, Officer in charge of the Woburn Experimental Station. The main work of preparing the bibliography of Dr. Mann's writings fell to Mr. Thorold.

We are deeply grateful to the co-workers, librarians, editors, and publishers who have helped to make possible this volume. Our work in locating the writings of Dr. Mann was greatly aided by Mr. D. H. Boalch, Keeper of Scientific Books, Radcliffe Science Library (Department of the Bodleian Library) Oxford; Shri N. G. Gokhale, Director, Tocklai Experimental Station, Cinnamara, Assam (Indian Tea Association, Scientific Department); Shri M. V. Sovani, Librarian, Servants of India Society, Poona; and Mr. G. V. Jacks, Director, Commonwealth Bureau of Soils, Harpenden.

It is a pleasure to acknowledge the help afforded to us by the Librarians, staff and collections of the British Library of Political and Economic Science, London; the British Museum Reading Room; the Indian Institute Library (Department of the Bodleian Library) Oxford; the India Office Library (Commonwealth Relations Office), London; the Rothamsted Experimental Station Library, Harpenden; the Royal Geographical Society, London; the Royal Botanic Gardens, Kew; the School of Agriculture, University of Cambridge; the Royal Institute of International Affairs, London; the Tata Library, Bedford College, London; the Library, India House, High Commission for India, London.

29th January, 1967. DANIEL THORNER.

## CONTENTS

## PART III

## TOWN STUDIES IN INDIA



## PART IV

## THE SOCIAL SETTING OF AGRICULTURAL DEVELOPMENT IN INDIA



## PART V

## THE USSR AND THE MIDDLE EAST

## PART VI

## AGRICULTURAL SCIENCE AND METHOD

# SCIENCE AND AGRICULTURE: THE CONTRIBUTION OF HAROLD H. MANN

Dr. H. H. Mann's career as a research worker was as remarkable for its range and diversity as for its duration. In its duration it must surely be unique, for he was an active and productive experimenter for 68 years, as is evidenced by the simple statement that his first paper, which described the effects of antiseptics on yeast, was published in French in 1894, and at the time of his death on December 2nd, 1961, he was preparing for publication the results of the "Market-garden" experiment at Woburn. The subject matter of his first and last papers also gives some idea, though an inadequate one, of the diversity of his research interests.

Mann graduated in chemistry from Yorkshire College (later the University of Leeds) in 1892, when he was twenty, and with the award of an "1851 Exhibition" went to work in Paris with Duclaux at the Pasteur Institute. There he showed considerable insight in applying his chemical training to studying the action of antiseptics, work that introduced him to biology and may well have had a decisive influence in determining the course of his career. Certainly, except for some work on osazones, published in 1895 and 1896, he turned his back on pure chemistry and worked increasingly on biological problems, using his chemical knowledge and techniques whenever they seemed likely to be profitable, but never otherwise. His introduction to agriculture came in 1895 when he was appointed assistant to Dr. J. A. Voelcker, then Chemist to the Royal Agricultural

Society of England, and even more directly in 1898 when he went as the first resident research worker at the Society's Experimental Station at Woburn. His stay there was relatively brief, for in 1900 he was appointed Scientific Officer to the Indian Tea Association, but it left him with a lasting affection not only for the Experimental Station but also for the countryside of Bedfordshire and the villagers living there.

*Research for the Tea Association*

It was in India that his versatility, industry and productivity became fully evident. Not only did his work provide much needed information for the development of the Indian tea industry, but it also established him as the world authority on the tea crop, whose advice was sought and given to many countries during the next 50 years. Mann was equally at home in studying all the many problems raised by the cultivation and preparation of tea. Although not a pathologist, he contributed much to knowledge about the pests and diseases of tea; indeed, within three years of going to India, in collaboration with G. Watt, he had written and published the second edition of the major book *The Pests and Blights of the Tea Plant.* But this was only one of his many activities, for he was also characterising tea soils, studying the manuring of the crop, its pruning, how to rehabilitate poor gardens, the factors affecting the quality of the tea leaf, the enzymes it contains and the processes involved in fermentation.

As his "swan-song" before leaving the post of Scientific Officer of the Tea Association, to become Principal of the Agricultural College at Poona and Agricultural Chemist to the Bombay Government, in 1907 he published a 284-page monograph entitled *The Tea Soils of North-East India and their Treatment.* For any other person this alone would have been more than an adequate reckoning for seven years' work and how, in addition to his very many other achievements, he managed to produce this authoritative work is difficult to imagine. It described, with a wealth of analytical figures, the physical and chemical characteristics of many different soils, and reports the results of experiments on many different factors that affect the growth of the tea crop, but perhaps its most important contents are Mann's shrewd conclusions made after his careful observations on the consequences of different practices. The monograph is rich in wise advice, much still relevant, and it provided the tea planters with a sound basis for the successful growing of tea.

*Principal of the Agricultural College*

While still with the Tea Association, he did some work with crops such as rubber and sisal, and in his later posts other crops and problems increasingly occupied his attention, although the tea crop still remained a prime interest, because he not only appreciated its great economic importance to India but was excited by the many scientific problems posed both by its growing and processing. His personal research inevitably diminished because of his teaching and administrative duties at the Agricultural College; even so he was more productive than most people would have been engaged in full-time research, and his Annual Reports contain many results and valuable recommendations on the growing, manuring and diseases of various crops. His versatility, too, is shown by his other publications of this period, which range from the chemistry and physiology of the leaves of the Betel-vine, through safflower as a drying oil, to the composition of the milk produced by cows and buffaloes. Outstanding in this period was his study of the "Rab" system of cultivating rice. In many ways this work epitomises Mann's outlook on and approach to research; it shows (1) his enquiring mind, which was never satisfied by simply establishing phenomena but always sought their explanation; (2) his insight in appreciating how many factors may be involved in a seemingly simple phenomenon; (3) his ability to design experiments to distinguish between the effects of different factors and to identify the most important; (4) it further shows that his desire to find explanations was not simply to satisfy his intellectual curiosity but with the added hope that explanation would allow practices to be changed for the better, so that yields of crops could be increased and the life of the cultivators improved.

In the "Rab" system, trees, vegetation or cow manure were burnt on the seed-beds, from which rice seedlings were later transplanted. The Forest Department was anxious to protect wooded areas by getting the practice forbidden, but rice cultivators protested that it was necessary. Mann, with N. V. Joshi and N. V. Kanitkar, not only confirmed the opinion of the cultivators that the practice was beneficial and showed that burning vegetation or cow manure increased yields more than applying the unburnt material, but analysed the effects in terms of the plant nutrients supplied by the ashes and of the chemical, physical and biological effects of the burning itself on the soil. This work predated the discovery of partial soil sterilisation by heating, and although Mann, Joshi and Kanitkar probably under-estimated the biological effects from burning, they went far to explain the benefits of the system and to show how, with a given amount of material to burn, the maximum benefits could be achieved.

*Director of Agriculture*

Mann's influence while Principal of the Agricultural College went far beyond what shows in his publications, for his enthusiasm, energy and knowledge made him an inspiring teacher, and long after he left India he continued to have an effect through his students. From the College he went, in 1918, to be Director of Agriculture for the Bombay Presidency. Despite his administrative duties, he continued to do valuable research on problems as different as diseases of the potato, the cultivation of fodder crops and genetic variability in cotton. Increasingly, though, his writings are concerned less with specific problems of individual crops and more with the general need to increase yields and to adopt agricultural practices that would not only maintain but increase soil fertility. From his social surveys he more than anyone else appreciated the poverty of the rural population and how critical were their conditions even in years favourable for agriculture, but if further point was needed to make him urge much needed reform it was provided by the famine in the Deccan districts after the drought in 1918.

He knew the benefits that irrigation could bring, but also its limitations. Not only that there were many areas where irrigation was impossible, but that irrigated crops often received all the manure in a district and thereby impoverished the nearby land carrying unirrigated crops. He stressed the need for water conservation and showed how to make the most of the limited rain in growing crops. He also started an analysis of rainfall records to see whether there was any truth in the belief that rainfall was diminishing in the Bombay Province and whether there was any way of predicting famine-causing droughts; the results of this analysis, which he continued for many years, were published in 1955 under the title *Rainfall and Famine.* His detailed knowledge of Indian agriculture, of the factors limiting yield and of what might be done to improve matters, are succinctly summarised in his article *The Agriculture of India,* published in 1929 in the *Annals of the American Academy of Political and Social Sciences.*

Mann's research aimed to establish agriculture as a science, with crops being grown and stock raised by methods developed from established facts instead of based on equivocal traditions, but he was well aware that extra knowledge from research could not alone improve things and that, rather than ignorance, lack of capital usually determines the way in which agriculture is conducted. His idea of what was needed, and what could be done to lessen the precariousness of farming in the Deccan districts is summarised in a masterly manner in the evidence he gave to the Royal Commission on Agriculture in India. Although he advocated great extensions of irrigation, he showed that most of the land must depend on rain and urged the need to develop methods of 'dry farming' to conserve the rain for crops and prevent soil erosion. He also stressed the need for insurance against famine years and for alternative work for the cultivators, who were solely dependent on agriculture although this could not occupy them for more than half the year.

*Woburn Experimental Station*

In 1927 Mann reached the official retiring age in India and returned to England. The Woburn Experimental Station had recently passed from the control of the Royal Agricultural Society to the Lawes Trust Committee, the Governing Body of Rothamsted Experimental Station, and Sir John Russell, then director of the Station, invited Mann to take charge of the experiments there. Mann willingly accepted this unpaid appointment and, in addition to supervising the experiments there, from 1936 to 1946 he also shouldered the responsibility for running the farming. In effect, this appointment began another research career about as varied and extensive as his previous one in India. Any other person of his age might well have been content simply to see that the experiments designed by workers at Rothamsted were properly conducted, but not Mann. He not only took a prominent part in deciding the programme of major field experiments but also engaged enthusiastically in many research projects solely initiated by him. Of special interest to him were problems in maintaining and enhancing soil fertility, the competition between weeds and crops, the reasons for crop failures, such as from clover sickness, when the same crop is grown repeatedly on the same land. The growing of unusual and exotic crops also attracted his attention and it is only because of him that we know the potentialities and limitations in England of such crops as soybean, maize, sweet lupin, serradella and the Jerusalem artichoke.

The list of his publications from 1927 onwards shows the range of subjects to which he profitably turned his enquiring mind while at Woburn. They extend from the role of organic matter in soil fertility and the residual values of manures, through the consequences of ley farming and the practice of irrigation, to the take-all disease of wheat. There is no need here to summarise his results, because fortunately he did this himself in an article that was published in the Rothamsted Report for 1958 and that article is reproduced later in this book. It not only shows the great width of his interest and his insight into agricultural problems, but must be unique as an account of experimental research by someone in years far beyond any official retiring age and unpaid for most of the time. He retired officially from his post at Woburn in 1956, when 84 years old, but this only meant that, relieved of his official duties, he could devote his whole time to research and writing, which he continued to do with undiminished enthusiasm to within a few days of his death.

Research into agricultural problems was far from Mann's only activity while at Woburn. Between 1930 and 1940 he often travelled abroad, to Russia, East Africa and other countries that sought authoritative advice on the cultivation of tea. The consequence of his three visits to Russia were such as to lead him to be referred there as "the father of Russian tea". As is evidenced by the list of his publications, he kept up his interest in other tropical crops than tea and in the agricultural and social problems of tropical countries.

Repeated visits to the Middle East, defined broadly to extend from the Caucasus to East Africa, left Dr. Mann with an abiding and insatiable interest in the agriculture of these regions and in the successive, sharply contrasting phases of their historical evolution. His knowledge of the crops in the region was such that Sir John Russell, the Director of Rothamsted Experimental Station, suggested to Dr. Mann that he write a book on Foodgrains in the Middle East and India. An illness and other difficulties prevented the project coming to fruition, but the draft outline he prepared is printed below (opp. 393-94).

At home in Bedfordshire he was an active member of many local organisations. His energy was prodigious and fortunately matched his reforming zeal. Few people can have lived fuller or more useful lives, and few can have been less self-seeking. He described himself as "always a rebel and reformer". This is true, but only half the truth, for it leaves out the facts that he was also compassionate and wise. What he rebelled against was ignorance and social injustice. His scientific work was aimed at discovering knowledge whereby agriculture could prosper and its future be assured. What he wanted to reform were systems in which poverty is based on ignorance, and his realisation that poverty could be abolished by knowledge rightly and justly applied was the driving force behind both his scientific and sociological studies.

F. C. BAWDEN, F.R.S.

# SOCIAL AND ECONOMIC STUDIES OF HAROLD H. MANN

Harold Mann's fame in agriculture and the natural sciences always overshadowed his remarkable achievements in the field of social study. He himself contributed to this process by his disarming habit of referring to his town and village studies as merely his hobby. With rare exceptions, his acquaintances in the agricultural world took these expressions at face value. Even the most summary review of his contributions to the social sciences will suggest, I believe, that his accomplishments in this field were probably as great as those which gave him an international reputation in the natural sciences.

## I

Harold Mann grew up and went to school in Yorkshire in the 1880's and early 1890's. This was precisely the period when Charles Booth was applying to London the detailed methods of studying urban poverty, slums, drink, etc., first worked out in France by Frédéric Le Play. In 1901, B. Seebohm Rowntree published an exhaustive enquiry into the extent and consequences of poverty in Dr. Mann's own home town, the city of York. Inspired by Rowntree's work, Dr. Mann decided to see whether the same methods could be applied to an agricultural village in England. With the aid of friends, he organised in October-November 1903 a study of Ridgemount, the village near which he had lived from 1898 to 1900, when he was working at the Woburn Experimental Station.

Through the intermediacy of Seebohm Rowntree, Dr. Mann's study was submitted to the newly-founded Sociological Society in 1904, and published in the first volume of *Papers* of that Society, under the title, "Life in an Agricultural Village in England: A Study in Economics". It was the pioneer application to the British *rural* scene of the Le Play-Booth-Rowntree methods and paved the way for many subsequent studies.

Dr. Mann's headquarters from 1900 to 1907 were not in England but in Calcutta where, as Mr. Bawden has indicated, he was Scientific Officer to the Indian Tea Association. Thus his field work at Ridgemount in the Fall of 1903 must have been sandwiched in while he was on home leave from India. Before and after that sojourn in England, Dr. Mann had been utilising his spare time in Calcutta to design a quite different type of field enquiry. On behalf of the Drink Problem Investigation Committee of that city, Dr. Mann, a dedicated advocate of temperance, conceived and carried out in 1903-04 a detailed investigation of the drinkshops and their clientele in Wards 6 and 13 of Calcutta. To the best of my knowledge, this study, parts of which are reprinted in Chapter 13 below, represents the first application of the Le Play-Booth inquiry method to an Indian *town*. Moreover, although the term "sample survey" was not current in those days, Dr. Mann in fact selected his subjects in such a way as to obtain a probability sample of the universe of liquor purchasers in the two wards which he covered.

In 1903, the same year that he launched his pioneer studies of village life in England and town problems in India, Dr. Mann also managed to issue his fundamental revision of Watt's basic manual on the diseases of tea plants. A glance at the bibliography of Dr. Mann given in the Appendix will show that he continued to exhibit this marvellous versatility right up to the time of his death in 1961.

While in Calcutta, Dr. Mann became friendly with Dr. N. G. Mukerji whose *Handbook of Indian Agriculture,* first issued in 1901, was the outstanding manual on the subject. Dr. Mukerji died in 1906, in the midst of preparing a thorough revision of his *Handbook*. Dr. Mann saw the job through; the new version appeared in 1907. Again in 1915 Dr. Mann brought the book up to date and considerably shortened it. On both occasions, he insisted on remaining anonymous so that the work would appear under Mukerji's name only.

Not long before the close of his seven-year stay in Eastern India, Dr. Mann found time to analyse the cost and adequacy of the diet of tea garden coolies in Upper Assam. (Chapter 6 below). His work for the Tea Association had often required him to visit the tea plantations in the hills. It was characteristic of the man to take a personal human interest in the Mundas, Oraons and other tribal peoples of east-central India who had migrated to Assam under indenture contracts to constitute the plantation labour force. We may note with some surprise that, as of 1906, Dr. Mann found that the normal diet of a working labourer amounted to 3,700 calories per day; and that the coolie's working wife, who performed similar tasks on the plantation, had the same food, viz., at the rate of 3,700 (sic) calories per day. Children were provided roughly half of this per head. At these levels of nutrition Dr. Mann found that the cost of feeding a labourer, his working wife, and three children amounted in 1906 to Rs. 10-10-0 per month.

## II

Dr. Mann moved from Calcutta to Poona in 1907 upon his appointment as Principal of the Agricultural College and Agricultural Chemist for Bombay Presidency. For the ensuing twenty years Poona, and, to a lesser extent, Bombay, remained his headquarters. After taking time to familiarise himself with the dry Maharashtrian setting of the Deccan—what a different world it must have seemed from the lush Ganges Delta and the hills and valleys of Assam!—Dr. Mann launched an intensive investigation of Poona as a characteristic Deccan town. He has indicated that his friend, Patrick Geddes, encouraged him to undertake this survey of Poona. Only a part of his results were ever published, and these concerned primarily the lowly or "Untouchable" classes, principally the Mahars and Mangs (see the articles of 1912 and 1916, reprinted as Chapters 14 and 15 below). In the same period Dr. Mann published a study entitled "The 'Mahars' of a Deccan Village", which dealt with Saswad, a little country place of a purely rural character, 16 miles from Poona (see Chapter 7 below). All three papers are detailed, precise studies in the Le Play-Booth tradition. They are based upon interviews with all the resident families of the castes studied in Poona and Saswad. They constitute, I believe, our first modern, quantitative studies of Untouchables. The study of Saswad was the first, to my knowledge, to apply the Le Play-Booth method to the Indian countryside.

At the time of the First World War, Dr. Mann's rural and urban enquiries had become well-known to social scientists in India and England, and his reputation led to his being recalled temporarily to Eastern India. The Tata Company, operators of India's first great steelworks at Jamshedpur, decided to have a social relations analysis made of their company town situated some 150 miles west of Calcutta. When they asked in England for recommendations as to who should undertake the study, Sidney Webb, Professor L. T. Hobhouse, and other distinguished personages in London referred them to Dr. Mann. He accepted the invitation from Tatas, obtained leave from the Government of Bombay, and carried out his investigation during the closing months of 1918. In his printed report dated January 1919, Dr. Mann showed that on the social side India's first steeltown had got off to a poor start. Lack of planning had led to grave problems in regard to housing, supply of food and other consumption goods, welfare of mothers and children, and health. As a comprehensive welfare analysis of a "new town" centered on a single great industrial complex, Dr. Mann's report on Jamshedpur is unique for India. It is a pity that the Tata Company has not yet released this report to the public, nearly half a century after it was first printed.

## III

Dr. Mann's most celebrated social enquiries were his two village studies in the Deccan. As early as 1911 Dr. Mann had suggested in print that it would be useful to make a detailed survey of a single village in India (see Chapter 2 below). In a talk which he gave in London in 1961, he has explained how he went about making his first survey(see Chapter 3 below); and he has indicated how much he owed to the remarkable group of young Indians who were then working with him in Poona. It took Dr. Mann and his half-dozen collaborators two years (1913-1915) to gather all the data they deemed essential. Their short but important book on the village of Pimpla Saudagar, was issued in 1917; Dr. Mann had earlier published an article in the *Indian Journal of Economics* summarising their work, which is reprinted below as Chapter 8.

When Dr. Mann's results became public knowledge, a storm broke loose. As a result of his enquiries Dr. Mann had concluded that only 35 percent of the families of the village surveyed could make do from agriculture alone. A finding of this sort by an Englishman, in fact a government official, was considered by many of Dr. Mann's compatriots as a bad case of "letting the side down." A more serious and thoughtful criticism put forward at the same time was that Pimpla Saudagar might be an unrepresentative village because it was located near the important government arms factory at Kirkee. Dr. Mann at once acknowledged the justice of this argument. With some of his original collaborators he soon set about the study of another and much more isolated village, Jategaon Budruk, situated about 25 miles northeast, as the crow flies, from Poona. The group of researchers had the satisfaction of discovering by the winter of 1917-18 that the economic data gathered in this far-off village yielded results which differed in no significant respect from those obtained in Pimpla Saudagar. From that time onward Dr. Mann's village studies began to gain acceptance, even among official circles. By 1927, Dr. Mann had the honour of being called as the first witness before the Royal Commission on Agriculture in India (the Linlithgow Commission). He was officially encouraged to submit to the Commission up to date evidence for 1926 on the two villages he had earlier studied, so that these could be compared with his initial results for 1916. (See Chapter 19 below, for Dr. Mann's testimony and the 1926 results).

Dr. Mann also extended in historical depth his examination of one of these villages, Jategaon Budruk. With help from his Maharashtrian collaborators, he managed to find in the Poona archives of the Peshwas records for Jategaon Budruk extending back to 1698. (See his paper, "A Deccan Village under the Peshwas", reprinted as Chapter 11 below).

In the course of conducting his social enquiries Dr. Mann came to formulate his views on methods of work appropriate for such studies. His reflections on this subject seem to me to be of such lasting value that I have placed them at the head of this volume, as Chapters 2 and 3.

Society, Dr. Mann wrote, presented the most complicated phenomena on earth. To decipher the meaning of a social system required a series of intensive studies of social conditions over comparatively small areas. Observation must be complete, small differences cannot be ignored—"inaccurate data are worse than no data at all". Naturally the observations which are made relate to specific hypotheses and theories. But how far in advance of knowledge can the investigator permit his theory to go? Here Dr. Mann was very firm. Theories, he stated, should go only a little in advance of knowledge, just enough to suggest further experiments. "If they go beyond this, they are often a hindrance".

In all this Dr. Mann was aligning himself explicity in the Le Play-Booth-Rowntree tradition. He considered Rowntree's 1901 study of York the most striking and perfect of such studies. Dr. Mann was so enthusiastic about Rowntree's later volume, *Land and Labour: Lessons from Belgium* (his review article on this book is reprinted in Chapter 2 below), that he adopted the same title when he published the results of his own Deccan village studies in 1917 and 1921.

As we have mentioned, almost all of Dr. Mann's village and town enquiries were made in time that he spared from his regular work as an agricultural scientist and later as Principal of the Poona Agricultural College. Dr. Mann enjoyed doing these studies and believed that they were worthwhile in themselves. Nonetheless, he was an exceedingly busy man. He would not have devoted all the effort that he did to these enquiries unless he thought there was a larger context in which they were important. That context was the poverty of India, in Dr. Mann's eyes, the unnecessary poverty. He believed that with the existing land and available manure the agricultural output of India could be raised by 50 per cent. (See Chapter 22 below). Furthermore, he held that, if the reserve of labour in the villages could be marshalled productively, then total production on existing land could be pushed up higher still. Such an increase in production might go a long way toward reducing poverty in the countryside and lifting up the quality of life. Why didn't that increase in output take place? What were the forces or factors operating to prevent it?

Dr. Mann had little patience with people who tried to answer questions like these on the basis of printed volumes to be found in libraries. The available data, he wrote, were scanty and unsatisfactory. No progress in understanding India's basic agricultural problems was possible, he felt, without much more evidence freely gathered by investigators sympathetic to the peasantry but yet determined to get information of a high degree of precision. His village enquiries were intended to show that such investigations were feasible in India, and to inspire other researchers to follow his example in other regions of the subcontinent. He drove himself relentlessly not only because that was his nature, but because he conceived the task as large and urgent.

## IV

Throughout Dr. Mann's twenty years of service in the Bombay Presidency and his later brief appointments in Hyderabad State, he had constantly to deal in practical terms with proposals for agricultural betterment. The first decades of the century saw a great expansion in the work of the Departments of Agriculture in the various provinces of British India. In the years from 1909 to 1914 four successive *Reports* were issued by the central Board of Agriculture of the Government of India under the title, *The Introduction of Improvements into Indian Agriculture by the Work of the Agricultural Departments* (1909, 1910, 1912, 1914). Dr. Mann worked on all of these and was, in fact, Chairman of the Committee which drafted the last three. These are administrative documents not intended for the general public, but a number of the main points from these *Reports* are presented in an interesting and forceful way in Dr. Mann's crisp paper of 1915, "The Lines of Development of Indian Agriculture" (reprinted as Chapter 18 below).

With regard to the agriculture of Western India, the area he came to know best, Dr. Mann was undoubtedly the most productive writer of the era. To my knowledge, his *Statistical Atlas of Bombay* (3rd Edition, 1925) remains the best agricultural atlas ever produced in India.

The views which Dr. Mann developed on the basic problems of Indian agriculture may be stated in very simple terms. There were, according to his analysis, many forces operating to make the peasantry stolid, inert, passive. At the same time there was a proved readiness on the part of the peasantry to adopt changes which would benefit them quickly, surely, and substantially, provided these involved small out-of-pocket cost. Most peasants were short on capital but long on labour. To get capital they had to borrow at high rates of interest. Hence their enforced preference for inexpensive improvements.

Even more crucial was the question of incentive. The peasants' fear, which Dr. Mann considered justified, was that little or none of the benefits from their extra exertions would accrue to themselves. Any fresh gains were likely to be taken away from them by the landlord, the moneylender, the crop merchant, or some petty local official. Thus the fundamental obstacle to the taking up of improvements by the peasants was not technical or narrowly agricultural. It was rather a question of the very foundations of village economy and society.

These views are set forth in three of Dr. Mann's outstanding contributions: his written and oral testimony before the Royal Commission on Agriculture of 1926-28 (see Ch. 19 below); his masterly article of 1929 entitled "The Agriculture of India" (Ch. 20 below); and his sober but not pessimistic survey in 1948 of the prospects for "Village Betterment in the New India and Pakistan" (Ch. 22 below).

Although optimistic by nature, Dr. Mann preferred not to blink at the massive roadblocks in the way of any genuine rural progress in India. He always emphasised that the crucial obstacle was the social one, the institutional framework of Indian agriculture. If any scheme of planned development was to get very far, he argued, it would have to bypass the normal channels of power in the countryside. The actual cultivators who worked out in the fields under the glaring sun would have to be given a major voice in all decisions.

Dr. Mann conceded that the passage of authority from established or vested interests to bonafide tillers would have revolutionary implications and that there was no precedent for this in the modern history of India. Without a change of this order of magnitude, however, he remained sceptical that Indian agriculture would be able to realize its admittedly great potentialities. There was a vast amount of work that was necessary to lift up output in India, and there was a large reserve of under-utilized labour-time in the villages. It might seem obvious, when viewed in abstract, that the spare labour-time and the work to be done should be brought together. But in everyday life it was not so simple. The excess labour-time was in the hands of the mass of the village poor; the main benefits of such works as road-building or clearing of irrigation channels were likely to accrue to the handful of well-off families in the village. Why, then, should the many bestir themselves for the benefit of the few?

Until this social problem was resolved, Dr. Mann had no great expectations for improvement of Indian agriculture through technical measures. While there was something to be said in each case for irrigation, fertilizers, land reclamation, consolidation, large-scale farming, and cooperation, he expected no panacea, no simple single solution. With regard to irrigation canals, for example, he felt that the most promising ones had long ago been built; the possibilities that remained were limited, and there were regions in which irrigation, unless introduced with proper precautions, could do more harm than good (see Chapter 25 below).

Trained as a chemist, Dr. Mann had looked closely into the value of natural and artificial fertilizers for India. Their use, he concluded, had to be proportioned to the availability and dependability of a restricted water supply. With regard to land reclamation, he wrote that India was unfortunately not so well situated as the Soviet Union; there were, alas, no substantial resources of good land waiting to be brought under the plow. As for land consolidation, which had been much talked of from 1900 to 1925, the actual results in the years after 1925 had proved disappointing.

## V

Dr. Mann was particularly interested by collective farming and by the cooperative movement—two quite different things which he was very careful to distinguish. Collectivization along Russian lines attracted his attention in the 1930's as a possible device to overcome the poverty of the peasants (see Chapter 27). By the end of the Second World War, however, he felt that collectivization had destroyed the Russian peasant economy as a structure, and turned the peasants into labourers working for the state (see Chapter 26).

Cooperation as a movement won enthusiastic and energetic support from Dr. Mann, particularly during his years in Bombay (1907-1927). He is, in fact, considered one of the major figures in the building of the cooperative structure of Bombay, the area of India in which cooperation has perhaps registered its greatest successes. Nonetheless, Dr. Mann did not let his enthusiasm blind him to the harsh reality that, on balance, the failures of the movement in India far outweighed its successes, and that the causes of these failures were deeply rooted, not only in the Indian soil, but in Indian society (see Chapter 23 below).

The sobering conclusion which Dr. Mann reached after long decades of anxious work and enquiry was that the agricultural transformation of India would have to come the hard way. The starting point would have to be the lands already long under cultivation, and the most difficult problem would be not how to obtain greatly improved yields from those lands, but how to do so in such a way that a substantial share of the increases in output would go to the actual cultivators.

Short of a genuine solution, Dr. Mann refrained from striking at easy targets, such as the village moneylenders. It was only too easy to attack them, Dr. Mann indicated, but *in the existing structure* there were essential functions which they were performing (at a high price, admittedly), and which no one else was in a position to fulfil. Until a better mode of supplying credit to poor peasants could be worked out *in practice,* it was no service to the peasantry to drive out the village moneylender (see Chapter 23 below).

Dr. Mann's defense of the much-abused village "*banias*" provides a good example of the scientific integrity that compelled him to study all relevant facts as objectively as possible, and then to express his conclusions without bowing to popular prejudice. With this stubborn fact-mindedness Dr. Mann combined a constant awareness of the larger human framework within which agricultural problems arose. All his work on Indian villages and cities is imbued with his deep concern for the welfare of the peasants and townsmen, at first within the context of a colonial dependency, and later, after a transition he had worked for and which he warmly welcomed, within that of an independent nation.

Dr. Mann was never afraid to issue the findings of his studies nor to point out their relevance to the broader issues of social justice and political freedom. No matter how unpalatable his results might be to the powers of the day, Dr. Mann spoke his mind and said what he thought had to be said. He paid for this—honours and distinctions went elsewhere—but he felt it was worth the price.

DANIEL THORNER.

**PART I**

**APPROACH AND OUTLOOK**

*Chapter* 1

# THE PHILOSOPHY OF A MAN OF SCIENCE*

It is impossible to exaggerate the effect which a study of science has on the mind of a young man who takes it up seriously. To me, as a school subject, science was intensely attractive: I could have spent the whole day or nearly so in connexion with it. But at school it stood on the same intellectual plane for me as other studies: it was more interesting perhaps, but it was in the last resort simply so much to learn, and to learn on the authority of a teacher and a textbook. I might make observations, I might collect specimens, but all these were to be considered in relationship to the same source of authority. One heard in the distance the rumblings of great scientific controversies, such as that in my young days on the theory of natural evolution, but whether it was to be accepted or not depended very largely on the one hand on the authority of the standard-bearers of each party, or on the arguments, often mere matters of logic, which they brought forward. But about the age of sixteen I went to college. and was launched on a career which from that time onward was chiefly if not entirely a scientific one. The atmosphere surrounding was different: authority of teachers and books of every kind became less insistent: without any definite

* Reprinted from the **Indian Interpreter** (Madras), Vol. VI (1911). No. 1, pp. 18-28.

instruction in the matter one began to require experimental proof for what was stated, and in examining controversies, it began to be more a matter of analyzing and examining the sufficiency of the data rather than considering the argument, from the data, of the protagonist on either side.

How quickly this change in attitude took place I cannot remember, but it certainly took place. And I find it has been noticed by others as one of the almost inevitable results of a genuine scientific training. And one of these almost certain consequences was to break down my conception of the certainty of some matters previously considered as beyond all controversy, and to bring other matters into a position of importance and assurance which they did not previously occupy. There is a striking passage in Drummond's *Natural Law in the Spiritual World* which puts in better and more forcible language than I can find, the effect of such a training as I received. He says: "No one can study modern Science without a change coming over his view of truth. What impresses him about Nature is its solidity. He is there standing upon actual things among fixed laws. And the integrity of the scientific method so seizes him that all other forms of truth appear to be comparatively unstable. He did not know before that any form of truth could so hold him: and the immediate effect is to lessen his interest in all that stands on other bases. This he feels in spite of himself; he struggles against it in vain, and he finds, perhaps to his alarm, that he is drifting fast into what looks at first like pure Positivism. This is an inevitable result of the scientific training. It is quite erroneous to suppose that science ever overthrows Faith, if by that is implied that any natural truth can oppose successfully any single spiritual truth. Science cannot overthrow Faith; but it shakes it. Its own doctrines, founded on Nature, are so certain that the truths of Religion, resting to most men on Authority, are felt to be strangely insecure. The difficulty, therefore, which men of science feel about Religion is real and inevitable, and in so far as Doubt is a conscientious tribute to the inviolability of Nature it is entitled to respect."

None but those who have passed through it can appreciate the radical nature of the change wrought by science in the whole mental attitude of its disciples. And this is often erroneously connected with the nature of the subjects usually classed as scientific, when it is really due not to the class of subject, but to the method of dealing with them. The subjects have the advantage of allowing the application of the scientific method easily, but that is all. It is the method of referring everything to observation and experiment which can be repeated by even the meanest disciple rather than to authority, which forms the essential part of the difference. And this method can be applied, perhaps with greater difficulty, to all *real* subjects, to all those which have to do with phenomena, whether of history, or literature, or even, as we are now learning, of conception. One begins in fact dimly to realise that in adopting the scientific method, one enters the study of a subject as *a master*, that one is not to be bent to the subject, but to bend the subject to oneself—and this sense of independence of all others, of being able to ascertain certainties for oneself, is characteristic of the scientific attitude.

It must be at once owned that the effect of this is to narrow one's outlook at once, but at the same time to make it very much more intensely felt. It narrows it, in fact, to phenomena, and makes all that is not phenomenal so remote as to be entirely on a different plane and to be at first sight so uncertain as to be scarcely worth considering. I will return to this a little later. Just now I want to consider what I have called the increase in intensity of the outlook.

By this I mean that the scientific training gives a far more rigid criterion of what is and what is not—of what conclusions are legitimate from certain facts and what are not, than we ever had before. Of course, perhaps I am a purist in these matters, but I often feel that half the difficulties of so-called science, and all the pseudo-science of our or any day is due to the non-realization of the rigidity of the scientific criteria of truth and the method of their application.

Of course, in searching for scientific truth as for all truth, the first thing is observation. But, says the scientific method, if you are to rely on observation alone, your observation must be complete. There must be no ignoring of slight differences, no being content with a sample of the type of thing you are studying. If you are to content yourself with observation, you must observe all and in every detail. But it is at once answered. This is ridiculous. No man can observe on this scale or in this manner. Quite so: I quite agree. Then observation is useless as a method of research. No—not useless, but rarely or never adequate, I at once answer. The realization of this is essential to the real true rigid scientific attitude. The lack of it has rendered the conclusions drawn from some of the finest observations ever made quite of temporary interest. The observation has not been deep enough, it has not been wide enough to justify the conclusions. All the mathematics and logic in the world will not make laws deduced from observations incomplete either in depth or width stand on anything but a very uncertain base.

May I say again here that in the recognition of this fact our scientific method differs from all others. To a logician, I suppose, to make sure of his axioms or to get them generally accepted is the primary duty—and then his results must be accepted as true. Anything of this sort must be equally strong throughout, and men can fight for the conclusion. And provided the axioms are undisputed, there can only be progress by more complete deduction. The same would be the case with science were the observation complete, but this is practically never the case. Hence the scientific theory—which is generally the part which appeals to the imagination—is generally only valuable as it is suggestive.

It will at once be said, perhaps, that I am apparently making scientific results less, and not more, certain than other truth. I do not acknowledge this, for I may now turn to speak of the second method of attack of problems, namely, that which is based on experiment. And this is after all the essential characteristic of the scientific method. So long as inquiries are in the stage of observation merely, they have not advanced, and cannot advance far. Arrangement, classification there may be, or principles which enable similar things to be connected together may be discovered, but the causes underlying phenomena are not likely to be discovered, except as brilliant guesses. In this stage are some of our sciences even now—sociology for instance, psychology in a less measure—and they cannot advance seriously until experiment becomes a regular method of attack of their problems.

By experiment I mean this—that one *makes* facts under known conditions—that you simplify conditions, and then when you know what these are, ascertain what takes place —that you follow the modifications of reaction which take place when conditions are modified. This may seem little, but it is what makes science unique. It is what takes a student from the library to the laboratory, from a study of what other people think to what he can make nature say. I use nature here in the widest sense of all phenomenal things. By defining your conditions, and then ascertaining the effect that changes in condition produce, one feels master of a situation. You think I speak perhaps in superlatives, but the sense of power that comes when one has forced nature to speak, so to say, to unlock a secret, however small, is one which once experienced can never be forgotten.

One of the most striking illustrations of what I mean is in the recent development of our knowledge concerning heredity or the transmission of characters from generation to generation of living things. For centuries little progress in the understanding of the subject was made, though on the practical side much was done. But a comparatively few years ago Mendel put forward as a result of observation and experiment in his own garden the idea that every living thing contains characters segregated from one another in units, and that these are transmitted as unit characters to a number of descendants in accordance with the mathematical theory of probability. This may or may not be true but it has indicated a line of experiment, of determining conditions of production of new generations of your organisms, and so forcing nature to give up secrets. The result is that in this matter there has been more progress in ten years in understanding the mechanism of heredity than in the previous thousand.

Probably one of the finest masters of the application of the experimental method in our times was Pasteur, who stands (at any rate from my point of view) in the same position from this side as does Darwin as a master of observation. No one who examines the manner in which he arranged his questions to nature, modified his conditions so as to ascertain how change of circumstance altered his results, or how he secured a rigid simplification of conditions can help seeing how the means which I have tried to describe bring about a power of ascertaining truth which is indeed astonishing to those who have not before realized what the scientific method is. I especially commend to you the extraordinarily powerful investigation which he made to ascertain whether the lowliest of living things could be produced without pre-existing life.

The result is a certainty not obtainable otherwise. Define your conditions: ascertain your result—and the only criticism of this result that is possible is that the conditions were not accurately enough defined. And it is a legitimate criticism, and leads to the continual repetition of experiments with the conditions more and more accurately defined. Sometimes the conclusions stand: sometimes not. But in either case progress in knowledge is made—and progress whose value is fairly exactly known. To quote M. Duclaux again. "La science est une quèteuse à domicile: chaque jour elle fait le départ des pièces fausses qu'elle a reçues et ne garde nulle rancune a ceux de qui elle les tient."

This makes science progressive, and progressive in a different sense from that in which so-called philosophy or even mathematics is progressive. In these, so long as the axioms are untouched, all is necessarily true provided the logic is correct. As I have already said, progress can only be made by carrying the logic further. But doubt the axioms, and all the superstructure falls like a house of cards. In science on the other hand we are continually revising: the past structure must be continually modified. Destroy our mutual agreements, even though they are as fundamental as the law of gravitation—and our school books might have to be rewritten, but you would not affect materially the position of the knowledge obtained.

I have tried to describe what it is that seizes me in the general intellectual position which science gives. I have rarely had a more trying experience than after a few years of a scientific atmosphere to return to language study. In learning a language one has to accept: one asks why—and is told that there is no reason, it simply *is* so. One gets restive—that authority not experiment should be the arbiter. In science one is a master; I try to make even the humblest student feel this. Nature is a servant who, properly approached, can be made to give up her secrets by the application of experiment. And this mastery, I always feel, embraces or will ultimately embrace the whole phenomenal world. The sciences at present so called are only a small corner of the sphere capable of being attacked by the combination of observation with the experimental method which I have tried to describe—with the deductions which can legitimately be made from them.

But it is an important point to notice that while I believe that the application of the scientific method in all its strictness will enlighten ultimately the whole range of phenomenal knowledge, yet a consequence of using it is a hesitation to express any opinion outside the range to which it is applied. *Because* I believe that the experimental investigation of the brain and the phenomena of thought will ultimately enlighten the whole dark places of psychology, it makes me hesitate, on the basis of deduction from the very few observations, to express any opinion on psychological questions or even on the laws of thought. *Because*

I believe that the best means of dealing with social evils and social difficulties will ultimately be found by the most careful observation and experiment, I hesitate to give adhesion to any social nostrum, whatever it may be. I may support something provisionally, but I feel justified in suspecting most profoundly theoretical systems worked out from some accepted social axioms. Whether the foundation be a supposed social contract as with Rousseau, or a theory of the increase of population as with Malthus, or a discussion of the necessary conditions of trade as with Adam Smith, or ideas as to the necessary relationship of men to one another as with Ruskin, I feel (though I may provisionally act upon any of them) that till each of them has observed and especially experimented under known and defined conditions, I cannot accept any save as a working hypothesis on which to base further experiment. I mention these sociological questions in order to show how far I conceive the scientific method is applicable, not because I wish to discuss these questions themselves.

But you will perhaps see whither I am leading. A scientific man, I conceive, so far as science is capable of leading must be an agnostic. I hesitate to use the term, but I know no other which meets the point. So far as phenomena are concerned there is a tendency (if you believe in the integrity of the method) to be driven to say 'I don't know' about anything in connexion with which observation and experiment have not gone far enough to allow legitimate conclusions to be drawn. From observation alone I doubt whether such legitimate conclusions can be reached, and the most is a working hypothesis: if combined with experiment something more than this may be attained, but it is always subject to revision. Useful bases for further inquiry you may have—but the true position is one of questioning and not of certainty.

I am aware that scientific men are often dogmatic and positive when they ought to adopt the attitude already suggested. Even the inventor of the term agnostic, Huxley, became violent and used argumentative methods worthy of the theologians of the dark ages in defending what was only a working hypothesis. There seems an inevitable tendency to rally round a theory and fight for it not always with the legitimate weapons of further experiment, but with those of the philosophical schools. It appears time after time, I could quote cases in the scientific world at the present moment. But it is unscientific, and the only way of settling a question is not to argue about it, but to experiment with it. To quote a French author again with some slight modification. "You must not ask argument or calculation to illuminate a mass of observation of whose inter-relations you know nothing. Argument, calculation, are like a musical box which only plays the tunes which have been put into it. Instead of asking for results, the materials for which have not been fully supplied, the only way is to proceed to experiment."

Applied to phenomenal matters, therefore, I can only take one position. I accept only that theory, that conclusion, which has been properly attacked and so far as it is deciphered by observation and experiment. I accept no authority, save as a recommendation to examine for myself. I cannot be a heretic, simply because there can be no heresy. My conclusions are subject to revision in the light of further experiment. In the meantime I can and do act on them, always on the look out for further evidence, always unconsciously experimenting to obtain further evidence which will lead me to a truer conception of the truth. And if in those ranges of knowledge which one cannot personally examine, I take the conclusions of others as bases for action, I only do so because I have confidence that they have applied the same criteria and are equally ready to modify them in the light of new data.

You will say that this is pure materialism. And so perhaps it appears at first sight. In the first half century in which the scientific position was first fully realized, the tendency of scientific thought was wholly or almost wholly materialistic. That is the effect on beginners in science almost always. *Here* your bases are so independent of authority, and so absolutely capable of verification continually that there is a tendency to feel that this is the only field of knowledge. But, as in so many other cases, the effect of more science is to alter the position somewhat. To quote a striking passage from Bacon: "This I dare affirm in knowledge of nature, that a little natural philosophy and the first entrance into it doth dispose the opinion to atheism: but, on the other side, much natural philosophy and wading deep into it, will bring about men's minds to religion."

Science, in fact, has become more humble in recent years because it recognizes that there are certain spheres which are *at present* beyond the sphere of scientific investigation. There are hosts of things subjective, which are undefinable in objective terms; there are intuitions which cannot be accounted for on any rational basis; there are influences which cannot be described in scientific terms; there are worlds which science has not touched and (though I doubt it) may never be able to touch. And here there is a tendency in present science, with which I am in entire and hearty sympathy, to stop. It is not that I deny the possibility of scientific investigation into these questions in the future. That may be. But it is not yet. And yet many of these questions are more real to many of us than all the objective truths revealed by science. We cannot ignore them: we cannot write over them *'agnosco'*: we cannot put them on one side and leave them until a future generation is able to submit them to scientific inquiry with the aid of weapons of which we are ignorant at present.

But, and here I may come in conflict with the philosophical students, I do not think these things can be unravelled by argument and by logic—any more than by science at present. The axioms of Descartes or of Kant afford no more guidance than the atomic theory or the law of gravitation, because in both cases we are attempting to compare incommensurable quantities. I try to keep this intuitive world, perhaps more real to me than any other, as small as possible,

I hesitate to assent to any creed which attempts to define it—and yet it determines my actions more than all scientific discussion. But it is subjective: it is mine. I cannot impose it on others. But it is real, and science is realizing that it cannot be ignored.

And apart from this there is a range of knowledge into which science makes no claim to enter, namely, into that examination of the true nature of things which has been the subject matter of philosophy from all generations. Our inquiries as men of science are concerned with how things happen—and not why they happened. This, you may say, is a narrowing of knowledge. So it is, but it is better to aim at a mark you can hit, rather than aim at something which it is very doubtful you can hit, at least with the weapons in hand. We look with contempt on the fields of metaphysics, discussing to-day precisely the problems of ten generations ago, and with no further real light on them. We have certainly progressed: we do know a little more as to *how* things happen than was known a hundred, or even ten years ago. Our object may be humble, but we are getting there. I, personally, and I speak in all humility, feel an impatience with those who would unravel the riddle of the universe by mental processes on the basis of a few axioms.

In recent years there has been a phenomenon which is somewhat striking. Some of our scientific leaders have entered the region of metaphysics, and tried to give the bases of a philosophy whose foundation was science. The first of these was perhaps Auguste Comte, but in this connexion he hardly counts. Then came Herbert Spencer, and his synthetic philosophy, mighty at least in bulk. In recent years he has been followed by a crowd, of whom I may mention Prof. Ernst Haeckel, and the latest, Sir Oliver Lodge. Now all their works in this direction seem to me to be fundamentally unsound. You cannot unravel questions by the scientific method unless you adopt it—unless you demand the same standard of observation and experiment as you do in your daily work, The observation is incomplete;

experiment has been hardly made. Hence, at present, the scientific method is inapplicable. But, is it not valuable to have theories in advance of experiment? Certainly, but only a little in advance, in order to suggest further experiment. If they go beyond this they are often a hindrance.

Hence, as soon as you enter the field of metaphysics, as soon as you begin to examine the *real* nature of things, as soon as you leave the sphere of phenomena, I cannot consider these scientific men as any better equipped than others who are not known in the realm of science. They are as logical, but not more logical. They are as ingenious in argument, but not more ingenious. They are as well equipped mentally, but not more so. And, I think, that their scientific reputation is apt to give them a false appearance of importance when they speak outside the range of the application of the scientific method.

I think I have spoken enough. But I must summarize my own position. One cannot study science without acquiring an admiration of the rigidity and certainty of the scientific method. That combination of observation and experiment under known conditions seems to give such certain results that it is apt to destroy confidence in every other method of inquiry. And so far as any other method of inquiry is concerned which attempts to deal with phenomena that destruction of confidence seems to me justified. I can only say 'I do not know' before the range of things capable of scientific investigation, but not yet investigated. All I can hope to do is to try and add my quota of knowledge, and to act *provisionally* on such information as I can get. But all objective phenomena are capable of rigid scientific investigation, and this will come—and in my mind this is the only investigation of any great value.

Outside the range of phenomena, I have interest, but little confidence, either in argument or speculation. I feel the scientific method will ultimately tell us *how* things occur, and that if ever we find *why* it will be as the ultimate end of the patient and steady inquiry as to *how*. There are no short cuts in knowledge, and argument on the basis of doubtful axioms is a very barren field of study. And this applies equally when scientists enter the field of such speculation as when others do so.

All our knowledge outside the range of phenomena (and I speak with the greatest diffidence here) seems to me to be subjective—but not on that account to be any the less real but perhaps even more so. This subjective knowledge influences me more than all the science I have acquired. It is different, I know, for me and for others, but for each of us it is a very real thing. I do not know how to define it in philosophical terms, and I do not like to speak of it here. But it is the knowledge that influences more, which I cannot argue with, which I cannot understand, which makes me something more than a materialist. These I try to define. I find them agree in a measure with others though not entirely. I find I can use this agreement to influence others when I cannot understand the process. I find that others can appeal to me by things which I cannot reason about. All this may come within the range of science ultimately: at present it seems more likely to me that there are other sources of knowledge than perception. And, feeling this, materialism seems to me to be a very partial knowledge, perhaps the least part of knowledge.

*Chapter* 2

# THE SOCIAL STUDY OF A PEOPLE[1]

It is now many years ago since I first read Herbert Spencer's *Study of Sociology,* and I took upon that reading as one of the important events in my life, or at any rate of my public life. And this is the case, not because of any of the facts detailed in the book, but because of the way in which Spencer insists on the extreme complexity of social life and organization, and the absurdity of basing conclusions with regard to it, as is so frequently done, either on a very superficial study, or on a series of general principles which have been evolved in the study. We have, I have felt more and more since I first read Herbert Spencer, in all social questions, the most complicated phenomena on earth. We know little about them and their relationship to one another beyond a most superficial acquaintance. Hence, we should spend our time in an attempt, at first hand, to obtain information as to what conditions really are and how they are inter-related, rather than on the basis of general principles, pass our days in proclaiming the virtue of some social nostrum which is at best but a half truth.

This recognition of the extreme complexity of social and public problems has been more and more generally felt in recent years, and has resulted in a series of intensive studies of the social conditions over comparatively small areas, which are, taken together, in my mind the

[1] *Land and Labour*: *Lessons from Belgium*, by B. S. Rowntree, London, 1910.

(This article is reprinted from the *Indian Interpreter*, Madras, Vol VI, 1911. No. 2, pp 79-83.)

most important contribution to Sociology in our time. The 'intensive' study of social conditions altogether is a comparatively new thing. The first man of any importance to apply it was the French student LePlay about the middle of the last century, but an immense increase in activity and in closeness of study of social problems has taken place in the last ten or twelve years, and nowhere more than in England and America. The immediate exciting cause was the wonderful study of social conditions in London made by Charles Booth about fifteen years ago. But the closeness with which the social and economic state of small groups of people has been examined has immensely increased since then. Rowntree's study of the city of York[1] is perhaps the best known and in some ways the most striking and perfect of these. In that study, he obtained information about the economic condition of every working-class family in the city, and by comparing it with recognized standards, was able to show that necessary irreclaimable poverty was the heritage of a far larger proportion of the people of urban England than had previously been recognized. His results indicated that nearly one-third of the people of a typical English town were below the 'poverty line', that is to say, lived below the standard necessary to produce physical efficiency. By a similar study, though not such a complete one of a typical English village, I was able to show[2] that the conditions of rural life in England were equally bad, if not worse.

These results have, on the whole, been confirmed by an immense amount of similar information collected since, and have led to a very earnest discussion as to why this state of things should be, and how it can be remedied. It is generally recognized that the terrible spectacle of one-third of the population of the richest country in the world living below the poverty line is a matter of extraordinary

[1] **Poverty: a Study of Town Life,** by B. S. Rowntree.

[2] **Life in an Agricultural Village in England**, by H. H. Mann. Sociological Papers, 1904 (Macmillan & Co.)

seriousness. It is also recognized that it is due to the unsatisfactory distribution of wealth. But how has this unsatisfactory distribution arisen? Can it be remedied? If so, the devising of a remedy depends on a knowledge of the cause of the disease.

And, here, come in the principal differences of opinion. Some blame the land laws, and the concentration of the land in the hands of a few persons. Some consider lack of mobility on the part of labour as the fundamental evil. Some think that the bulk of those who lie in the submerged strata of the country are chiefly inefficient degenerates, and should never have been born. And there are many other explanations. The din of the discussion is heard everywhere among those to whom social evils are a matter of real concern. As a discussion, it will do little good; if it induces further investigation which will lead to action on a sound and certain basis, it will be of immense value.

I have before me a book which embodies the results of one of these special investigations, undertaken in order to get a clearer insight into the question of poverty under western conditions, and its cause. Mr. B. S. Rowntree, to whom the raising of the question in its present urgency is largely due, was anxious to find out its relationship to the question of the tenure of land. Many of the social and economic evils revealed by Rowntree and others have been put down, as I have already noted, by a very large school of thinkers to the restricted ownership of land in England. How far this is true can only be settled by examining the social conditions in other countries where the system of land tenure is different. Hence, Mr. Rowntree has devoted himself for five or six years to a close study of the position in a country, near to England, sufficiently small to be studied in detail, where conditions vary sufficiently from those in England to enable a satisfactory comparison to be made. Belgium was chosen for this purpose. Small, compact, easy of access, definitely modern in every sense of the word, it was admirably suited for the matter in hand.

And yet such an inquiry was laborious enough. Belgium is a small country, but it contains more than eleven thousand square miles and has a population of seven millions. Its conditions do not vary as widely as in large countries, but they are widely enough different to make it a matter of impossibility to argue from one part of the country to another. The study which has been made involved first, an inquiry as to the number of landowners and the size of their holdings throughout the country and the amount of their debts, a set of figures not available in any European country; then it meant a detailed investigation of the wages paid, the hours of work, and the conditions of labour in the chief industries of the country. Then again a personal and first-hand knowledge was required of the way in which the people engaged in agriculture live, how far they have secondary occupations, how far they own their own land, how much the condition of those who rent the land differs from that of those who own it. Still further Mr. Rowntree found that he needed to know the price and rent of land and its changes in recent years throughout the country. And similar investigations were made with regard to the facilities of transport or, in other words, the mobility of labour, the system of taxation, the general level of education, the general standard of life, including the character of the housing, and many more subjects.

Reading a programme like this, one is tempted to exclaim that an inquiry of this size could only be made by Government, and yet a Government *will* never make such an inquiry. If they did, I am afraid the results would be of small value, because the agents employed by a Government are not such as to be able to obtain much more than mere mechanical data. If it is worth while to find out these matters, it must be by private agency, assisted, as in the present case, by the Government so far as was possible, but essentially, nevertheless, conducted by private workers. Its cost, however, is enormous. Only rich men like Mr. Rowntree or a society organized for the purpose could do it. But the results are bound to be of immense value in attempting the task of deciphering the meaning of the social system and its defects in a country, a province or a city. The work in the present case took four years, and though the mass of data obtained is enormous, and the discussion masterly, yet the author would be the first to recognize that it only represents the beginning of the study of social and economic conditions even in a small and compact country like Belgium.

I wish that time and scope allowed me to consider and discuss the conclusions to which Mr. Rowntree's study leads. That Belgium is a country of small holders, that co-operation has developed as perhaps nowhere else in the world, that the mobility of labour has been made almost complete, that the land produces larger crops per acre than in any other country in Europe—and yet that the price of land is very high, that wages are low, that the standard of living is beneath that of several adjoining countries, that poverty abounds and the relief of the poor is unsatisfactory —all these and many others are conclusions which lie on the surface. The greatest value comes in when one considers the relative effect of causes in producing results, in other words, when the comparative study of the social conditions in Belgium and elsewhere is taken up. And here I have no space to follow Mr. Rowntree, but would commend the increasing number of those to whom these questions are of vital importance to study the book now before me.

But, before concluding, I want to insist on one thing, because it is important in India. We shall never, in this country, get beyond vague generalizations with regard to the social and economic conditions of the people unless we face the problem and go and get the facts at first hand for ourselves. The atmosphere has been ringing for many years with the battle cries of opposing economic views. The Government, in dozens of official publications, maintain that India is getting richer and that the economic condition of the people is improving. On the other hand, the popular leaders and such responsible men as the late Mr. Romesh C Dutt maintain that the mass of the people are becoming poorer, and that India is being slowly drained of its wealth. Which is true? There is evidence on both sides. I have never seen Government completely answered. I have never seen the arguments of the critics fairly met. The only way to get evidence of real value is a first-hand study of preferably a small area, on somewhat the lines indicated in the work under notice. Up to the present practically nothing has been done but to pore over the pages of Government reports and come to conclusions—often to absolutely opposite conclusions—from data collected for an entirely different object. Even a study like that of Theodore Morrison[1] on the industrial conditions in the United Provinces is extremely slight, and savours rather of the study than of the field.

But we must, I repeat it, face matters. Take a taluka, a town, a village even, and work up the actual living conditions there, and you do more to ascertain what really exists and to bring about beneficial changes, than perhaps anything which can be done at the present moment. Government with its mechanical arrangements, almost wooden in their lack of elasticity, cannot do it. It must be the work—very monotonous work—of those who—with a vision set on a future far ahead, but a future in which India shall be the best that India can be—are content to toil personally and with enthusiasm in finding out the present condition of all, aye, even the lowest classes of the people.

[1] Theodore Morrison. *Indian Industrial Organisation*, London: J. Murray.

*Chapter 3*

# HOW TO CONDUCT AN ECONOMIC SURVEY[1]

I feel that it is rather presumptuous on my part to attempt to speak to-day on how to conduct an economic survey when my own experience of making such a survey is a generation or more old. But I have ventured to speak to you on the subject precisely because I came into the carrying out of such surveys when they were much less common than they are to-day, when there was little guidance as to how to carry them out or how to draw conclusions from them when they had been carried out by others. At the turn of the last century very few such surveys had been made, and, when I found that this was the only way in which a real understanding of the actual condition of the population could be obtained, I started on the first serious study of this sort that I had undertaken.

But before I tell you about my early attempts and failures, and the conclusions to which they led me, may I enlarge for a few minutes on certain principles which have nothing to do directly with economics, but apply equally to all forms of scientific enquiry, though perhaps more importantly to economic and social studies than to any others. After I graduated at Manchester, I was awarded a post-graduate scholarship which enabled me to spend a year at what was perhaps the greatest research centre in Europe, in Paris. That year, in contact with some of the greatest research workers in the world, emphasized three principles which have guided all my work ever since, and which seem

[1] Reproduced from the manuscript of a talk to the Seminar of Dr. Anstey at the *London School of Economics*, Spring, 1961.

to be perhaps more essential to economic and social research than to any other, though it was a group of biologists who gave them to me. These were as follows:

(1) It is wrong, as well as unscientific, to start a study with a definite purpose of proving any theory which may have impressed itself on you beforehand. One of the greatest of my Paris mentors put it this way: "We forecast no results. In every hundred thousand experiments one succeeds. We anticipate nothing. We simply report what we find." You will see the point. In so many studies in both arts and science one feels that the worker has already decided the results he wants to get before he starts. Macaulay wrote history to prove that the Whigs were always right, and, to a less extent, Gibbon wrote history to prove that the second century A.D. was the happiest time in European history. In economic matters so many studies have been made to prove that the only method of advance is by co-operative methods; it may be so, but it is a fatal fault to start work in a survey to prove it.

(2) Whatever results you obtain in any study, which appear to have a simple explanation, it is wise to doubt that explanation. It is much more likely that the results obtained have a more complicated explanation. As one of my Paris mentors stated: "If there seem to be two explanations of your results, one of which is simple and the other more recondite, always follow up the more complicated one first". That reminds me of another saying, namely, that inexperienced workers at scientific problems often make the biggest discoveries, but it is unfortunate that the discoveries are so often not true.

(3) It is wrong to be swayed by the fashion of the moment or by what the official opinion is about any matter. I remember discussing with a colleague, certain scientific results—(they were not of an economic nature)—in which I was sceptical about what was the current opinion about a certain scientific theory. He said to me, after I had expressed my scepticism: "Did the authorities who appointed you to your present position know hat you held

these views?", evidently meaning that, if this had been known, I should never have been appointed. In another case, and this was an economic one, I was asked by an eminent economist, whether it was wise to publish results which seemed to be against the tendency of government policy. It would perhaps have been more politically wise if I had pigeon-holed my results, but political advantage should not, I think, affect conclusions which seem to follow from results that one has obtained.

From these general principles which apply to research in all science and all surveys, I will now pass on to the economic and social surveys which I have had the opportunity to do in the last sixty or so years. Though I was a scientific worker in a very different field, namely the borderland between chemistry and biology, yet I have been always very interested in social and economic questions, and was inspired at the turn of the century by Rowntree's study of the City of York from an economic point of view. Rowntree, you will remember, gathered data with regard to the economic condition of almost all the inhabitants of that city, which had, at that time, about 60,000 inhabitants. Rowntree showed, as a result, that about 40 percent of the population of a prosperous English city must be living below a very meagre standard of life, and were, hence, in what he called "primary poverty". His work inspired me to see how far these results would also be obtained in an agricultural village and in 1903, with the help of several friends, I organised a similar study of such a village. In doing so, I came up against a couple of what I have since found to be very essential difficulties in such a study. The first of these is the failure of the so-called "census" method of getting accurate data except for the simplest facts, such as the relative number of males and females, the size of families, or the age structure of the population. When one attempted to get data as to economic position, a census enumerator would be apt to get answers to his questions which would be far from an accurate presentation of the facts. In other words, it is only when the investigator is closely in touch with the people, and is quite familiar with their background, that the data obtained are likely to be correct. Inaccurate data are worse, I feel, than no data at all.

The second point which I realised at that early period was that one can only get the data that one can collect by using people who are well known to the people themselves, and in whom they have confidence. In the survey which I made in 1903 in an English village, I was lucky enough to have as colleagues two friends who knew the village and its people even better than I did, and I was thus able to get material very much better than I could have done myself. I was thus able to get an approximate budget for practically every working-class household in the village, and I was able to compare these with a very austere budget based on that used a year or two previously by Mr. Rowntree in York. As a result I showed that, at that period, the proportion of the population living in "primary poverty" was not very widely different from what Rowntree had found for a city, that is to say that 40 percent of the population living in an agricultural village in midland England were below the poverty level. The results of my study were published by the Sociological Society and I do not think that they have ever been controverted, though I myself was strongly criticised for what I concluded in my report.

My next efforts were in India. It was a good long time before I felt I knew conditions well enough to do any survey work. There was, too, the language difficulty which I never completely got over. But I did, while in Calcutta, make a very interesting survey of the drink shops of Calcutta, in which I was able to make what I fear was a very superficial study of the types of people who visited and used these shops. But my next serious survey was, after I removed to Poona, a study of the conditions, economic and other, under which the depressed or untouchable classes of Poona City lived. In order to do this work, it was necessary to get into fairly intimate touch with the people I was intending to study. It took me several years to get that intimate touch, but, ultimately,

I had the co-operation of several members of the untouch able classes themselves. Without them, I could not have got the data that I needed, for I found the people very suspicious of strangers. In the areas where they lived in Poona City, very few outsiders ever penetrated, and then they were usually officials who were there on official business and went away as soon as possible.

I did this as an individual and not as the envoy of any society or institute, though I had the patronage of the so-called Depressed Classes Mission, and of its then head, the late Mr. V. R. Shinde. I rather want to insist on this, for I doubt whether, as the envoy of an institute, I could have got the co-operation of the people that I certainly did have. I was able to get data for 1400 households who formed perhaps 80 or 90 percent of the depressed population of Poona City. In a few cases, I was able to get details of expenditure of a family, though with a largely illiterate body of people, it could only be done in a very limited number of cases. In one, fairly representative case, where three members of a family (two women and one man) with at that time an income of Rs. 17/1/- per month, I found that they would be short, on their own estimate, about Rs. 6 per annum, which would be saved chiefly on the amount to be spent on clothes. As regards food, I made the following notes at the time, and I am wondering how far they still remain true nearly fifty years later:

"The simplicity of the food, in which the price of grain is a dominant factor to an extent almost inconceivable elsewhere: the gradual introduction of tea as a luxury for very special occasions: the large amount of expenditure on sugar (i.e., *gur*) which is, nevertheless, I believe, one of the most economical parts of the expenditure: the absence of butter or ghi from the diet, and its replacement purely by a vegetable oil; these are all factors which may be regarded as almost universal. The large expenditure on alcoholic liquor is particularly noticeable, but known to exist among those where the religious influence does not prohibit, though it discourages indulgence . . . The indulgence in liquor is the weakest economic factor in the budget, and makes what would otherwise be a fairly satisfactory economic position into one which is certainly not satisfactory."

Soon after I had completed the survey to which I have just referred—in the year 1914, to be exact—I had gathered round me a group of men who were very anxious to do something for the improvement of the rural conditions in the Deccan, and I suggested to them that a preliminary to any such work was to gather information as to what were the actual conditions in the villages which were accessible to us in Poona. Nobody seemed to know, for the information on almost every aspect of village life was very meagre. Nor was the information regarding the yield of crops, the effect on productivity of well irrigation, the rotation of crops and the varieties cultivated much more than superficial. On some matters the information was good. The character of the soils of almost any village was very carefully studied in connection with revenue settlements, and I fancy that there are few areas in the world where the quality of the soil of every field is known, if trouble is taken to look it up. As regards the people, considered either as individuals, or as households, very little seemed available beyond that collected for the several censuses, and even this was usually obtainable in a summarised form.

So we set to work, and I cannot speak too highly of the work done by my colleagues. We had no money and no economic institute behind us, and all I was able to offer was to liberate all the members of the group from their usual work on Saturday mornings. We decided to work on a village near Poona, but which was not materially influenced by the neighbourhood of the city, which was near enough to be accessible to a party whose sole means of transport was the bicycle. I do not think we thought of it as necessarily typical but we knew that we could find dozens of others which would not differ in general character. My colleagues set themselves to make friends of all the people of the village including the headman and other influential members of the village society, and then the work of study was divided up between the party. One studied the soils, the water supply, the wells and the water in them, another considered the crops and the rotations both as to yield and as to their economic advantage, another got into special close relations with people of every class and caste, and they were particularly successful in gaining such confidence that they were able to get reliable information on matters which are usually not revealed to strangers.

With all this it took us about two years to gain the information that we wanted, and I remember that I began to write up the whole results in the middle of 1916.

*Chapter 4*

# THE INVESTIGATION OF ECONOMIC CONDITIONS IN UNDERDEVELOPED COUNTRIES[1]

What is an underdeveloped country? The answer that springs to mind is that it is one in which the production, whether agricultural or industrial, is less than is easily possible. But this answer is far too simple. Production in many of the so-called advanced countries is far below what might easily be obtained—in fact, in this sense, many of the most prosperous areas of the world are hopelessly underdeveloped. In practice, however, the term "underdeveloped" is employed almost entirely as descriptive of regions where there is a peasant economy; where the primary purpose of land cultivation and of local industry is subsistence; and where production, whether by cultivation or otherwise, for a market, i.e., for sale, has a subordinate position in the minds of most of the population. In many regions, including those where the largest part of the world's population live, the usual assumptions of Western economic thought are by no means dominant. The greater part of what is produced is for use and not for sale, and so never leaves the local area; and a money economy, though it exists for certain purposes and at present is extending its scope almost everywhere, nonetheless forms a subordinate part of the local economy.

1 Reprinted with permission from *Economic Development and Cultural Change*, Chicago, Vol. X, No. 1, October, 1961. pp 102-104.

Most of India is typical of the areas where these characteristics prevail, although similar conditions obtain in much of South and Southeast Asia. Hence, any serious attempt to study economic conditions in such a region is very welcome. Obviously, the methods used in such a study will be considerably different from those employed in inquiries about areas where the assumptions of Western economic thought are applicable. The book here under review[1] is an attempt to formulate appropriate methods for the former kind of investigation. It has been issued by the leader of a group of economic workers in Delhi who have applied these methods to a number of villages in northern India. The methods may be criticized as far as the collection of data is concerned, as I shall show below; but they do form an honest attempt to devise a scheme for such study. This is by no means easy, for the data required for any reasonable conclusions are very difficult to obtain in villages whose populations are largely illiterate; where answers to even the most carefully arranged questionnaires are likely to be very far from the real facts; and where no written records are kept, except by the minor officials responsible and by the few merchants and money-lenders resident in the village or its neighbourhood.

Even if these difficulties are partially overcome, another is inherent in the study. The investigators come from a town centre and have been brought up on the views of more or less orthodox economics; they will find it very difficult to see with the eyes of a villager, to whom their ideas are quite strange and very remote from his conception of the situation. Although praiseworthy attempts have been made to overcome this obstacle by living among the people in an Indian village and trying to grasp their outlook on life, a tendency generally remains to think in terms rather of production for a market and the advantages of growing crops which yield a larger monetary return, than of those which are required for subsistence.

[1] P. K. Mukherjee, *Economic Surveys in Under-Developed Countries: A Study in Methodology*. Bombay: Asia Publishing House, 1960. Pp. xii + 263, $ 5.50

This seems to me to be the real difficulty in getting to know the economic condition of an Indian village. The author of this book hardly seems to face the problem seriously enough. He mentions it, of course; but he appears to consider it possible to train the field staff who would collect the data a good deal more easily than I have found to be the case.[1] Dr. Mukherjee does state that it took the Indian Statistical Institute eight to ten years to create a pool of workers competent to collect data from villagers; and he quotes with approval Beatrice Webb's requirement of special preparedness in the mind of any investigator of social and economic conditions. However, I doubt that he has insisted enough on this. In India, the urban-minded educated classes are more widely separated from the peasants in the villages than in almost any Western country with which I am acquainted. It would be a rare urban individual in India who could forget his town outlook and really perceive conditions in a village as the villager himself sees them.

All this suggests that the collection of economic and social data in a peasant economy like India's is, except for the very simplest items, not only difficult but also very prone to extremely large errors. Even in an inquiry about the total agricultural production of a village, the results obtained from answers to any questionnaire may be very wide of the mark. The average cultivator himself knows only very approximately what the yield has been of any but cash crops grown for sale; and any inquiry tends to be vitiated by the suspicion that it may be connected, even very indirectly, with the assessment of government revenue. Any estimate of debts and other similar liabilities, however carefully checked, may have little relation to the facts. Even in such matters as the birth and death rates, the figures supplied can rarely be trusted. Those for the death rate are better than those for the birth rate; and the latter, in my experience, are most unreliable in India.

1 *Ibid.*, p. 195.

If we turn to two examples of the application of the methods described by our author in two other recent works, we find that the difficulties of collecting material are very great. It is clear that some of the conclusions are much more limited in scope than one might have hoped. Moreover, the questionnaires on which they are based contain many questions to which it is impossible to get correct answers, largely because the informants do not know the answers. As a result, the answer—if any—given to a question may be one which has been suggested by the questioner. In Gupta's account of the village of Shamaspur, over 200 questions are listed, many of which could not be answered by the majority of peasants in an Indian village.

My own criticism of the methods employed or suggested in this book is chiefly concerned with the collection of the data. Unless the investigator can identify himself with the fundamental ideas of a peasant community, and unless he can limit his inquiries to matters on which the village people themselves have the information required, the conclusions drawn are very likely to be erroneous. They can thus lead to extremely questionable legislative or other action.

Having obtained the facts by field study, the author deals with the examination of the results by the usual statistical methods. These, as described, are strictly orthodox; and, if the data obtained are correct, the general conclusions would be justifiable—subject, of course, to the estimation of errors. There can be no criticism of this part of Dr. Mukherjee's book. But the estimation of errors is so unsatisfactory that rigid conclusions may be neither possible or justified.

The picture of an Indian village as a dynamic unit—the goal of studies like those suggested in the present book—*can*, I believe, be obtained by closely concentrating workers of very special character on a very limited area. The census

[1] P. K. Mukherjee and S. C. Gupta, *A Pilot Survey of Fourteen Villages in U. P. and Punjab* (Bombay: Asia Publishing House, 1959) and S. C. Gupta, *An Economic Survey of Shamaspur Village* (Bombay: Asia Publishing House, 1959).

method of going over very large regions can yield satisfactory results only when one deals with very simple questions. Any attempt to apply it to the collection of intimate details about the life of an Indian community is doomed to be disappointing, if it is used to get any idea of the social conditions in a peasant village. Such an effort has been made in the very elaborate Rural Credit Survey, organized some years ago by the Reserve Bank of India. The results have, I fear, been discredited in both theory and practice, in spite of the large amount spent in obtaining them. Other large-scale studies on similar lines also seem likely to be condemned and to lead to mistaken conclusions.

However, I can congratulate Dr. Mukherjee on having given, in this work, the best account of methods which can be applied successfully in many rural areas in India—if the men to conduct them can be found. These men must be able to put themselves into the minds of the people they are studying. With a body of such workers, and repeated examinations at intervals of the same areas, we shall be able ultimately to obtain a dynamic picture of villages in India as economic units and as social centres, and of changes in their position, as the years go by.

**PART II**

**VILLAGE STUDIES IN INDIA AND ENGLAND**

*Chapter* 5

# LIFE IN AN AGRICULTURAL VILLAGE IN ENGLAND:

## A STUDY IN ECONOMICS*

During the past few years the whole question of the economic condition of the population of the cities and towns in England has come before the public in a measure far greater than at any earlier time. This result has been largely due to the admirable investigations made by Mr. Charles Booth in London, and, in an even more thorough manner, by Mr. Seebohm Rowntree in York. To complete the chain of evidence as to the economic position of the people it seems of great interest and importance to follow the methods used by these investigators in obtaining evidence as to the life in our agricultural villages in different parts of the country. The following is an attempt to obtain, in so far as my spare time and that of my coadjutors allowed, and to set out information obtained during the autumn of 1903, as to the condition of the population of a Bedfordshire village, which lies in the centre of one of the largest purely agricultural districts in England.

Ridgmount, the village in question, lies from one to two miles from the station of the same name on the Bletchley and Cambridge branch of the London and North-Western Railway. It lies, moreover, twelve miles from the country town of Bedford, and sixteen from the

* Reprinted with permission from the Sociological Society, **Sociological Papers**, 1904 (London, Macmillan 1905), pp. 163-193.

industrial centre of Luton. The village is bounded on one side by the Woburn Park of the Duke of Bedford, who is the greatest landowner, house-owner, and employer of labour in the district. Until a few years ago, however, a very considerable amount of freehold land existed in the village, but by slow stages this is being bought up, and apparently, if the present process of absorption goes on, in a very short time the whole parish will be in the hands of the Duke of Bedford.

The parish is an old one. A pre-reformation church exists, but has been replaced during the last century by a new building, and the old one is only now used for burials. There is also a Baptist Chapel, founded, so it is said, by John Bunyan himself, and also a Wesleyan Chapel. There are three farms in the parish, and several more just adjoining the border and even partly contained in it, as well as a portion of the Duke of Bedford's Park. I have not, however, dealt in my inquiries with the whole of the population of the parish, as about one hundred people live in farms and cottages far removed from the village itself, and it seemed neither possible nor advisable to complicate the data obtained by the introduction of these outlying members of the community. The figures and arguments which follow are therefore strictly limited to the population resident in the village itself, and are thus concerned with a population of 467 residing in 127 houses.

The village was chosen for investigation partly because of its purely agricultural character, and partly because it was the one most typical of the character of the surrounding districts with which I was acquainted. The lower land in the parish is a heavy clay, but as one ascends to the higher sections it becomes much lighter in character. The farms, on the whole, are about equally divided between arable and grass land, and from an agricultural point of view there is little to distinguish the place. From two of the farms, I understand, a little milk is sent to London. The markets attended by the farmers are Bedford

(Saturday), Leighton Buzzard (Tuesday) and, occasionally, Ampthill (Thursday).

Almost all the village people are, therefore, employed directly or indirectly in agricultural pursuits. The sole exceptions of any importance consist in the residence of two railway signalmen in the village, and of one man and three lads who work in a printing works at Aspley Guise, two-and-a-half miles distant. At one time the village was a large centre for pillow lace-making, and for straw-plaiting. Of these I shall have more to say a little later on, but, for the present, it may be pointed out that the latter has entirely and the former nearly died out.

Before any conclusion could be drawn as to the economic condition of the people, it was necessary to find the minimum standard, on the basis of local prices, of life, consistent with the physical efficiency of the population; and I have, thanks to Mr. Rowntree, been able to fix this with much greater ease than could have been done prior to his investigations. It will be wise to divide the consideration of this question, as he does, into three parts: (1) the necessary cost of food, (2) the necessary cost of rent, (3) the necessary cost of clothing and household sundries.

*Food.*—To take first the question of food. The local prices for materials in the autumn of 1903 are set out in the following list, and, alongside, are placed the prices found by Mr. Rowntree to be current in York in the autumn of 1899.

In this list I have set out Flour and not Bread, in order to bring it into comparison with the prices at York: all bread in the village is however bought from the baker; none is made at home. Owing to the presence of the County Council Farm School in the village (founded by the Duke of Bedford), separated milk is obtainable easily and cheaply. The remainder of the articles of food are bought usually from Luton firms who send vans round several times each week to these villages. No meat is specified in this list, my object being to get a minimum ration, so far as cost was concerned, and yet one which would be sufficient for physical efficiency. Meat however costs quite as much as in a city like York.

| | Ridgmount, 1903. | | | York, 1899. | | |
|---|---|---|---|---|---|---|
| Flour ... | 1s. 8*d.* | per | stone. | 1s. 4*d.* | per | stone. |
| New Milk ... | 1½*d.* | „ | pint | 1½*d.* | „ | pint |
| Skim or Separated Milk | ½*d.* | „ | pint | ¾*d.* | „ | pint |
| Oatmeal ... | 2*d.* | „ | lb. | 2*d.* | . | lb. |
| Bacon ... | 8*d.* | „ | lb. | 6*d.* | „ | lb. |
| Cheese ... | 7*d.* | „ | lb. | 7*d.* | „ | lb. |
| Sugar ... | 1½*d.* | „ | lb. | 1¾*d.* | „ | lb. |
| Potatoes ... | ½*d.* | „ | lb. | ½*d.* | „ | lb. |
| Margarine ... | (Lard) 6*d.* | „ | lb. | 8*d.* | „ | lb. |
| Butter ... | 1*s.* to 1*s.* 4*d.* | „ | lb. | 1*s.* | „ | lb. |
| Biscuits ... | 4*d.* | „ | lb. | 4*d.* | „ | lb. |
| Cocoa ... | 1*s.* | „ | lb. | 1*s.* | „ | lb. |
| Tea ... | 1*s.* 4*d.* | „ | lb. | 1*s.* 5*d.* | „ | lb. |
| Coffee ... | 1*s.* | „ | lb. | 1*s.* | „ | lb. |
| Treacle ... | 3*d.* | „ | lb. | 1¾*d.* | „ | lb. |
| Onions ... | 3½*d.* | „ | lb. | 3½*d.* | „ | lb. |
| Currants ... | ½*d.* to 1*d.* | „ | lb. | ½*d.* | „ | lb. |
| Suet ... | 6*d.* | „ | lb. | 8*d.* | „ | lb. |

The figures given approached so near to those given by Mr. Rowntree that I decided at once to adopt his standard of necessary minimum cost for food, for the maintenance of physical health, which is as follows:—

| | | |
|---|---|---|
| Man ... ... ... | 3/- per week | |
| Woman ... ... ... | 3/- „ | |
| Young person over 16 ... ... | 3/- „ | |
| Child 8 — 16 ... ... | 2/3 „ | average 2*s.* 3*d.* per week. |
| Child 3 — 8 ... ... | 2/1 „ | |
| Child 0 — 3 ... ... | 2/1 „ | |

In this it is supposed that all necessaries are bought, but allowance is made later on, in considering revenue, for materials produced on allotments or obtained otherwise, so far as was possible, in each individual case.

*Rent.*—The lowest rent is about 1*s.* per week, but this is for very wretched cottages. As a general rule, there are two standards for rent. The first is that of the Duke of Bedford's cottages, in which a good four or even five roomed cottage is obtainable for 1*s.* 6*d.* per week including rates, and in fact, there is hardly any higher rent. In these cases the Duke has recently compelled the tenants to pay the rates, and reduced 1*s.* 6*d.* rent to 1*s.* 4*d.*, as compensation. In the present paper I have, however, taken the old rent, which included rates, as the actual rent. The error, if any, will make the rent slightly too low.

The other class of houses are those *not* belonging to the Duke of Bedford, which are alone open to those whom he will not have as tenants for any reason whatever. They are generally higher in rent for similar accommodation, not much of any value of more than three rooms being obtainable for 1/6. On the whole I have thought it fair to take the following as the minimum adequate rents, merely noting that accommodation is increasingly limited, as houses are being pulled down, thrown together, or closed when the land passes into the Duke of Bedford's hands, and no new ones have been built for a good many years.

For one person it is possible 1*s.* per week might be a sufficient rent.

For a family up to man, wife and 6 children, 1*s.* 6*d.* week would be a minimum.

For a family exceeding this number 2*s.* per week would, at least, be necessary.

*Household Sundries.*—On inquiry I found that Mr. Rowntree's standard of 6*d.* per week for a man or woman, and 5*d.* per week for a boy or girl under 16 years of age, for clothes, was regarded by the people as an absolute minimum, and I have therefore retained these figures.

As far as fuel is concerned, it is considered that if the amount of wood, brush, thorn, &c. which could be picked up,* be taken into consideration, 1*s.* per week throughout the year would probably be enough to allow as the necessary expenditure for coal per household, independent of the number of its members.

Beyond this, 2*d.* per head per week has been allowed for other sundries such as soap, light, furniture, crockery, and similar articles.

Taking the above expenditure together we have, as the minimum necessary expenditure per week, as follows:—

| Family. | Food. | Rent. | Household Sundries. | Total. |
|---|---|---|---|---|
| 1 Man, or 1 Woman | 3/- | 1/- | 1/8 | 5/8 |
| 1 Man, and 1 Woman | 6/- | 1/6 | 2/4 | 9/10 |
| 1 Man, 1 Woman, 1 Child | 8/3 | 1/6 | 2/11 | 12/8 |
| 1 Man, 1 Woman, 2 Children | 10/6 | 1/6 | 3/6 | 15/6 |
| 1 Man, 1 Woman, 3 Children | 12/9 | 1/6 | 4/1 | 18/4 |
| 1 Man, 1 Woman, 4 Children | 15/- | 1/6 | 4/8 | 21/2 |
| 1 Man, 1 Woman, 5 Children | 17/3 | 1/6 | 5/3 | 24/- |
| 1 Man, 1 Woman, 6 Children | 19/6 | 1/6 | 5/10 | 26/10 |
| 1 Man, 1 Woman, 7 Children | 21/9 | 2/- | 6/5 | 30/2 |
| 1 Man, 1 Woman, 8 Children | 24/- | 2/- | 7/- | 33/- |

If a child be replaced by an adult in this table, 10*d.* must be added to the allowance (9*d.* for food, 1*d.* for clothes) per week.

If we now compare these figures with those in York, it is here found that a man, wife, and three children will need 18*s.* 4*d.* per week at least to keep them in physical efficiency, while in York 21*s.* 8*d.* was found by Rowntree to be required, the difference being accounted for by lower rent and lower cost of fuel.

*"Primary"* Poverty is here taken to be that poverty caused by an insufficiency of earnings, even when most economically applied, to provide for physical efficiency. *"Secondary"* Poverty is here taken to be that due to an uneconomical application of earnings.

* There is no right to a certain amount of faggots yearly in Ridgmount such as exists in several adjoining villages.

Before giving the figures of the proportion of poverty of either kind in the village, I will indicate the additional possible sources of income, over and above the actual wages earned, obtained by the people of the village. In all the cases cited below, I have indicated every family or household by a number which will serve to identify them in my lists.

The principal of these additional sources are as follows:—

(1) *Money sent home by children.*—This occurs in a few cases only in which the amount estimated to be received has been allowed for. The following are some of these.

*No.* 12 is a family in which one daughter is away at service, but has quite enough to do to keep herself. I know the family well, and the amount of income to her parents from this source is negligible.

*No.* 46 is an old widow. In conversation during December 1903 she told me that her parish allowance just paid for rent, fire, and light, but that her children were good to her, though she found it hard to get along.

*No.* 52 is a tailor and wife with two daughters away in service. I know the family well, and it is practically certain that nothing is received from this source.

*No.* 64 is a widow and daughter, who are raised substantially out of poverty by children living away in the town. The amount varies, and cannot be exactly got at.

(2) *Allotments.*—In view of the fact that a great deal has been made of the profits to be obtained by holding an allotment, it is interesting to have obtained the views of some of those who hold them in the village. This particular village is, however, not quite a typical case, for most of the allotments lie too close to the Duke of Bedford's park, where game is strictly preserved, and the result is that havoc is usually wrought among the crops sown. Corn of any sort is, in fact, rarely grown here, and the crops are limited to potatoes and a few other vegetables. Potatoes

are considered the most profitable crop, and hence I have taken these in measuring the actual value of an allotment.

Taking these special conditions into account the following estimates of allotment value may be considered as perhaps typical, the first of a piece of better land, the second of a piece of very poor land near the park and its game. In each case the acreage is twenty poles or one-eighth of an acre.

No. 1 (fairly good land).

*Average crop.*—27 bushels potatoes at 2/- per bushel £2 14*s.*

*Expenses.*

| | | | |
|---|---|---|---|
| Rent (1/8 to 3/4) ... ... | say | 3*s.* | 4*d.* |
| Manure ... ... ... | say | 10s. | 0*d.* |
| Seed ... ... ... | say | 3s. | 0*d.* |
| Digging land and potatoes ... ... | say | 10s. | 0*d.* |
| | £1 | 6*s.* | 4*d.* |

This leaves a profit on the allotment of £1 7*s.* 8*d.* per annum, or say 6*d.* weekly throughout the year.

No. 2. (bad land).

*Average crop.*—14 bushels potatoes at 2/- per bushel £1 8*s.*

*Expenses.*

| | | |
|---|---|---|
| Rent ... ... ... ... | 3s. | 2*d.* |
| Manure (2 loads every 2nd year) ... | 5s. | 0*d.* |
| Seed ... ... ... ... | 3s. | 0*d.* |
| Digging land and potatoes ... ... | 5s. | 0*d.* |
| | 16*s.* | 2*d.* |

This leaves a profit on the allotment of 11*s.* 10*d.* per annum, or say less than 3*d.* weekly throughout the year.

It may be stated that these estimates were given to me by actual allotment-holders of many years' standing.

The following notes may be made with regard to several of these items of expenditure.

*Rent.*—This varies very much with the class of land and may run, for twenty poles, from 1*s*. 8*d*. to 3*s*. 4*d*. per annum. It will be noticed that this rate (even the higher one) is lower than the average for agricultural land in the neighbourhood, but as a matter of fact much of the allotment land could not be let agriculturally. Some choice pieces go up to 6*s*. 8*d*. for twenty poles.

*Manure.*—It is commonly stated that if a cottager keeps a pig, this will manure his allotment. In this village however pigs are not encouraged by the principal landlord, and, in any case, as the cottagers have to buy the straw for litter, they gain little in this way. See below however. Most of the allotment-holders buy the manure, and for a full crop two loads at five shillings per load will be required for a twenty-pole allotment. Less than this will mean a smaller crop of potatoes.

*Seed.*—Most of the seed is got by exchange or the old seed is used over again. But a little seed always must be bought, and my informants put this as needing about 3*s*. per annum for one-eighth of an acre.

*Labour.*—Again and again I was informed that one man cannot work an allotment, and do full time at his regular work. If he does do it, he has to take two or three days off twice a year at least. The allowance of 5*s*. given by my second informant is regarded as a minimum to be paid for hired labour, or lost by stopping away from regular work.

Taking all estimates received into account, I have allowed 6*d*. per week as profit on a 20 pole allotment, and less or more proportionately for smaller or larger ones respectively. To show that the game trouble is a real one, I may mention that one field near this village was set apart by the Duke of Bedford for allotments just outside the park. The number taking up allotments became, however, less and less, and it has now been attached to the adjoining farm, and taken out of allotment culture. On enquiring the reason, I found that the cottagers with one voice declared that owing to game the crops were so damaged that they did not pay the cost of working.

(3) *Harvest Money*—To labourers on farms in these districts the harvest money (both for the hay and corn harvest) has been in years past a great stand-by. This has been paid in two ways—(1) by a special payment for the corn-harvest month of double the ordinary wages, and allowance for overtime at hay-harvest, or (2) by piece-work during the harvests.

The rate for harvest-money on several farms round Ridgmount at present is as follows:—

*Corn Harvest.*—£5-10*s.* for the month, inclusive of ordinary wages, beer, overtime, &c., &c.

*Hay Harvest.*—3*d.* per hour overtime, giving from 12 to 20 hours overtime each year.

I have calculated that these extra payments to the men getting them are worth, as a maximum, 1*s.* 6*d.* per week throughout the year.

Where piecework is paid instead of harvest-money, the amount extra earned by this means has a tendency to get smaller. The men themselves working at farms where this is in vogue, tell me that they must not earn as much as they could, or the rate would be immediately lowered, and that really their piecework gives them little or no extra pay.

There is only one farm in the parish affected by this piece-work system, and I have allowed special harvest-money at the same rate as at the other farms, as I could get no reliable estimates from the labourers.

The men who work in the Duke of Bedford's park are quite different. Only those in certain departments get extras at all, but in these departments they receive 1/- per day extra in hay-harvest, and this includes beer-money and overtime. I have calculated this to be worth 9*d.* per week throughout the year.

Some of the men also, during the shooting season, are employed as beaters, and get 1*s.* extra per day for this; and one or two, specially employed game-watching on Sundays, occasionally make, by this means, an extra full day's pay.

(4) *Parochial Charities.*—The income of the parish for charitable purposes is, now, £72 12*s.* per annum. Of this one-third is devoted to church purposes, one-third to school purposes, and one-third to directly charitable uses. Of these the principal are—(1) an annual dole of 3*s.* 6*d.* per head to applicants belonging to the village at Christmas; (2) £10 paid occasionally for the apprenticeship of a boy. This is under the control of two trustees, and is only available once in several years.

(5) *Home Industries.*—Ridgmount used to have a considerable industry done in the cottages in *lace-making* and *straw-plaiting.* The latter has entirely disappeared and the former is now only carried on by a few middle-aged or old women. One woman (No. 42) maintains herself by this means, but, working ten to twelve hours per day, can only get about 3*s.* per week from the dealers. I was talking to some of the old women in the village early in December 1903, and was told that there was not perhaps one woman in the village who could earn, by working all the time, a shilling per day. 6*d.* was the usual maximum, and this would mean very hard and constant work. A poor worker will only make 2*d.* to 3*d.* per day, and the work is very trying. All the lace now made is Maltese: old point lace can only now be made by two people, and as it requires very close work it is never done. Lace-making as a supplementary source of income has however been taken note of in my tables.

*Carving* and *Carpentering* have been taught in the village for several years, at considerable sacrifice by the public-spirited teacher. There are several who make a little by this means, but only one perhaps anything material, and in this case it has been allowed for.

(6) *Pigs and Fowls.*—There are seven households of the working classes who keep a pig in the village. As a rule it is not allowed in the Duke of Bedford's cottages, unless special structural provision exists for the purpose.

The value of a pig to a household is so intimately bound up with the growth of barley on the allotment at the same time, and the provision by this means of barley-meal and litter, that in a village like the present, where the allotment-holders rarely grow barley (as, if they did, it would be destroyed by game), the profit of pig-keeping is quite different from that in one where circumstances, in this respect, are dissimilar. One case came to my notice in January 1904, where a man on killing and selling his first pig only received from the butcher six pence more for the pig than he had paid for the pig originally and the barley-meal and straw required.

As a rule the men in the village estimate the profit per pig at about ten shillings, and as two can be fattened per annum, this would amount to about one pound for the year, or say five pence per week. The value of the manure is little, if any, more than the cost of the straw-litter; that is to say, if the manure had to be bought it would cost practically the same as the straw-litter of which it is made.

Comparatively few fowls are kept by the working-classes, as here again the Duke of Bedford, I understand, does not encourage the practice. Those who do keep fowls declare the income from this source to be negligible, when they are kept as a cottager keeps them.

(7) *Odd Jobs.*—Quite a number of people supplement their income by outside work, and the boys before leaving school do odd jobs in several cases. The following is a complete list.

*No.* 17 is Parish Clerk and gets £4 per annum.

*No.* 23 helps with cleaning the church and earns about 1*s*. 6*d*. per week. The amount, however, varies.

*No.* 33 cleans the reading-room. This is often closed, and then payment is only made for special meetings. It cannot amount, regularly, to more than 1*s*. per week.

*No.* 87 acts as caretaker of the Baptist Chapel, which brings in £7 10*s*. per annum, or with additions for special occasions, to about £9 per annum.

As to boys and girls the following are the only ones, not in regular employment, who earn anything. There is practically no absolutely casual earning for boys and girls in the village.

*No.* 37.—One girl (12) is employed to help at the School and gets 6*d.* per week.

*No.* 41.—One girl (12) is employed paper-carrying and gets 1*s.* per week.

*No.* 90.—One boy (13) is employed as messenger and gets 2*s.* per week.

(8) *Fruit Farm.*—During the height of the season a few women and girls have been employed picking fruit on the Woburn Experimental Fruit Farm. This however is getting less, and less, and I was repeatedly informed that in 1903 it was negligible as a source of income in the village.

(9) *Pensions.*—Quite a number of old people are pensioners, principally of the Duke of Bedford. These have all been noted, and allowed for. The fact of the existence of such pensioners increases very considerably the number of old people in the village.

(10) *Poor Relief.*—According to a list supplied in January 1904 by the clerk to the Board of Guardians (Ampthill), there are 25 households in receipt of parish-relief in the village (and one in the outlying part of the parish, which is, hence, not included here). Three of these are on account of children, the remainder being principally old or feeble people. The amount varies from 2*s.* 6*d.* to 5*s.* per individual.

The households obtaining poor-relief contain 75 persons with a total receipt of £ 5/6/-* or an average of 4*s.* 3*d.* per family. Of these families 12 were *principally* dependent on the parish-relief with a population of 21 and a weekly receipt of £2 17*s.* This gives an average per

* This included one family with two members obtaining 3/-, who does not occur in the table supplied to me. The person here obtaining parish-relief had however died in December 1903, and I have retained the household here as my inquiries were made in October and November, 1903.

family of 4*s.* 7½*d.*, or 2*s.* 8½*d.* per individual member of these households.

I have now set out all the sources of income available in the village to ordinary working-class people, and making every allowance for all the additional sources here set out, the following table sets out the final result with regard to the primary poverty in the place:—

| | |
|---|---|
| Total population ... ... .. ... | 467 |
| Total working-class population ... ... | 390 |
| Total number of families ... ... ... | 127 |
| Total number of working-class families ... | 104 |

| | Number | Percentage on Total | Percentage on Working Class |
|---|---|---|---|
| Families in Primary Poverty | 40 | 31.5% | 38.5% |
| Population in Primary Poverty | 160 | 34.3% | 41.0% |

It may also be remarked that the average number of members to each household was as follows:—

| | |
|---|---|
| Number of members per household in total population ... | 3.68 |
| Number of members per household in working-class population ... ... ... ... | 3.76 |
| Number of members per household in total population ... | 4.00 |

The conclusion to which we come from a consideration of the figures, after every allowance has been made for subsidiary sources of income, is that no less than 34.3 per cent. of the population of a typical agricultural village in Bedfordshire do not obtain the necessary amount of money to enable them to remain in physical health. This percentage rises to no less than 41.0, when the working class alone is considered.

If we consider those just above the primary poverty line, by taking the number of those whose income does not exceed that required for physical efficiency by (*a*) two shillings per family per week (*b*) six shillings per week, (the latter being the average amount given by Rowntree and Sherwell as spent in England per family per week in drink), we have still more striking figures as below:—

| | No. of persons. | Percentage of the Wage Earners. | Percentage of total Population. |
|---|---|---|---|
| Persons below primary poverty line ... ... ... | 160 | 41.0 | 34.3 |
| Persons belonging to families with less than 2s. per week above the primary poverty line ... ... ... | 213 | 54.6 | 45.6 |
| Persons belonging to families with less than 6s. per week above the primary poverty line ... ... ... | 284 | 72.8 | 60.8 |

We will divide the causes of primary poverty into the following sections which appear to include all the cases.

(1) Death or desertion of wage-earner.

(2) Illness or old-age of wage-earner.

(3) Wage-earner out of work.

(4) Irregularity of work.

(5) Largeness of family, *i.e.,* more than four children.

(6) Lowness of wages.

Though in dividing the cases it is often difficult to say to which class a family properly belongs, yet the following table shows the *immediate* causes without much room for doubt:—

| No. | No. of Families | Immediate cause of Poverty | No. of children affected | No. of adults affected | Total umber affected | Percentage on Population in Primary Poverty |
|---|---|---|---|---|---|---|
| 1 | 5 | Death or Desertion of wage-earner. | 11 | 8 | 19 | 11.9 |
| 2 | 14 | Illness or Old-age of wage-earner. | 3 | 23 | 26 | 16.2 |
| 3 | ... | Wage-earner out of work. | ... | ... | ... | ... |
| 4 | 6 | Irregularity of work. | 13 | 14 | 27 | 16.9 |
| 5 | 4 | Largeness of family. | 25 | 8 | 33 | 20.6 |
| 6 | 11 | Lowness of wages. | 31 | 24 | 55 | 34.4 |
| | 40 | | 83 | 77 | 160 | 100.0 |

Let us now carefully consider in detail each section of those in primary poverty.

### Section 1

**Poverty Caused by Death or Desertion of the Wage-Earner**

The following are notes of some of the cases which come under this heading.

*No.* 3.—In this house the family has been deserted by the husband probably owing to the immorality of the wife. In any case it is now a squalid abode of drink and immorality. The female head of the house makes a variable amount from unknown sources, but the only certain earnings are made by a boy of 13 who has recently gone labouring (4/-).

*No.* 10.—Another bad case. All the children in this house are illegitimate. The female head of the house earns about 1/- per week by washing, but the principal earnings are by a grandson of 21. I understand that the children are the grandchildren of the head of the house, being the illegitimate children of a daughter who is away. The husband is dead.

*No.* 12.—The husband in this case died during the last three years, but was little loss to the family. The female head of the household gets her living chiefly by carrying goods in a perambulator to Woburn. She also does a little charring (about 10/- per week, altogether). One of the boys has recently gone to work.

The following are statistics with regard to this Section.

| | |
|---|---|
| Total number ... ... | 19 |
| Number of families ... ... | 5 |
| Average size of family ... ... | 3.8 |
| Average earnings ... ... | 8*s.* 10¼*d.* |
| Average rent ... ... | 1*s.* 9*d.* |

Of the total earnings 24*s.* 3*d.* or 4*s.* 10*d.* per house is contributed by the head of the house, while 20*s.* 2*d.* or 4*s.* 0½*d.* per house is earned by other members of the family.

## Section II

### Poverty Caused by Illness or Old-Age of the Wage-Earner

Details of some of the families in this section are here given.

*No.* 2 is an old spinster, chronically ill, and somewhat demented, who subsists entirely on Parish-relief.

*No.* 6 is an old couple, the man being deaf and quite incapable. The woman works at lace-making, and tells me she cannot do much more than about 2*d.* per day. Parish-relief counts for practically all the income.

*No.* 7 an interesting old couple, who have lived together for 20 years, but who were recently compelled to get married by the guardians, as otherwise the parish-allowance would be stopped. The man is quite senile and useless: the woman does odd jobs round the neighbourhood.

*No.* 18 is distinctly a case of premature old age and break-up of constitution through drink. The wife takes in washing.

*No.* 27.—An old man and two daughters, both of whom are semi-invalids. They are lace-makers, and thus supplement the parish-relief.

*No.* 43 is an old labourer who has had a very little money. Owing to not being able to pay his way he has drawn on this and is still drawing, but it is nearing exhaustion. Very respectable and no debt.

*No.* 46.—A widow who has recently lost husband and daughter. One of the best lace-makers in the village. She can do old point lace, but says it is too hard on her eyes now. She is helped by her children, for she told me the parish-allowance only covered rent, fire, and light.

The statistics of this section are as follows:—

| | |
|---|---|
| Total number of persons ... ... | 26 |
| Number of families ... ... | 14 |
| Average size of family ... ... | 1.9 |
| Average earnings ... ... | 6*s.* |
| Average rent ... ... ... | 1*s.* 6*d.* |

Of the total earning 63*s.* 4½*d.* or 4*s.* 6*d.* per house is supplied by the head of the household, while 20*s.* 6*d.* or 1*s.* 6*d.* per house is earned by other members of the family.

Taking sections 1 and 2, and opening a debtor and creditor account for them together we have—

| Income. | | £ | *s.* | *d.* | Expenditure. | | £ | *s.* | *d.* |
|---|---|---|---|---|---|---|---|---|---|
| Weekly income | (1) | 3 | 8 | 10 | Weekly rent | (1) | 0 | 8 | 9 |
| of 19 families. | (2) | 4 | 3 | 10½ | | (2) | 1 | 0 | 11 |
| Deficiency per week | | 2 | 0 | 11¼ | Weekly minimum cost of food, &c. | | 8 | 4 | 0 |
| | | 9 | 13 | 8 | | | 9 | 13 | 8 |

This shows a deficiency per head on 31 adults and 14 children of 11*d*. per head, or of 2*s*. 1¾*d*. per family.

Of those householders earning money other than parish-relief we have as follows:—

| | | | |
|---|---|---|---|
| Washerwomen | ... | ... | 1 |
| Charwomen | ... | ... | 3 |
| Pensioners | ... | ... | 2 |
| Labourers | ... | ... | 2 |

## SECTION III

### Poverty Caused by Wage-Earner Being Out of Work

In this class should be included all where the head of the family is out of work, due to inability to obtain employment. In a country village the occurrence of an excess of employment for the labour at hand, or an excess of labour for the work to do, is largely a matter of season. If these figures had been obtained a month or two months later, it might, and would probably, have been the case that quite a considerable number would have been out of work. As it is, there were none. All decent men, who wanted work on the land, could find it; but it must not be supposed that this is a condition of affairs which exists all the year round, even in the village we are considering.

## SECTION IV

### Poverty owing to Irregular Work

This section includes those whose poverty is caused by irregularity of work, whether due to illness, to vice, or to any other similar cause. The following are details as to some of these families:—

*No*. 1 is an old and incapable cobbler, who lacks regular work chiefly because he is a very bad workman. I have no information as to why the elder daughter is not earning anything.

*No*. 9 is a case of irregular employment through drink and vice. Otherwise the house would, for the present, be out of poverty.

*No*. 16 is an ordinary case of uncertain work, no information as to the cause of the irregularity being attainable.

*No*. 20.—Another case of irregular employment through drink, though this is only partly the cause of the poverty.

*No*. 24 is one of the worst cases in the village. The man is said to be a regular poacher, but, at any rate, with drink, vice, and laziness this was the worst case with which I have come in contact.

*No*. 4.—The man is often ill, but is also a confirmed drunkard, the drinking probably causing, in large measure, the illness.

The statistics of this section are as follows:—

| | |
|---|---|
| Total number of persons ... ... | 27 |
| Number of families ... ... | 6 |
| Average size of family ... ... | 4.5 |
| Average earnings ... ... | 11*s*. 9*d*. |
| Average rent ... ... ... | 1*s*. 8½*d*. |

Of the total earnings, 55*s*. 5*d*. or 9*s*. 3*d*. per house, is supplied by the head of the household, while 15*s*. or 2*s*. 6*d*. per house is earned by other members of the family.

Taking section IV and opening a debtor and creditor account for it, we have—

| Income | £ | s. | d. | Expenditure | £ | s. | d. |
|---|---|---|---|---|---|---|---|
| Weekly income of 6 families. | 3 | 10 | 5 | Weekly Rent | 0 | 10 | 3 |
| Deficiency per week | 1 | 14 | 1 | Weekly minimum cost of food, &c. | 4 | 14 | 3 |
| | 5 | 4 | 6 | | 5 | 4 | 6 |

This shows a deficiency for each of 27 persons of 1*s*. 3¼*d*., or, for each of 6 families, of 5*s*. 8*d*. per family per week.

In this group four of the six householders are agricultural labourers, one is a cobbler, and one a hawker.

SECTIONS V AND VI

**Poverty Caused by Largeness of Family or Lowness of Wages**

These are the groups in which poverty is caused by low wages, for the group (V) in which it is attributed to largeness of family is merely a variant of the same cause. This lowness of wages or largeness of family, in some of the cases cannot be regarded as the sole, but only as one of several causes.

The following are notes on some of the families. In section V:

*No.* 37 is a blacksmith, who drinks badly.

*No.* 51.—The amount of income is a maximum. Drink is a cause of a certain amount of uncertain employment.

In section VI the following notes could be made.

*No.* 16.—The head of this house is one of the only cases where a man has benefited materially by learning carving, from a financial point of view. He makes about 1/- a week, and possibly a little more by his work in this way.

*No.* 42 is a very bad case. A spinster makes her living entirely by lace-making, and works 10 to 12 hours per day for an income of about 3/6 per week. The guardians refuse poor-relief as she is a middle-aged woman; but heavy work is impossible to her, and as her sight is failing, it is not likely she will be able to keep up the present close work long.

Statistics relating to these two classes are here given.

| | Section V. | Section VI. |
|---|---|---|
| Total number of persons ... ... | 33 | 55 |
| Number of families .. ... | 4 | 11 |
| Average size of families ... ... | 8.2 | 5.0 |
| Average earnings ... ... | 21*s.* 4d. | 15*s.* 6½*d.* |
| Average rent ... ... | 1*s.* 10*d.* | 1*s.* 8¼*d.* |

Of the total weekly earnings of Section V, 69*s*. 9*d*. or 17*s*. 5*d*. per house is supplied by the head of the house, while 15*s*. 3*d*. or 3*s*. 10½*d*. per house is earned by other members of the family: similar figures for Section VI show a weekly income of 152*s*. or 13*s*. 9¾*d*. per house supplied by the head of the house, and 1*s*. 8¾*d*. per house by other members.

Taking sections V and VI together, and opening a debtor and creditor account for them we have:—

| Income. | | £ | s. | d. | Expenditure. | £ | s. | d. |
|---|---|---|---|---|---|---|---|---|
| Weekly income | (5) | 4 | 5 | 3 | Weekly Rent | 1 | 5 | 10 |
| of 15 families | (6) | 8 | 11 | 0 | | | | |
| | | | | | Weekly minimum necessary cost of | | | |
| Deficiency per week | | 2 | 11 | 7 | food, &c. | 14 | 2 | 0 |
| | | 15 | 7 | 10 | | 15 | 7 | 10 |

This shows a deficiency for each of 88 persons of 7*d*. per week, or for each of 15 families of 3*s*. 4¾*d*. per family.

In these groups the householders have occupations as follows:—

| | |
|---|---|
| Labourers ... ... ... | 12 |
| Blacksmith ... ... ... | 1 |
| Lace-maker ... ... ... | 1 |

Now taking all the five sections of the primary poverty prevalent in the village, the deficiency works out according to the table here given:—

| | Sections I & II. | Section IV. | Sections V & VI. | Total |
|---|---|---|---|---|
| Per family | 2*s*. 1¾*d*. | 5*s*. 8*d*. | 3*s*. 4¾*d*. | 3*s*. 8*d*. |
| Per individual | 11*d*. | 1*s*. 3¼*d*. | 7*d*. | 11*d*. |

One of the most interesting and striking results which can be obtained from a consideration of my figures is that relating to the proportion of the population in primary poverty at various ages. These figures are set out below:—

| Ages. | Sections I & II. | Section IV. | Sections V & VI. | Total. |
|---|---|---|---|---|
| Under 16 | 14 | 13 | 56 | 83 |
| 16—25 | 1 | 4 | 7 | 12 |
| 25—55 | 7 | 6 | 23 | 36 |
| Over 55 | 23 | 4 | 2 | 29 |

This table gives the percentage of those under 16 to the total in primary poverty of 51.9: between 16 and 25, of 7.5: between 25 and 55 of 22.5, and finally of those over 55 of 18.1. The significance of these figures will become clearer when similar statistics are shown for the population not in poverty.

*Secondary Poverty.*—In a village lying in the heart of the agricultural districts of England, and which the bulk of the place belongs to landlords who are particular as to the character of their cottage-tenants, the amount of secondary poverty is necessarily limited. But it does occur, though in much less proportion than in town life. In deciding whether a particular household is to be classed as poverty-stricken, the personal feeling of the enumerator also affects the results considerably. But the following statistics represent very closely the amount of that poverty which is not due to lack of means, but to waste of means.

| | |
|---|---|
| Persons in Secondary Poverty ... ... | 33 |
| Families in Secondary Poverty ... ... | 10 |
| Percentage of working-class population ... | 9.0% |
| Percentage of total population ... ... | 7.1% |
| Number of individuals per family ... ... | 3.3 |

*Causes of Secondary Poverty.*—The general cause of secondary poverty is the unwise and vicious expenditure of the money or the deliberate avoidance of work when it can be obtained. It is manifestly impossible to apportion satisfactorily the immediate circumstances which cause such poverty. They not only overlap, but also vary at different times. With every reservation, however, for such causes of uncertainty, we may roughly divide the poverty of the ten families thus enumerated as being due in five cases to drink and its associated vices; in three cases to bad management of the home, sometimes through over-work of the head of the household in getting a living; and in one case to uncertainty of work. The last is one of those cases where the man is a good workman, but will not work regularly.

The crying cause of secondary poverty is the drink habit. Further it has a great deal to do as a contributory cause of much primary poverty. In the village there are three public-houses, so that, considering that the through traffic is not now of much importance, the allowance is very excessive, and while this does not altogether account for the amount of drinking, and hence of poverty, the large number of opportunities has probably a very considerable influence on it.

The occupations of those householders marked as being in secondary poverty is as follows:—

| | | | |
|---|---|---|---|
| Labourer | ... | ... | 6 |
| Carpenter | ... | ... | 1 |
| Cobbler | ... | ... | 1 |
| Charwoman | ... | ... | 1 |
| Pensioner | ... | ... | 1 |

The following figures show the total amount of poverty both primary and secondary.

| | Primary. | Secondary. | Total. |
|---|---|---|---|
| Families in poverty ... ... | 40 | 10 | 50 |
| Population in poverty ... ... | 160 | 33 | 193 |
| Percentage of working-class population | 41.0 | 9.0 | 50.0 |
| Percentage of total population ... | 34.3 | 7.1 | 41.4 |

| | |
|---|---|
| Families *not* in poverty ... ... | 77 |
| Population *not* in poverty ... ... | 274 |

There remains now only to consider the portion of the population which is above both the primary and secondary poverty line, and the inhabitants thus included, as has just been stated, number 274, comprised in 77 households. They may be divided into those who belong to the wage-earners and those who being to the trading and upper classes. Let us take these two separately.

*Wage-earners not in poverty*

| | |
|---|---|
| Total number of persons ... ... | 197 |
| Number of families ... ... | 54 |
| Average size of family ... ... | 3.71 |
| Average family earnings ... ... | 23*s*. 7½*d*. |
| Average rent ... ... ... | 1*s*. 6½*d*. |
| Per cent. of population ... ... | 42.1 |

The amount of these earnings contributed by the head of each household and by others is as follows:—

| | Total. | Per family. |
|---|---|---|
| Contributed by householder ... | £41-8*s*-4½*d*. | £0-15*s*-4¼*d*. |
| by others ... | £22-7*s*-0*d*. | £0-8*s*-3¼*d*. |

When classified with respect to occupations we have these 54 heads of households showing as follows, as regards the principal source of income:—

| | |
|---|---|
| Agricultural foremen ... ... ... | 4 |
| Engineers and machine-minders ... ... | 2 |
| Bakers ... ... ... ... | 1 |
| Painters ... ... ... ... | 2 |
| Tailors ... ... ... ... | 1 |
| Signalmen ... ... ... ... | 2 |
| Labourers ... ... ... ... | 26 |
| Pensioners ... ... ... ... | 6 |
| Charity. Parish Relief and supported by lodgers | 8 |
| Policemen ... ... ... ... | 1 |
| Lodge-Keepers ... ... ... ... | 1 |

The economic position of this group is shown in the following table:—

| Income | £ | s. | d. | Expenditure | £ | s. | d. |
|---|---|---|---|---|---|---|---|
| Weekly income of 54 households. | 63 | 15 | 4½ | Weekly Rent | 4 | 2 | 10 |
| | | | | Weekly minimum cost of living | 36 | 2 | 6 |
| | | | | Excess income over what is absolutely necessary | 23 | 10 | 0½ |
| | 63 | 15 | 4½ | | 63 | 15 | 4½ |

The excess over what is absolutely necessary works out at 8*s.* 8½*d.* per family and 2*s.* 4½*d.* per head.

It must be remembered that this satisfactory result has only been obtained by the inclusion of every penny earned by every member of the household.* Thus for instance in No. 68 there are four young men all earning a full labourer's wage and living at home with their parents.

* Except mere lodgers.

Of course it is certain that the whole of their earnings are not available for the household expenses of the family in which they live, and therefore these figures show a much more favourable economic state of affairs than is really the case. That the position of the families in this group is largely due to the number of young wage-earners living at home with their parents is shown by the following age-table, the figures in which though not absolutely reliable in every case are very nearly so, and the figures are certainly accurate to the unit place. It shows the number at various ages in the families of wage-earners *not* in poverty and the proportion they bear to the whole number in such families.

| Age | | Number | | Percentage of Group |
|---|---|---|---|---|
| Under 16 | ... | 53 | ... | 26.9 |
| 16—25 | ... | 28 | ... | 14.2 |
| 25—55 | ... | 70 | ... | 35.5 |
| Over 55 | ... | 46 | ... | 23.4 |

If we compare this table with those already given for the population living in primary and in secondary poverty, the point on which I am trying to insist becomes very much more striking as follows:—

| Age. | Primary Poverty. Percentage. | Secondary Poverty. Percentage. | Wage-earners not in Poverty. Percentage. |
|---|---|---|---|
| Under 16 | 51.9 | 30.3 | 26.0 |
| 16—25 | 7.5 | 15.2 | 14.2 |
| 25—55 | 22.5 | 39.4 | 35.5 |
| Over 55 | 18.1 | 15.2 | 23.4 |

One of the causes of poverty, as shown here (provided resources are not wasted as with those in secondary poverty), seems therefore to be the removal of the young wage-earners from the home, or from the children not having arrived at the wage-earning stage. That the latter consideration does enter into the calculation seems to be shown by the fact that in the group we are at present discussing (the families *not* in poverty), the head of the household alone would only just manage, *on the average,* to earn sufficient to maintain the family above the primary poverty line. If, however, one more child appeared in each family, the heads of the households would no longer be in a position to do this.

I say, "on the average." however, for a vast proportion of households are quite dependent on the supplementary earners, or in other words, usually, on the children of the house, to be enabled to live out of a poverty-stricken condition. Out of the 54 families in the present group only forty would be above the poverty line, and fourteen below it if only the head of the household was earning money: in other words, 25.9 per cent. of the families above the poverty line would be in poverty but for the supplementary earners.

I have only now to speak of the trading and upper classes in the village, and a very few words will suffice with regard to them, as they are presumably all well above the poverty stage. The following data will however be of interest:

*Trading and Upper Classes*

| | |
|---|---|
| Number of persons ... ... ... | 77 |
| Number of households ... ... | 23 |
| Percentage of total population ... ... | 16 5 per cent. |
| Average size of family ... ... | 3.35 |

The heads of these households pursue the following occupations:

| | |
|---|---|
| Engineers | 2 |
| Clerks, &c. | 2 |
| Shopkeepers | 3 |
| Publicans | 3 |
| Ministers of Religion | 2 |
| Wheelwright | 1 |
| Baker | 1 |
| Butcher | 1 |
| Farmer | 1 |
| Carrier | 1 |
| School-teacher | 1 |
| Blacksmith | 1 |
| Gentlemen, or undefined | 3 |

There are three farms in the parish, and part of the Duke of Bedford's Woburn Park. If the farmers and their dependents in the outlying parts of the parish were included, the proportion of classes would only be very slightly different from that here presented, but the object of this investigation being to obtain the data for the village, I have not thought it necessary or even advisable, (nor, indeed, have I had the opportunity) to complicate it by including outside cottages and houses.

The final question remains. What is the present economic status of agricultural labourers in districts where this has not been very materially modified by the near approach of manufacturing or mining centres. For determining this point there are a considerable number of data in existence; the best and latest of which are those given by Mr. Wilson Fox in a report to Government (in connection with the recent Agricultural Commission) on the condition of agricultural labourers. For the county of Bedford, he there gives as the normal wages (without extras) for June and December in various years as follows:—

| | | June. | December |
|---|---|---|---|
| 1851 | ... | 9*s.* | 9*s.* |
| 1861 | ... | 11*s.* | 11*s.* |
| 1871 | ... | 11*s.* | 11*s.* |
| 1881 | ... | 13*s.* | 13*s.* |
| 1891 | ... | 13*s.* | 13*s.* |
| 1899 | ... | 13*s.* | 13*s.* |

While with extra payments, he quotes the average of agricultural labourers in the county as being 16*s.* 2*d.* at the times of his investigation (1899-1900). Unluckily he does not give analytical details showing how the last figure was obtained in Bedfordshire,—the nearest specific cases cited being at Newport, Bucks,—where various classes of labourers earned as follows: —

(1) Shepherd.—15*s.* per week, (£36/-); Harvest £6/-; Lamb Money £1 7*s.* 8*d.*; Cottage £6 10*s.*; or a total of £49 17*s.* 8*d.*, equal to 19*s.* 2*d.* per week.

(2) Cattleman.—13*s.* per week (£31 4*s.*); Harvest £5 4*s.*; Cottage £6 10*s.*; or a total of £42 18*s.*; equal to 16*s.* 6*d.* per week through the year.

(3) Horseman.—16*s.* per week (£41 12*s.*), and casual allowances equal to 9*s.*; or a total of £42 1*s.*, equal to 16*s.* 2*d.* per week.

(4) Ordinary Labourer.—12*s.* per week (£28 16*s.*); Harvest £4 19*s.* Cottage £6 10*s.*; or a total of £40 2*s.*; equal to 15*s.* 5*d.* per week throughout the year.

(In each case beer was given at hay and threshing in addition.)

How do these figures compare with those obtained at Ridgmount? I have gathered together the wages and extras of the fully adult labourers, working on full wages (that is to say, I have excluded those who were earning little owing to age. and who were earning little owing to youth), and I have calculated at full time. those whose wages were irregular. With all this the wages in Ridgmount are distinctly less than the average given by Mr. Wilson Fox for

Bedfordshire when extras are included, though distinctly more when the bare wages themselves are taken. In fact, as I have indicated in an earlier portion of this paper, certain extras, common elsewhere, are here not given, the most important of these being that cottages are nearly always paid for by the labourers, and rarely, if ever, supplied free by the employers. The average amount of *extra* pay earned by the labourers in this village (exclusive of beer which may be supplied at hay-harvest, corn-harvest, and threshing, which I have neglected) is 8½*d*. per head per week throughout the year, or a total for the year of £1 6*s*. 10*d*. This may appear to be very small, but the smallness is accounted for very largely by the fact that the Duke of Bedford, in many departments of the employment he gives, has less extra payment than the farmers round,—but considerably better wages,—a state of affairs which appears on the whole to be preferable from the labourers' point of view, and to be preferred by them.

If foremen be excluded, the average wages paid in the village amounts to 13*s*. 7½*d*. per head per week for pure agricultural labourers, sixty-five in number, who are working at full rates. The Duke of Bedford's standard is about 15*s*. per week; the standard of the other farmers 12*s*. to 14*s*., though, as has been said, the latter usually carry more extras than the former. This gives an average weekly wage, including extras, of 14*s*. 4*d*. per head.

On the other hand a foreman makes as a rule from 20*s*. to 30*s*., and if these be included as agricultural labourers, the average amount without extras touches 14*s*. 3*d*. per week, and including extras reaches 14*s*. 11*d*. per week, a sum very considerably below that obtained by Mr. Wilson Fox. After a very careful examination of Mr. Fox's figures, I cannot help thinking that in working out his averages, he has not allowed enough for the enormously greater number of the lower grade of labourers over higher grades, and I think if this were taken into account his figures would not be very different from mine. But taking the actual figures obtained, it appears clear that a man earning the average rate of wages, and the head of a household, must descend below the primary poverty line so soon as he has two children, unless he is able to supplement his income by an allotment, by fattening and breeding pigs, or by other means. It is also clear that he will remain below the poverty line until the eldest child leaves school and begins to earn money, and that, even if he has no more than two children, his only chance to save will be in his later life when the children are grown-up, and are earning money or have left home. This is the most favourable case: if there are more children the period of poverty is longer and the chance of saving less. In any case during life it is a weary and continual round of poverty. During childhood poverty conditions are almost inevitable. As a boy grows up there are a few years intermission till, as a young man, he has two children; then poverty again till these children grow up, and, finally, at best, a penurious old age barely lifted above the poverty line.

I do not wish to draw conclusions in the present paper, but one thing I must say. The cry of "back to the land" has a curious commentary in the results I have obtained. As at present existing, the standard of life on the land is lower than in the cities; the chances of success are less and of poverty are greater; life is less interesting; and the likelihood of the workhouse as the place of residence in old age is greater. It is evident that the outcry against the depopulation of the country and the concentration of population in the towns must remain little more than a parrot cry until something is done to raise the standard of life, and hence the standard of wages in our purely agricultural districts, to increase the chances of success in life, to make life more interesting, and to bring about a more attractive old age than at present, when under existing conditions, the workhouse is apt to loom too large on the horizon of the agricultural labourer.

**N.B.** When this paper originally appeared in the **Sociological Papers**, 1904, it carried the following footnote by the Editors of those papers:

The inclusion of this paper—offered by the author through the intermediacy of Mr. Seebohm Rowntree—in a volume professedly sociological, calls for a word of explanation.

The distinction between what properly belongs to Sociology and what is internal and technical to the several social sciences, whether economic or other, is not always easy to make. But in this case, one may admit that Mr. Mann's investigation was made from a point of view mainly economic, and yet claim for it a place alongside those "Monographies Sociales" of the Le Play investigations which mark an important phase of transition in studies alike sociological and economic. The change from the more abstract and general writings of the older economists to the more concrete, geographical, and particular investigations of Le Play, Booth, Rowntree, etc., is itself indicative of a growing re-orientation of Economics in a sociological direction, i.e. its reorganisation as a specialized department of a generic sociological science. Mr. Mann's paper is noteworthy as continuing and developing this sociological re-orientation of Economics, and it gains further importance from being—we believe—the first effort to apply the Le Play-Booth method in a comprehensive way to English village-life. —EDITORS.

*Chapter* 6

# NOTE ON THE DIET OF TEA GARDEN COOLIES IN UPPER ASSAM AND ITS NUTRITIVE VALUE*

In the past few years quite a large number of investigations have been undertaken by Church, Leather, Hooper and others with regard to the composition of the foods commonly used by the people of India. Little, if anything, however, has been done to ascertain, not the value of individual foodstuffs, but that of the diet of which they form a part, except with regard to the minimum required. This last has been worked out with considerable thoroughness in connection firstly with jails, and secondly with famine conditions. It seems, however, that a knowledge of the food value of Indian diets, not under special conditions like those of famine or scarcity, but in the regular course of daily life, will be of considerable value, and the present is a preliminary note designed to introduce the subject to the notice of those who are interested in the question, and to enable me to ask for cooperation in extending dietary studies to a large number of the races, castes, and groups living in India.

The fact that I have been able to get together the details which follow is due to the system under which the coolies are employed in Upper Assam. It is very important that every coolie who *can* work should be kept in condition for work, and hence it is customary when any man or woman is noticed to be becoming weak, anaemic, and unfit to do the regular daily task, to provide, under the

* Reprinted with permission from the **Journal and Proceedings of the Asiatic Society of Bengal (New Series)**, Vol. III, No. 2. 1907, pp. 103-108.

supervision of the garden doctor, properly cooked food for them at what is called a 'hotel,' to which they go, obtain and eat their rations before going to work in the morning and after returning from work in the evening. The existence of this so-called 'hotel' enabled me to obtain exact information as to the nature of the diet, which was regarded as sufficient and suitable by the coolies, and which was closely similar to that which they provided for themselves under normal conditions. It has been found that the coolies usually improve in health under the diet of which I now give the amount and composition.

I have obtained data from two very large properties on the Sadiya Road in Upper Assam. The figures are, however, so closely similar that one set of figures will be quite enough to represent the facts. It must be noticed, however, that the amount of food calculated for men and women is the same, whereas in Europe and America it is customary in similar studies to only allow a woman four-fifths of the quantity provided for man. This is explained by the fact that in Assam, during a large part of the year, both women and men are doing similar work, and hence will require the same nutritive material. A child is reckoned, on the average, as consuming half the food required for a man.

This being the case, the ration allowed per adult man or woman per day is as follows:—

| | | | | | | |
|---|---|---|---|---|---|---|
| Rice | ... | 14 | chitaks | or | 815 | grammes. |
| Dal | ... | 2 | ,, | ,, | 115 | ,, |
| Salt | ... | 1/3 | ,, | ,, | 19 | ,, |
| Spices | ... | 1/16 | ,, | ,, | 3½ | ,, |
| Oil | ... | ⅛ | ,, | ,, | 7½ | ,, |
| Potatoes | ... | 2 | ,, | ,, | 115 | ,, |
| Onions | ... | ¼ | ,, | ,, | 14½ | ,, |

The elements of this diet, as used, need some description.

*Rice.*--The rice normally employed is that known on the Calcutta market as coolie rice of the quality 'Kazla No. 1.' Occasionally, and on some gardens, the lower priced 'Kazla No. 2' is employed, but this is not usual in Upper Assam. It is a brown or red rice, containing a considerable proportion of the inner coats of the husk attached to the grain, and is hence richer as a nutritive material than if polished white. One occurrence which has come under my notice illustrates this. A tea company recently obtained a higher type of rice, known on the Calcutta market as 'Jhabra,' for their coolies. This was white and polished, but shortly afterward the coolies complained of it as not being as satisfying as the commoner grade. A higher grade rice is employed in the hospitals known as 'Kalchitu.' Analysis of samples of Kazla No. 1 and Kalchitu rise gave the following figures:—

| | | Kazla No. 1. | Kalchitu Rice |
|---|---|---|---|
| | | % | % |
| Moisture | ... | 7.75 | 5.75 |
| Oil | ... | 1.11 | 1.53 |
| *Albuminoids | ... | 8.25 | 10.06 |
| Carbohydrates, &c. | ... | 81.22 | 81.46 |
| Crude Fibre | ... | .15 | .10 |
| †Ash | ... | 1.52 | 1.10 |
| | | 100.00 | 100.00 |
| *Containing Nitrogen | ... | 1.32 | 1.62 |
| †Containing Sand | .. | .56 | .11 |

All the water in which the rice is cooked is eaten with it, so that the nutritive value actually absorbed is that above given.

*Dal.*—The Dal, the principal source of nitrogen in the food of the people of Eastern Bengal and Assam, in the present case consisted of equal proportions of *Arhar* (Cajanus indicus) and *Masuri* (Lens esculenta). Both of these are leguminous seeds of very considerable nutritive value. Both are grown, in small quantities, in the neighbourhood of the garden where these were used, but by far the greater amount is imported from Bengal. On analysis they gave the following figures:—

| | | Arhar Dal. | Masuri Dal. |
|---|---|---|---|
| | | % | % |
| Moisture | ... | 3.50 | 5.66 |
| Oil | ... | 3.03 | 1.02 |
| *Albuminoids | ... | 24.69 | 25.12 |
| Carbohydrates, &c. | ... | 62.57 | 65.37 |
| Crude Fibre | ... | 1.79 | .39 |
| †Ash | ... | 4.42 | 2 44 |
| | | 100.00 | 100.00 |
| *Containing Nitrogen | ... | 3.95 | 4.02 |
| †Containing Sand | ... | .21 | .29 |

During the present season (1906) the prices of both rice and *dal* have been extraordinarily high, ranging in May, when these data were obtained, from five to six rupees per maund for rice, and from four to five rupees per maund for either of the above dals. In the present instance, however, rice was always supplied at Rs. 3 per maund under the Inland Emigration Act under which the coolies are taken to Assam.

*Spices*—These consist of a mixture of Black pepper, Caraway and Coriander, which gave on analysis the following figures:—

| | | Black pepper. | Caraway. | Coriander. |
|---|---|---|---|---|
| | | % | % | % |
| Moisture | ... | 6.76 | 6.48 | 4.21 |
| Oil | ... | 7.86 | 14.35 | 17.68 |
| Albuminoids | ... | 13.69 | 19.25 | 15.19 |
| Carbohydrates | ... | 56.66 | 40.54 | 37.31 |
| Crude Fibre | ... | 9.77 | 7.55 | 14.63 |
| Ash | ... | 5.26 | 11.83 | 10.98 |
| | | 100.00 | 100.00 | 100.00 |
| Containing Nitrogen | | 2.19 | 2.67 | 3.18 |
| Containing Sand | | .20 | 3.08 | 2.42 |

*Mustard Oil*.—All the fatty matter employed in cooking and eating these rations is in the form of oil from Mustard or Rape (*Brassica* sp.), a seed very extensively cultivated in the Assam valley. *Ghi* or clarified butter is not used there.

*Vegetables*.—The garden, where the results under discussion were obtained, is situated in a district where potatoes are grown in large quantities. The onions shown in the ration are replaced as necessary by other vegetables, potherbs and whatever is obtainable of a similar sort. It was impossible to submit the potatoes or onions to analysis at the time, and I could not keep them. In calculating nutritive value of the diet I have, therefore, used the following figures, which will be certainly very close to the actual fact:—

| | Potatoes. % | Onions. % |
|---|---|---|
| Oil ... ... ... | .15 | .15 |
| Albuminoids ... ... | 1.99 | 1.60 |
| Carbohydrates, etc. ... | 20.86 | 11.00 |

Combining all the analyses given it will be seen that the daily ration as set out will contain the following maximum weights of nutritive materials:—

| | Albuminoids. Grammes. | Fat. Grammes. | Carbohydrates. Grammes. | Fuel Value Calories. |
|---|---|---|---|---|
| Rice | 67.2 | 9.0 | 661.9 | 3073 |
| Dal | 28.7 | 2.3 | 73.6 | 441 |
| Spices | .6 | .5 | 1.6 | 14 |
| Oil | — | 7.5 | — | 70 |
| Potatoes | 2.3 | .2 | 24.1 | 110 |
| Onions | .2 | 0.2 | 1.6 | 8 |
| Total | 99.0 | 19.52 | 762.8 | 3716 |

In Europe the minimum food value of the diet necessary for the maintenance of physical health and efficiency has been laid down by numerous observers, and in recent years a very close agreement has been reached among investigators in Europe and America. The best figures are, undoubtedly, those of Atwater in America[1] which follow. In accordance with modern ideas of the replaceability of various food constituents, only the albuminoids (protein) and Fuel-value are indicated, it being considered of minor importance whether the latter be obtained from albuminoids, carbohydrates or fats.

| | Albuminoids. Grammes. | Calories. Fuel Value. |
|---|---|---|
| Man with little physical exercise | 100 | 2700 |
| Man at light physical work | 112 | 3000 |
| Man at moderate physical work | 125 | 3500 |
| Man at active, hard physical work | 150 | 4500 |

The figure obtained for the diet under discussion is, it will be seen, equivalent in albuminoids to that of Atwater's standard of a man with little physical exercise only. In the present instance the work done would be comparable at least with Atwater's 'moderate work' standard, and is probably higher. If we argue direct from these figures, the standard of our Assam coolie diet is very deficient in albuminoids, but not otherwise a low one.

But it seems that we are not justified in so calculating. On the whole a man of the coolie classes, which in the district we are considering are chiefly composed of Mundas, Oraons and other Chota-Nagpur tribes, is considerably lighter in weight than a European or an American. Hence the amount necessary for maintenance, as distinguished from energy expended in work, will be less than that required in America. What correction should be, however, introduced on this account is unknown. The correction, in terms of food, necessary to bring the food value under the higher temperature conditions of India into relationship with that of America or Europe is also quite unknown.

[1] Investigations in the Chemistry and Economy of Food by W. O. Atwater, U.S. Department of Agriculture, Bulletin No. 21.

I have explained that this is a diet which is given to coolies who are weakly, and it does not hence quite represent that normally consumed by the people. In the first place the amount of rice is greater; a coolie in full work in Assam eats normally about 12 chitaks of rice per day only, or 22 seers 8 chitaks per month of thirty days. In other respects the diet is fairly closely followed except that potatoes are only used at certain times of the year, their place being much more normally taken by potherbs, or other vegetables which are often available. Thus it may be said that the above dietary represents, except in the matter of rice, almost the normal standard of food for a coolie in Upper Assam in respect to quantity. Under usual conditions, however, in their own houses, the coolies only eat one cooked meal each day, this being in the evening. In the morning only parched rice is consumed.

There is only one seriously doubtful factor in the above diet as applied to Upper Assam coolies. This is the amount of fish which may be obtained. I am assured, however, that among tea garden coolies, for by far the greater part of the year, the amount is negligible. Milk is not used, nor *ghi*.

The actual cost of the diet I have given above per head per month will alter very considerably with the time of year and the circumstances. The present year has, even apart from rice (which as I have said is by law supplied at Rs. 3 per maund), been one of notoriously high prices. Though not perhaps of the same permanent value as the remainder of the figures, it may be interesting to give the prices at which the materials were actually bought in the bazar at the time the present notes were made. They were as follows:—

| | Rs. | A. | P. | |
|---|---|---|---|---|
| Dal (both kinds) ... | 4 | 12 | 0 | per maund. |
| Salt ... ... | 3 | 13 | 0 | " |
| Spices ... ... | 0 | 6 | 0 | per seer. |
| Mustard Oil ... | 18 | 0 | 0 | per maund. |
| Potatoes ... ... | 3 | 8 | 0 | " |
| Onions ... ... | 3 | 13 | 0 | " |

At these prices and rice at Rs. 3 per maund, the cost of a man's diet per month in May 1906 would be:—

| | | Rs. | A. | P. |
|---|---|---|---|---|
| Rice | ... | 1 | 15 | 6 |
| Dal | ... | 0 | 7 | 2 |
| Salt | ... | 0 | 1 | 0 |
| Spices | ... | 0 | 1 | 0 |
| Mustard Oil | ... | 0 | 2 | 0 |
| Potatoes | ... | 0 | 5 | 3 |
| Onions | ... | 0 | 0 | 8 |
| **Total** | ... | 3 | 0 | 7 |

A family of man, wife and three children, which is usually considered a normal one, will require for the 'hotel' diet here described, a minimum of Rs. 10-10-0 per month in order to provide it, under the conditions of prices prevailing during the early summer of 1906.

*Chapter 7*

# THE 'MAHARS' OF A DECCAN VILLAGE* (SASWAD)

In the organisation of a Deccan village the *Mahars* play an essential part. They are outcast. They are not allowed to reside inside the village. But they are necessary and are found with their rights and duties, recognised as part of the village, both by the people and the rulers everywhere. It has long been my wish to study more closely than has yet been done the actual conditions of life which prevail among them as found in actual village life, and some time ago my friend Dr. N. Macnicol offered to make for me a house to house inquiry, as to their condition, among the *Mahars* of Saswad. This is a small country town, with a population of a little over six thousand, sixteen miles from Poona in the Bombay Deccan. Though a town, it is yet of such a purely rural character that almost any conditions prevailing there may be considered as also applying to the villages in the Poona district.

The manner in which the *Mahars* live in a Deccan village is, in general, very well-known. They have their quarters apart from the rest of the people, outside the village, in the *mahar wada*. The right to live there belongs to the *Mahars* of the village: it is one of their *vatan* or hereditary rights. In addition to this, they have a certain amount of land, the produce from which is their own, and which is divided among the families of *Mahars* belonging to the village. In return for this land and for the right

* Reprinted with permission from the **Social Service Quarterly**, Bombay, Vol. II (1916), No. 1, pp. 1-8.

to live in the *mahar wada*, the *Mahars* have definite duties which they must perform. If some members of a *vatandar* family go away to Bombay or Poona, the remainder must arrange for the duties to be performed. These duties consist of acting, under the orders of the village *Patil* or headman, as village messengers, as guides to strangers, and often as village watchmen. They erect the tents of strangers, and they have the duty of removing and disposing of dead cattle and other animals, the flesh and skins of these belonging, however, to themselves. There are, in addition, a few other minor duties which they must perform.

In recent years, all over the Deccan, a change has taken place in the position of the village *Mahars*. The fact that work is available in the large towns, and especially in Bombay, has taken very many away. There they get high—almost unheard of—wages, and though the cost of living is also high yet many of them do send money regularly, as we shall see, to their relatives and families left behind in the village. For, as I have insisted elsewhere, although a *Mahar* may go, or one branch of a family may go away, yet they will rarely, if ever, think of sacrificing their connection with the village where they hold *vatan* rights. To do this would make them landless outcasts, and hence they return,—many of them very frequently, but almost all at intervals—to the old home of the family.

In addition to this, the village *Mahars* in recent years have taken to work as artisans, and are tending more and more to do so. This takes them away often, for a mason or a carpenter cannot be always on the spot in a village,—but they return and often the family remains while the head of it goes away for work.

In the present article I wish to try and indicate the nature of the families, the sources of income, the conditions of life and the extent to which disintegrating influences have affected life in a definite village or small town,—Saswad being taken as an example.

So far as my records go, there were, at the time of the investigation about four years ago, fifty-nine *Mahar* households in the village of Saswad, and the actual number of souls in these fifty-nine households was two hundred and eighty-eight, or an average of 4.8 persons per household. I speak of households and not of families, because the fact that so many men are away in Bombay prevents those left behind separating into different households, however meagre the accommodation may be for the number present. The households thus formed consist largely of the wives and families of those remaining. This is illustrated by the relative number of men, women and children* which was as follows:—

| | |
|---|---|
| Men ... ... ... ... | 60 |
| Women ... ... ... ... | 112 |
| Children ... ... ... ... | 101 |
| Not ascertained for two households ... | 15 |
| Total ... | 288 |

Leaving out the two households, where the number could not be ascertained, it will be seen what an enormous excess of women and children exists. A more striking evidence of the extent to which the towns are denuding the countryside of its labouring population so far as the *Mahars* are concerned, could not be given.

But though the drawing influence of the towns on the *Mahars* of Saswad vilage is great, yet large numbers of those away send money home for those who remain. Of the fifty-seven households for which the information is fairly complete, no less than twenty-four (or forty-two per cent.) had members of their families away who, whether regularly or only occasionally, remitted money to those who were left at home, while in four more cases representatives had been away and had returned only a short time

* The distinction between men and boys and between women and girls is not perfectly definite. I have tried to take men and women as those over about 16 years.

previously for various causes. There were still other cases in which members were in the habit of going away for part of the year. We shall not be far off the truth if we say that sixty per cent. of the *Mahar* households of a country town like Saswad have, either occasionally or regularly, some of their members away working, and that most of these remit money home when they are away. In the case we are discussing fourteen households had one representative each away from the village for work, seven had two representatives, and three had not less than three members each away.

The people who leave the villages are mostly artisans or those who can work as labourers in mills and other places. Saswad is rather famous for its masons, and an unusually large proportion of those who go away ply this trade. To this extent the figures for Saswad are not typical,—but they are as follows. I have considered all as labourers whose trade was not specified.

| | |
|---|---|
| Masons ... ... ... ... | 19 |
| Labourers ... ... ... ... | 15 |
| Butlers ... ... ... ... | 2 |
| Mistri ... ... ... ... | 1 |

Of these, twenty wage-earners were in Bombay, seven in Poona, six at Nasik, and only four in all other places. The superior attraction of Bombay to artisans and labourers is obvious from these figures.

If we add all these to the number of males in the households of the *Mahars* of Saswad, the disproportion of the sexes becomes very much less than seems the case at first sight. We have in fact:

| | |
|---|---|
| Men in village ... ... ... ... | 60 |
| " away at work ... ... ... ... | 37 |
| Total ... | 97 |
| Women ... ... ... ... | 112 |
| Children ... ... ... ... | 101 |

The large number of women is made up because the number of old women is large, of old men comparatively small. The average number (including the wage-earners away) per household is:

| | | | | | |
|---|---|---|---|---|---|
| Men | ... | ... | ... | ... | 1.7 |
| Women | ... | ... | ... | ... | 1.9 |
| Children | ... | ... | ... | ... | 1.7 |

The condition under which the people, who remain in the village, live must next be considered. All the households living in the *mahar wada* of Saswad were *vatandar Mahars* of the place, except one. This ensures them a right to erect their houses in the *mahar wada*, and practically every household lived in a house which belonged to some member of the family, and hence no rent was paid. This is a great advantage and practically means that apart from the expenditure for clothes,—very small under village conditions—the whole income is available for food and other direct personal uses. Of course the houses are exceedingly primitive and generally exceedingly small. But where the people live so largely out of doors as is done in a village this has less importance than it would have in a town. Very often the people sleep outside the hut or house, and for a large part of the year the house may be considered largely rather as a shelter from sun and rain and a storehouse for food than as a sleeping-place. Nevertheless, the actual figures or house area per head of the population is as follows:—

| | |
|---|---|
| Number of **large** houses, with more than 50 square feet per head ... ... ... ... | 15 |
| Number of **medium** houses, with 25 to 50 square feet per head ... ... ... ... | 9 |
| Number of **small** houses, with 12 to 25 square feet per head ... ... ... ... | 19 |
| Number of **very small** houses, with less than 12 square feet per head ... ... ... ... | 14 |

These figures show one of the very marked features of many *mahar wadas* that I have visited. Most of the houses are primitive, small and low. A few are large, well-built, often with two storeys, and usually represent the houses of members of the community who have been away and been successful, and have come back to spend their money and life in the village home. Considering that the people are all of one community I fancy the obvious difference (as indicated by the kind of house) between the condition of life of various members and various families is more marked in the villages among *Mahars* than among most other castes.

This difference is shown in other ways. Take for instance education. Among most of the families, the children are left quite uneducated though a number of them send their boys to school when small. No less than seventeen boys and one girl were at school at the time of the inquiry or about eighteen per cent. of the total number of children. One boy was in the high school. This represents a much better condition of things than I expected, but it must be repeated that these children came from the better-off part of the community and from a comparatively few families.

I have already described the occupation of the men belonging to the village. Of the inhabitants left behind, each household supplies some one to assist in carrying on the village duties as described above. But nearly every one does other work besides. Of sixty adult males actually residing in the village, twenty-nine may be classed as labourers, either in the field, as coolies, or in other capacities; ten were masons, one was a carpenter, two were clerks of a sort (one being a petition-writer) and one was an independent gentleman. Eight lived by charity in one form or another, though among the men the charity was not pure begging but was usually of a religious character, the persons being *gosavis* or something similar. Of these for whom I have records, fifty two per cent. were labourers of one sort or another, and twenty five per cent. carried on skilled work. The latter proportion is larger than would usually be expected, and shows how the *Mahars* are extending their activity among the skilled industries. I expect this will go on more and more, and they will drive competitors in the future from these trades.

Of the women, no less than thirty-one depended principally on money sent by husbands, brothers, or other members of the family working away. Among the rest, fourteen depended on coolie work of one sort or another, and six were beggars. It was curious to find one *Mahar* woman carrying on a trade in eggs, collecting them in Saswad and the surrounding villages and carrying them to the Poona market.

Of the amount of income per household or even per person, it is somewhat difficult to speak, largely because the custom of paying in kind, in the form of grain, persists still to a considerable extent. All I can do will be to give some typical examples of the rates (as they were four years ago), rates which, however, from all accounts since, are continually tending to increase. A man's labour, among the *Mahars,* at the time of this inquiry, was paid at the rate of four annas a day when money was given. Women were paid at little more than half this rate. Thus in one household consisting of two men and four women, the men and the women worked on the roads, the former receiving four annas and the latter two annas a day. Women employed in sweeping in the streets of the little town received Rs. 3-8-o per month, or about again two annas per day. When paid for in money, the rate was similar for field labour, but often it was paid for in grain.

A skilled mason who stayed at home was able to make Rs. 10 per month when full work was obtained, but this was very uncertain owing to weather conditions, and the local need for the work. This is, as will be seen, about six annas a day. At Nasik or Poona a mason might, at the time of this investigation, get eight to ten annas a day if there was work, and one mason who had just returned from Satara had there received fourteen annas per day. A *Mahar* carpenter seems to be paid rather less in Saswad than a mason, but the work is more constant.

I regret that I have been unable to find out so far the scheme under which the *vatan* land belonging to the *Maharas* is cultivated and the produce divided. This matter will need further inquiry which promises to yield exceedingly interesting results when made for several villages.

A few, but only a few, of the *Mahars* of Saswad supplement their income by keeping fowls, or goats, or occasionally a she-buffalo for milk. It is rather astonishing that a large number of the households in a caste, which has no prejudices against keeping fowls, do not have some. If kept in the villages they cost very little, for they find their living largely among the garbage of the village, while there is an excellent market for eggs in Poona for which they are regularly collected. Only eight households out of fifty-seven, however, according to any records, keep them. Three households have one or more goats, while two possess a she-buffalo for milking purposes. Beyond this there seem to be no secondary occupations. Attention to village duties, at the necessary seasons attending to the *vatan* lands, where these have not been rented out, and the work as labourers or artisans absorb the whole remunerated activity of the *Mahars,* men and women, of Saswad village.

As to the general material condition one would perhaps describe it as being in a transitional state. The old isolation is gone. The people know that there is highly-paid work to be had by intelligent young men, either as labourers or as artisans at the large centres of population. And as a result, I feel, there is a tendency here, as there has been in several of the countries of the west and notably in England, for those who remain in the village to be the weak and the unintelligent. So long as the families and children remain, perhaps this is not a serious matter, but the men who have gone tend more and more to have their families away with them, merely returning occasionally for short periods to the village. This will lead to an exceedingly serious problem in the future, for the village *Mahars* will tend to become duller and duller from generation to generation, and the possibility of improvement more and more remote.

Such are the general results of the present study of the *Mahars* of Saswad. They are fragmentary and by no means complete, and especially the data which we have do not permit me to attempt to deal with the deeply interesting question of the relation of the village *Mahars* and their families to their *vatan* lands, *vatan* rights, and *vatan* duties. But they present to us a picture, partial no doubt, of the position of the *Mahars* as they are in the present period of rapid change. The old condition of a down-trodden class who could not escape from their bondage is rapidly disappearing. The fact that they can migrate to large centres where their outcaste character is much less continually obvious, where they can secure higher wages, and where a good workman can obtain his value whatever be his caste, and finally where he can be his own master in a sense which is not the case in his own village, all this is altering their outlook on life, and tending to the growth of independence of character, but at the same time, to the increase of vices, such as excessive drinking, against which caste rules give no protection. At these results one can be glad, and yet sorrowful. But the increase of independence will go on without anyone's help. The evil results will continue and increase, unless more means be taken than at present to prevent them in the great centres of population.

*Chapter 8*

# THE ECONOMICS OF A DECCAN VILLAGE*

## I

Among the questions which those who are interested in the future of India must most frequently ask themselves are—How far is the condition of the rural population improving or the reverse? and—How far are changing economic conditions in India affecting rural life both in its economic and agricultural aspects? It has hence long been my ambition to try and ascertain by careful and close inquiry in a comparatively few rural centres in the Deccan what is the actual present condition of the people. I have wished to do so because I believe that it is only by such intense inquiry in a comparatively limited number of centres that these questions can receive an adequate answer. The publication of Mr. Keatinge's book on "Rural Economy in the Bombay Deccan" very markedly increased my wish to conduct such an intensive investigation as I have mentioned. In that book he drew, largely from his own personal experience in Deccan villages and largely from Government and other records, a considerable number of conclusions with regard to the condition and tendencies of rural life in this part of India. The reasons which he gave for these conclusions seemed rather of a nature to stimulate further inquiries than to make one feel they could be accepted as they stand.

* Reprinted with permission from the **Indian Journal of Economics** (Allahabad), Vol. I (1916), Part 4, pp. 409-433.

This being the case I undertook with the help of a number of my friends and assistants to conduct a survey of a single Deccan village, and wish to lay before the readers of this journal a few of the more obvious results of my inquiries. Space will not allow of more than this, but those results which I can present will, I think, be found to be at least of a character to stimulate discussion and to add considerably to the data existing for an understanding of the condition of the rural population in this part of peninsular India.

Just one more point by way of introduction. The village which I shall proceed to describe represents one type only of the villages in the Deccan. It lies distinctly in the zone of fairly assured rainfall, and it is also a village in which practically no irrigation takes place. In order to get a full picture of village life even in the western Deccan two other types of village would have to be considered, namely, that in which a village possesses a large number of wells for irrigation, and in which the prosperity or otherwise of the village depends on the existence of these wells; and that type of village which is a creation of irrigation canals, in which cultivation is rich with an intensity such as is hardly to be seen in any other large area in India. The eastern Deccan again, with its different distribution of rainfall, the extreme uncertainty of its moisture, and the depth and heaviness of its soil, represents a different series of conditions which would need again a further and distinct investigation.

## II

The village with which I am to deal represents the dry cultivation of the western Deccan. It is named Pimpla Soudagar and lies nine miles from Poona to the west, and about two miles to the south of the Bombay Road, separated from it, however, by the river Pawana. It was chosen for this inquiry for three reasons: (1) It was sufficiently near to Poona to make it possible for us to investigate it; (2) It was considered to be sufficiently far from Poona to prevent the influence of an adjoining city being paramount in its life, and (3) It was sufficiently small to make it possible for an outsider to get a clear idea of the village as a whole. The area of the village amounts to one thousand and sixty-five acres. All is held under the ordinary *ryotwari* tenure of the Deccan. *Inam* rights, that is to say, the right of the whole or part of the land assessment, of two hundred and eleven acres are held by private inamdars, while there is a charge to another inamdar against the revenue of the whole village dating from the commencement of British rule in the Deccan.

This village lies on the banks of the Pawana river and is typical of many hundreds. The bank of the river is rocky and on a rocky ledge, parallel with the river, lies the village site. The outer boundaries of the village on every side but one are also high, rocky and uncultivable or with very shallow and poor soil. Between the comparatively rocky river bank and the high land round the village lies a hollow filled up during many centuries with the black soil characteristic of the trap region of Western India. The depth of this soil depends on the level of the rock below the surface. Where the rock is far below, there the soil is deep; where it approaches the surface there the soil is shallow. Below the soil there is either hard rock or the *murum* which represents disintegrated rock and which is of very little use from an agricultural point of view. The depth of the black soil, however, is the feature which determines the agricultural character of any particular field in the village. Round the edge of the village area, as I have already said, the soils are shallow. Sometimes there is no black soil at all, and the material which exists for the growth of plants is the disintegrated trap rock or *murum* which represents one of the driest and most 'hungry' soils I have ever seen. The main part of the village is better. The soil, though not rich, is retentive and capable under ordinary conditions of rain of giving fair returns with good cultivation. The rainfall amounts to between thirty and thirty-five inches per annum.

As I have already said, the village owes its existence to the rocky ledge on the banks of the river retaining the black fertile soil in the hollow behind, which has been deposited there by long-continued washing from the surrounding hills. As the hills are near, the soil is not—for black cotton soil—very heavy, for heaviness in soil in the trap area depends largely if not entirely on the distance to which it has been washed by water. The village derives its supply of drinking water from a long reach of the river which lies in face of it. As in so many other cases, the river current almost entirely fails after January in each year, but a ridge of rocks at the extreme east of the village causes the formation of a long pool which retains water throughout the year. The river being almost entirely composed of surface water is very soft, and if it escapes contamination higher up its course, forms an admirable drinking supply. Beyond this, the village contains twelve wells all of them with water which partially or entirely fails at least in April or May in each year. This being the case, the village is essentially one which depends on dry cultivation. Two or three of the wells are occasionally used for irrigated crops, and formerly even sugar cane was grown over a few acres every year. This has tended to disappear in recent years, though in the present season (induced perhaps by the present very high price of *gur*) several acres are again under the sugar cane crop.

The wells depend, as do most in the Deccan, on fissure water. They are a most doubtful, and a very expensive speculation. If a well taps a fissure of perennial water, the fortune of the man who digs it may be made. If the digging either reaches no water (and at least forty per cent of the wells sunk in the Deccan are failures), or only reaches water running for a few months in the year, it is a failure which usually ruins the man who undertakes it. Of the twelve wells in Pimpla Soudagar one only has a fairly constant supply, five have a supply which sinks to half its quantity in the hot weather, in four other cases the supply sinks to less than a quarter, while the last practically dries up each year in March, April and May. Such a condition has not tempted the people to experiment further, and one cannot blame them. Five hundred rupees invested in a well, unless there is a greater probability of success than has hitherto been reached with the available appliances, would be nothing but a gamble.

As in all Deccan villages, the cutting up of the land by rush of surface water has been the cause of the ruin of a considerable area of the village land. The damage is obvious, and some at any rate of the land near the river round about where the nalas break through the rocky ridge which borders it, is so badly cut up as to be spoilt. More important however is the constant wash of surface soil in small quantity from the fields—against which, except in a few places where irrigation has been or is carried out, very few precautions are taken. The limitation of embankment to such sites and the almost complete abandonment of the rest of the land to surface wash is, I fear, characteristic of very large areas in Deccan villages similarly situated.

I have already stated that the village consists of 1065 acres. Of this land 1020 acres is held by various holders. The balance consists of roads, *nalas* and the village site. The division of the village land can perhaps best be shown up by a table.

| | | | |
|---|---|---|---|
| Total area of village | ... | 1065 | acres |
| Area kept for public purposes | ... | 44 | acres |
| Area held by private owners | ... | 1021 | acres |
| Area held for cultivation | ... | 1006 | acres |
| Area fallow in 1914 | ... | 91 | acres |

This table at once indicates the characteristics of cultivation in a Deccan village. Its features are first the very large area held for cultivation and hence assessed which is not actually cultivated in any year and which only bears a very thin crop of grass and is used as inferior grazing. The existence of a large fallow area is very striking and shows how modern methods of maintaining the fertility of land have hardly reached the outlying unirrigated villages in the Deccan.

## III

As in all *ryotwari* tracts the revenue assessment of all the land goes direct to Government unless a definite grant to some one else, that is to say to an *inamdar*, has been made. The history of the growth of such grants in the later Maratha period is very interesting, and in the present case we have been able to trace it from 1699 to the present day. At various periods these *inam* grants were as follows:—

| | | 1699 | 1770 | 1779 to 1819 |
|---|---|---|---|---|
| | | acres | acres | acres |
| (1) | For temples ... | 84 | 84 | 84 |
| (2) | For village **patel** ... | 42 | 42 | 42 |
| (3) | For village **mahars** ... | 28 | 28 | 28 |
| (4) | For individual inamdars | 84 | 112 | 112 |
| (5) | For village **Kulkarni** | — | — | 12 |

The origin of the first or temple *inam*, devoted to a wellknown temple in the adjoining village of Chinchwad is a very old one and we have not got any record of it. Of the others only one now remains in its original form namely that for the *mahars*. The total area of the *inams* has been reduced and only 211 acres now remain under this privileged tenure. Only 35 per cent of the ordinary assessment is payable by the holders of these lands.

In 1819 a new factor was introduced in the grant to one man, who had assisted in the British occupation, of the whole government revenue of the village for his own life and his sons', and then of half of it to his heirs for ever. The original grantees wasted their substance and by mortgage or otherwise the actual amount received had sunk from Rs. 910 to Rs. 211 by 1840. In 1856 the grant was converted into a definite money payment, based on the revenue in 1817-18, and this is now distributed among six members of the family. This *inamdar* is thus in no sense a landlord—he is merely a person who has a definite money charge against the revenue of the village.

## IV

The history of the land revenue of this village is very interesting; and previous to the coming of the British it affords a very striking record of the disturbed state of the country. The actual amount of revenue obtained and the expenses at different dates are shown as follows:—

| | 1770<br>Rs. | 1791<br>Rs. | 1797<br>Rs. | 1811<br>Rs. | 1829<br>Rs. |
|---|---|---|---|---|---|
| Assessed revenue collected | 801 | 792 | 1037 | 1158 | 889 |
| Expenses ... ... | 74 | 202 | 697 | 310 | 186 |
| Net Revenue ... ... | 727 | 590 | 390 | 848 | 703 |

Nothing could illustrate in a more striking manner the increase of leakage of revenue during the later periods of the Maratha rule. The revenue assessed and collected was higher by 45 per cent, but the amount actually received by the Government was exceedingly variable. New local officers like the *deshmukhs* (of whom there were two) and *deshpandes* made a charge on the village revenue, the village officers took much more than their *inam*, the extra expenses charged increased enormously. In one year (1791) which I take as an example, the expenses charged against the Government in the records include *solatia* to Government officers, festivities at the time of visits of officers and others, money spent on wandering groups of *gosavis* and large amounts for which details are not even given. As far as one can judge, it would appear as if at least about Rs. 85 from the Government revenue was paid in moneys to greedy local officials of one sort or another beyond fees which they seem to have been entitled to charge. The extreme of this leakage is however seen in 1797 when out of a revenue of Rs. 1,087 only Rs. 390 reached the treasury. We do not know how much the people themselves were compelled to pay, but the time was that of the passage of the marauding armies of Shinde and Holkar. These leaders charged the village revenue with Rs. 179, and demanded a feast to their officers costing Rs. 14. Beyond this, presents to the messengers who brought the news of the Peshwa's court were thought to be legitimate charges against the Government revenue. Finally we get a glimpse of the unsettled condition of the countryside in the fact of money paid to a village servant who did not get his usual fees on account of a riot.

In 1811-12 conditions were a little better but the local leakage of revenue continued on an even greater scale than before. The assessment was higher, and the amount collected higher, and reached a maximum in 1817-18 when Rs. 1,226 were collected, an amount which fell to Rs. 889 ten years after British rule was established. After this we enter the modern period, the time of regular settlements and regular revenue returns. The actual figures for three dates are as follows:—

| | Rs. |
|---|---|
| 1849-50 | 1,115 |
| 1886-87 | 1,128 |
| 1914-15 | 1,660 |

We have thus a large increase in the land revenue during the later Maratha period, some reduction in the early British period, followed by a large rise. The gross amount now obtained for Government from the village is almost exactly double what was obtained in 1770. This does not mean that the assessment on the land is double, as some of the land whose revenue had been alienated at that time has since been resumed and its value now flows into the regular treasury.

## V

In no matter is the study of this village more interesting than in connection with the holdings. When we first find detailed records in 1770, there were only 24 holders of land, including the five inamdars. The number at various dates since that time has been as follows:—

| | Number of land-holders | Average size of holding |
|---|---|---|
| | | Acres |
| 1771 | 24 | 44 |
| 1791 | 41 | 25 |
| 1797 | 34 | 31 |
| 1811 | 54 | 19 |
| 1817 | 48 | 22 |
| 1829 | 58 | 18 |
| 1840 | 60 [1] | 17½ |
| 1914 | 156 | 7 |

During the Maratha rule the holdings were large and the holders few, and moreover, it almost appears as if the number declined in periods of disturbance and rose in periods of peace and good government. The same results have followed in the long British period, and now we are found with a reduction in the size of holdings which would have been almost inconceivable a hundred years ago. It is evident in fact that in the last sixty or seventy years the character of the holdings has altogether changed, until now the average holding is below what is necessary to maintain a cultivator's family. The number of holdings of various sizes now is as follows:—

[1] Previous to 1856 I have counted the inams as being in possession of single holders. We cannot get certain data about this but it makes the figures before that date and after not quite comparable.

| Size of holdings | Number of such holdings |
|---|---|
| More than 40 acres | 1 |
| 30 to 40 acres | 1 |
| 20 to 30 acres | 9 |
| 10 to 20 acres | 18 |
| 5 to 10 acres | 34 |
| 1 to 5 acres | 71 |
| Below 1 acre | 22 |

The excessive subdivision which has progressively increased during British rule is recognised as a very great evil. Mr. Keatinge has suggested that an economic holding of good dry land such as most of this village consists of, in the western Deccan and with an Indian *raiyat's* standard of life would be about ten to fifteen acres. Even therefore if each holding were held in one block, it is evident what a large proportion of them (81 per cent) are below this size. It means, therefore, that by far the larger number of holdings cannot under the most favourable circumstances, maintain their owners, but that they must rely on other occupations, either at home or away to support their families, or that they must sublet their holdings.

The conditions are worse even than this makes out because the land held is not only small in area but is divided into a large number of fragments. That is to say that when under Hindu law landed property must be divided among the members of a family, the division is made by partitioning each piece of land, and not by the various claimants taking the whole of various sections of it. The evil result is very apparent in the village now under consideration.

In fact out of 156 landholders in the village only 28 hold all their property in a single survey number and in a single piece. The land is split up into no less than 711 separate plots and the largest proportion of the plots are under one acre in size. The following table shows the extreme condition to which the subdivision of lands has gone:—

| Size of plots | Number of plots of this size |
|---|---|
| Over 20 acres ... | 1 |
| From 10 to 20 acres ... | 7 |
| From 5 to 10 acres ... | 21 |
| From 1 to 5 acres ... | 266 |
| From ½ to 1 acre ... | 211 |
| Under ½ acre ... | 212 |

So far nothing has been done by the public authorities to stop this excessive fragmentation of land which is an evil of far greater moment than that merely resulting from the subdivision of holdings themselves. It has in fact all the evil of very small holdings in that it prevents the use of machinery and labour saving methods, and on the other hand it has all the evils of large holdings in that it hinders the adoption of really intensive cultivation by any holder, which is the great advantage of small holdings. I do not want to discuss here the methods by which public authorities might attempt a way out of what is an obvious difficulty, but the matter has been very seriously tackled in two or three countries of Europe and also in Japan where a similar condition had grown up.

At first sight it would appear as if the people were settling this by natural methods, namely, by the abandonment of village life (though not of village land) by an increasing number of people, and by subletting a large proportion of the land. The extent to which the landholders have left the village without giving up ownership is shown by the fact that only 64 per cent of the registered landholders cultivate their holdings. The remainder have become labourers either in the villages or away from it, and though they still hold their land they have ceased being cultivators in the ordinary sense. The land is thus sublet, and the extent to which this is taking place is evident, for our investigations show only 109 actual cultivators in 1915 as against 156 landholders, or 140 if joint holdings really held by one of the joint holders are excluded. We have then the introduction of cultivators who are not landholders who form 17 per cent of the total number of actual cultivators and some of whom deal with larger areas of land than most of the actual landholders.

I had expected that these new non-landholding cultivators would be chiefly outsiders. This is not however the case. Practically all of them are members of the village who had not formerly cultivated land but who had official connection with the village, such as village mohammedan, the village barber, and so on, together with some members of the leading families who for some reason had no share in the land. It will thus be seen that the average size of the area cultivated by one man is greater than the area owned by one man. If we take 10 acres as being the smaller limit of an economic holding then we find that while only 17 per cent of the holdings are above this limit 23 per cent of the areas cultivated are larger than this. There is hence a tendency for the area of cultivation to be larger than the area owned. In spite of this tendency, however, the bulk of the areas cultivated are still incapable of supporting the families in ordinary village conditions.

If instead of considering the total area cultivated by one man we turn to the extent to which this is cut up into fragments we find an unexpected state of affairs. The number of fragments in a single man's cultivation is even greater than the number of fragments in a single holding. The number of separate plots cultivated separately in the village is not less than 729, and again by far the greater proportion of these are less than one acre in size. All the remarks which I have made regarding the evil effect of the minute subdivision or of the cutting up of holdings into minute fragments is of very much greater force when applied to the cutting up of cultivation into such fragments. In a village such as we are considering without any material amount of irrigation, it seems to be a disadvantage without any mitigating quality. If the plots are close together the evil is modified, but in other circumstances (and these are by far the most frequent) it means endless loss of labour and time without any compensating advantage.

I am not desirous of discussing here remedies for this state of affairs; but two results of this excessive fragmentation of holdings should I think have attention drawn to them. (1) It prevents effectually any outsider with capital from entering on cultivation on a large scale in this village. When a man may have to deal with 20 or 30 or even more landowners in order to get a stretch of land of 30 or 40 acres, and any one of those can spoil the continuity or self-contained character of his cultivation, nine men out of ten will refuse to have anything to do with such a scheme. (2) The impossibility of introducing outside cultivators with more enterprise makes the introduction of new and better ideas in agriculture exceedingly difficult. Small holders, when conditions favour the introduction of new ideas, make the most progressive cultivators in the world; when the land is so excessively cut up as in this village, which I believe is a typical one in the western Deccan, the people form a body as hard to move from tradition as any we can conceive of.

## VI

This leaves us to consider the character of the people, their caste, their material condition, their sources of income, and in general, the manner of life which results from the conditions in which they are placed. The village with which we have dealt is a small one, and from a recent house to house census which we have made it contains 112 families and a population of 556. By far the larger proportion of these are, of course, Marathas by caste. There are in the village, however, as in practically all other villages, a few families of village servants such as a single Mohammedan butcher, a chambhar for making shoes, a barber, four families of *mangs* whose principal duty is to make ropes, and in addition nine families of *mahars*.

Housing in the village is astonishingly good, considerably better than in many, if not most, villages I have visited. Each house, in fact, has an average area of about 200 square feet. There are five people per household—a rather large number at first sight. But many of the families are joint and the number of children is exceedingly small. Though there are nearly two men and two women per household, yet the number of children, both boys and girls, is only 164 in the 113 households of which the village consists. This is to me somewhat extraordinary and, if found to be the case in other villages, would seem to be an exceedingly serious matter. Some years ago I drew attention to the small number of children in certain communities of the depressed classes. The number here, in a village composed almost entirely of Marathas, presents the same condition of things in an even emphasized form.

By far the larger proportion of the people in the village work on the land, but the place is just near enough to Kirkee and to Poona to cause a considerable number to be attracted to work in these places. Of course recently, at the time when our census was taken, the ammunition factory at Kirkee was working at extreme pressure; but I must own that I was astonished to find no less than 87 men and boys from this village were going there every day to work. Considering that there are only 287 males in the village this means that about thirty per cent of the male population go five miles every morning to work in the Kirkee factory, and five miles back again in the evening.

Beyond this, eight people from the village carry milk daily into Poona City for sale. Each one conveys from five to six seers and may be considered to obtain about six annas per day for this service. Thus practically the total time of eight people is taken in carrying 100 pounds of milk to Poona City for sale in the city. There are a few others who work at jobs other than cultivation. Some who possess bullocks do carting on the roads. One or two stone quarriers live in the village, and there are besides the village servants of whom I have already spoken. So far hardly anyone has gone from this village permanently to work in Bombay. It has its representatives, however, at Bombay, Poona and at a few other large towns, who take temporary labour there and return to the village when needed. At the time of our inquiry there were six men away in this manner.

Thus what would be purely an agricultural village if the subdivision of land was less complete than it is, has developed into one from which nearly a third of the male population is practically compelled to work outside. One can say with confidence that if the village were further away from a labour centre, a large proportion of the people would have to go to reside in Bombay or elsewhere for work. The going is not a choice; it seems to be an economic necessity. Here the Kirkee factory takes the place of Bombay, but the principle is the same, and it seems the direct result of the fact that most of the holdings are now too small and too scattered to maintain their owners.

Such going away would probably be only the second effort on the part of the village people to retain a sound economic position. The first would be the borrowing of money either on mortgage of land, or on personal security with a guarantee of selling the crop through the money-lender who advances the money. I have been fortunately able to get what I believe is a fairly accurate statement of the debts of the people of the village. Debts on the security of land are now officially recorded; the remainder is generally so well known that a false statement would soon be detected. The debts are Rs. 5,820 on the security of land under one of the various forms of mortgage and Rs. 7,495 on personal security, or a total of Rs. 13,315 on the village. This is an excessively large amount, and as a result the interest is very high, varying from 12 per cent to 72 per cent. The average interest charged is 19½ per cent. This means an average indebtedness per family of Rs. 118,—and if we take all the debts as a charge on the land, an average charge of nearly Rs. 13 per acre.

Considering that the average sale value of land in the village is probably not more than from Rs. 70 to Rs. 90 per acre, this would mean that the whole of the land is involved to the extent of one sixth of its value at least.

This assumes, of course, that the land is the only capital, and this is very nearly the actual state of affairs. Beyond the land the cattle, numbering 284, represent the chief asset of the people. The value of implements may almost be neglected, and that of houses is very small. I hope later to get more accurate figures, but for the present I think the total sale value of the village and its property cannot be much more than one lakh of rupees. If this is the case its indebtedness is over 13 per cent of this sale value, and entails an annual charge on the village of Rs. 2,600.

This load of debt seems to me to be the first result of the extreme subdivision of the lands, or in other words it is due to the increase of the population tied to the land without the increase in the intensity of the cultivation necessary to support them. When by incurring debt the position became no clearer, then a certain number of the members of most families have gone away for work. In the case of seventeen typical cultivators' households for which we have data on which, I think, we can rely, no less than 33 per cent of the total income came from such outside sources. I cannot help thinking that this is likely to increase, and that either a complete separation from the land of a considerable proportion of the young male population will occur, or else while retaining an interest in the land as owners, they will in a much larger measure than at present become labourers elsewhere and sublet the land which they hold. The land in the village has still a rental value of Rs. 7 per acre on the average, or from four to five times the amount of the Government assessment.

The only alternative to this it would seem, would be a very considerable increase in the intensity of the cultivation by more careful treatment of the crops the people now have, or else by the introduction of more paying crops than they grow at present. We must therefore now consider their crops and methods.

## VII

The cultivation at Pimpla Soudagar is absolutely typical of the dry villages of the Western Deccan. Very little advance in implements and methods seems to have yet been made, though some new crops have been introduced in recent years. As to implements of cultivation a complete set such as is used by the people would cost about Rs. 40, and they are all made locally from beginning to end. The wood is grown in the village, the carpenter is a servant of the village, and except for the very small quantity of iron required for the tip of the plough-share and the blade of various other cutting implements, the village would be quite independent. All repairs to these implements are paid for in *baluta* or a fixed charge on each crop produced,—but the carpenters are paid in cash for new implements made. The use of the modern iron plough is only just beginning, and the hire of these from Poona shows signs of being taken up.

I have already alluded to the very large area of fallow. This is always a sign of backward agricultural methods, of insufficiency of manure, and generally of inadequate cultivation. As a matter of fact the only manure available in Pimpla is that from the cattle, and some of this is at least preserved in a series of primitive manure pits outside the village site. There are no cattle sheds in the village whatever. The animals either remain outside or else are tied in the verandahs of houses. No use whatever is made of the cattle urine and two thirds of the dung is used for making into dung cakes for the Poona market. All around the year, except in the rainy season for three or four months the process of preparing cakes for burning goes on on the rocky parts of the river bank. The sole manure in the village is therefore at most one third of the dung alone from under three hundred cattle, most of which are away the greater part of the day where the dung cannot be utilised for manure at any rate. I estimate that not more than 400 cart-loads or 150 tons of cattle manure, mixed with house refuse, is available each year in the whole village.

It might at once be asked why outside manures are not purchased. The people of this village, and of all others where irrigation facilities do not exist, maintain, and I believe that they are right, that unless you can secure an adequate and certain water supply, it does not pay to lay out money in buying manures. The water is so uncertain, that the risk of wasting this money is too great. The only manures which it has paid to apply, so say the people, are the cattle manure which costs them nothing, and that obtained by allowing shepherds to feed and keep their sheep on the land which it is desired to manure. The crops are grown, therefore, with very little manure at all, for 150 tons of cattle manure per annum, and say 40 acres on which sheep have been allowed to stay, makes a very small quantity. The yields of crops are hence very small. The small yield is probably still further diminished by the fact that rotations seem very little attended to. Only forty acres in the whole village were under leguminous crops (except peas, of which more later), or four per cent of the cultivated area, and these are recognised as the crops which keep the soil in good condition.

The cause of this seems to be that the people will not cultivate a crop unless they consider it is fairly certain, and unless it is either needed for the food of their own families and animals, or else brings cash into their hands. The idea of using crops to build up the fertility of the land has not entered the minds of the people at all. And this is the case in spite of the ease with which it could be done. There seems to me a very large field of work in villages similar to Pimpla in showing to the cultivators how, at a minimum of cost, the fertility of the soil can be built up by a green manuring crop in the rainy season.

The actual cropping of the village is very typical. *Jowar*[1] occupies by far the greatest area, giving, as it does, both grain and fodder, while *bajri*[2] either alone or with *tur* (pigeon pea) takes its place on the lighter land. These crops form the centre of the cropping scheme in most dry villages of this part of the Deccan. The produce of neither of them is usually sold, and they form the staple food of man and beast.

Side by side with them are the money crops—those which are sold and for which cash is received. I do not know why cotton is not grown here,—but there is none. Formerly the money crop used to be wheat, rotated with *gram*, and twenty years ago we were told the village was largely a sheet of wheat in the cold weather. This has now been largely abandoned owing, the people say, to the repated failure of the late rain in October and November which is essential to the successful cultivation, and to the increasing need of fodder for cattle, of which wheat furnishes very little as compared with *jowar*. There is no doubt that in the last fourteen years the late rains have been very much less certain than they were thirty years ago. Between 1877 and 1890 adequate rain for wheat fell in Poona in October in twelve years out of fourteen, from 1900 to 1914 it fell only eight times out of fourteen. In November in the former period the rain was sufficient in nine cases, in the latter only in three cases. Whether this is a cyclic variation and a period is coming when wheat will be again a suitable crop we do not know, but the change in the last twenty-five years is as clear as any such change can be, and the cultivators are quite right in their contention that the rain has become less suitable for wheat grown without irrigation.

It is curious to find that wheat has been replaced by two crops, largely grown together, whose disposal depends purely on the nearness of the Poona market. We were astonished indeed to find that wheat had to a considerable

[1] **Sorghum vulgare.** [2] **Pennisetum typhoideum.**

extent been replaced by carrots and peas as a money crop in the village. These crops are now grown extensively, and the crop is stated to need far less rain than wheat required. Carrots give a cash crop in Poona, so do green peas; the tops of the former and the vines of the latter are excellent fodder.

The rest of the land is devoted to less important crops. Niger seed or black *til* is an oilseed which forms the favourite crop of light and shallow land. Gram for some reason which I cannot understand is now a minor crop on the same class of land. Groundnut which has been such a great success elsewhere in Western India, is not welcomed here, and is said not to pay nearly as well as carrots and peas.

Let us consider the returns from these crops, for it must always be recognised that each crop, in the cultivators' minds, must pay its way. We can calculate these in two ways. One is that usual here, when we give the return supposing all labour to be paid for. In the other we suppose the cultivator's labour is his own, and that it should not be calculated in the return. The results with each of the crops, under both methods of caculation are as follows:—

| | Profit per acre paying for labour | | Profit per acre with cultivators' own work |
|---|---|---|---|
| | Rs. | as. | Rs. |
| **Jowar** ... | 12 | 6 | 18 |
| Bajri and tur ... | 7 | 1 | 13 |
| Carrots and peas | 13 | 0 | 33 |
| Wheat ... | 12 | 0 | 15 |
| Gram ... | 13 | 10 | 15 |
| Niger seed ... | 4 | 0 | 6 |
| Groundnut ... | 7 | 12 | 27 |

The first column gives the return to a capitalist farmer; the second to a working cultivator. The difference is greatest with those crops which require the greatest labour, and one can easily see why after the food for the household has been produced, a cultivator prefers a crop like carrots and peas to one like wheat, apart altogether from the question of the climatic difficulty in recent years. The latter crop (wheat) would be the natural resource of a capitalist; the former is naturally preferred, and is likely to be more preferred in the future by working cultivators.

The average net return per acre of land cultivated in the village to a working cultivator is about Rs. 14-8. Ten acres will therefore bring Rs. 145 or about Rs. 12 per month. It will be seen that this is little enough for the maintenance of a family; a less quantity of land will make such maintenance impossible.

VIII

With this I must stop. I have only been able to sketch in the slightest manner some of the more obvious results of our inquiry. But we have seen how, in a typical dry village in the Deccan, the population has increased, the number of land holders has increased, and the holdings have become so split up into fragments, that not only are the areas now held too small in the vast majority of cases to maintain the family which hold them, but also they now exist in the most awkward form for economic cultivation. We have seen how this was first met by the incurring of debt, which now hangs round the neck of the village to the extent of 13 per cent of the total value and pays about 20 per cent on the average in interest. The cultivation of a dry area can never pay for this, and the next step has been the going out of nearly 30 per cent of the whole male population to non-agricultural work. This enabled the balance to be kept for the moment,—but the next step seems to have been the modification of the cropping so as to make the village more self-sufficing for food and fodder and the devotion of more area to a crop which at least pays a working cultivator better than that formerly grown. What will be the next? I can see no hope unless one of three things happens. Either the intensity of cultivation must be increased, and to do this, the removal of some of the present load of debt and hence more easy financing seems to be needed. In addition better use must be made of the soil resources either by the demonstration of the paying character of green manuring and other methods, or else new crops must be found which will yield better and more certain returns in the soils of this village. Or finally some change must be made which will prevent, if not the subdivision, the excessive fragmentation of the land, and will remove from an interest in the village land a large number of those who now have it. I should myself regret to see this removal of population, and the creation of a large landless proletariat, but I do not wish to impose my views in the present paper. I have tried to present a cold-blooded sketch of actual conditions,—and I leave others to thrash out the many problems which our study of this village raises,—itself one of many thousand similar communities in the Bombay Deccan.

*Chapter* 9

# THE EFFECTS OF RISE OF PRICES ON RURAL PROSPERITY*

The effect of a rise in prices on the condition of the rural population in India has been a matter of very serious difference of opinion. Some authorities have held that, seeing that the produce of the land was the source of the prosperity of landholders and village people generally, the fact of their being able to obtain a higher price for it would obviously lead to their general economic advantage. Others, equally well-informed, have stated that whatever would happen in theory, in practice every rise in price has meant more difficulty and more poverty in the rural areas. The matter is really of great importance, because, among other reasons, in a country of periodic land settlements such as occur in India changes in price have always been considered a valid factor in determining changes in land assessment.

There are many methods of making a study of the effect of a rise in prices on rural prosperity. But the series of village studies in Western India which we have made in recent years seem to give us data for tackling the question at close quarters. The method of study in each case has been similar. The people of the village have been studied family by family. The sources of revenue have been ascertained for each family, as well as the necessary expenditure, including the payments of interest on debts

* This paper, originally presented by Harold Mann and N. V. Kanitkar to the Madras Conference of the Indian Economic Association, 1919-20, is reprinted with permission from the **Proceedings** of the Conference Madras, 1920, pp. 30-40.

which they had incurred. To complete such a balance sheet, family by family, naturally involved a certain number of assumptions as to the gross crop per acre, and hence the gross revenue from land, and also the minimum cost of living and clothing at the time when the study was made and at the prices prevailing. If the assumptions made be correct, then the survey affords a means of ascertaining the economic position of every family in the village, and also the effect which any change in prices or other conditions would make on this position. It only requires, in fact, a recalculation of the family data which we already possess, on the basis of a new level of prices, to enable a view to be obtained of the economic effect of any rise which may have occurred.

In attempting to carry out this scheme of utilising our family data for the study of the present problem it may be stated that we have used the figures in hand for two Deccan villages. One of these is Pimpla Soudagar, an account of which has been already published, and which is situated ten miles west from Poona. The other is Jategaon Budruk, a somewhat larger village about twenty five miles east from the capital of the Deccan. In the case of nearly every family in these two villages we have considered what would be the effect (1) of a general fifty per cent rise in prices without rise in wages, and (2) of a general fifty per cent rise in prices with a corresponding (fifty per cent) increase in wages at the same time.

In determining what meaning was to be attached to the term 'a general rise in prices', we have made it include a rise involving all types of materials used by the village. Two general expenses had, however, to be excluded. On the one hand, land revenue, being fixed by a thirty years' settlement, cannot rise and must be considered to remain constant. On the other hand, the interest charged on loans in the village has not altered materially as a result of the rise of price of articles or of money in the great centres. The only reason we can think of for this is that the rate of interest was already so high—the average for the two villages is 19¼ and 23 per cent respectively—that it can hardly rise further and is hence independent of moderate rises in the value of money elsewhere. But except for land and money, the price of every commodity was assumed to rise to an equal extent, and to stand at a level fifty per cent higher than that previously occupied. The results have been exceedingly interesting, and we venture now to lay them out.

To take the village of Pimpla Soudagar first, we may say that when the original study was made we divided the 103 families in the village, for which we had data, into three groups Group I. consisted of families which were able to maintain themselves, at a standard of efficiency and respectability which they considered satisfactory, from the income from land and cultivation alone. Group II. consisted of those who were, at their own standard, still in a solvent condition, though income from land had to be supplemented by income from their labour to make them so, while Group III. included all those families who could not maintain themselves, at their own standard, even from income from land combined with income from labour, without going into debt. The economic position, in other words, of Group I. and II. was sound; that of Group III. was unsound.

The economic position of the village families under these groups was as follows in 1916:—

| Group | No. of families | No. of persons | Income from Land | Income from Labour | Total Income | Neccssary Expen-diture | Interest |
|---|---|---|---|---|---|---|---|
| | | | Rs. | Rs. | Rs. | Rs. | Rs. |
| I. | 8 | 25 | 1.699 | 232 | 1.931 | 1,254 | 54 |
| II. | 28 | 136 | 1,740 | 6,710 | 8,450 | 6,890 | 662 |
| III. | 67 | 349 | 4.899 | 6,169 | 11,068 | 16,987 | 1,809 |
| Total | 103 | 510 | 8,338 | 13,111 | 21,449 | 25,131 | 2,515 |

Taken group by group we have in summary:—

1. In Group I. (solvent from land only) there was an excess of Rs. 677 per annum on eight families, or Rs. 84.6 per family per annum.

2. In Group II. (solvent from land and labour) there was an excess of Rs. 1,560 per annum in twenty-eight families, or Rs. 55.7 per family per annum.

3. In Group III. (insolvent) there was a deficit of Rs. 5,919 per annum in sixty-seven families, or Rs. 88.4 per family per annum.

On the whole on the 103 families for which we have data there was a deficit of Rs. 3,682 or Rs. 35.7 per family per annum.

Let us see what would be the effect of a rise in the cost of commodities of 50 per cent. This means, of course, a rise in the income from land, and a rise in expenditure on all items except labour and interest. The immediate result is, dividing the population into the same groups as before, to give us results as folows:—

| Group | No. of families | No. of persons | Income from Land | Income from Labour | Total Income | Necessary Expen-diture | Interest |
|---|---|---|---|---|---|---|---|
| | | | Rs. | Rs. | Rs. | Rs. | Rs. |
| I. | 10 | 30 | 3,233 | 418 | 3,651 | 2,423 | 318 |
| II. | 6 | 32 | 1,683 | 1,080 | 2,763 | 2,303 | 238 |
| III. | 87 | 448 | 7,608 | 11,667 | 19,275 | 31,056 | 1,959 |
| Total | 103 | 510 | 12,524 | 13,165 | 25,689 | 35,782 | 2,515 |

Taken group by group we have in summary:—

1. In group I. (solvent from land only) there was an excess of Rs. 1,228 per annum on ten families, or Rs. 122.8 per family per annum.

2. In group II. (solvent from land and labour) there was an excess of Rs. 460 on six families, or Rs. 76.7 per family per annum.

3. In group III. (insolvent) there was a deficit of Rs. 1,178 per annum on 87 families, or Rs. 135.4 per family per annum.

On the whole on the 103 families for which we have data there was a deficit of Rs. 10,093 or Rs. 98 per family per annum.

These results are very striking. The general effect of the rise of prices on the village as a whole is disastrous, but it has benefited a few families, which depend very largely on their own labour on their own land, and also *have very large debts.* The two families which have risen to the first group, show the following figures before and after the rise in prices.

| | Income from Land | Labour | Total Income | Necessary Expen-diture | Interest | Excess of Income |
|---|---|---|---|---|---|---|
| | Rs. | Rs. | Rs. | Rs. | Rs. | Rs. |
| Before rise in prices | 457 | 186 | 613 | 540 | 264 | 73 |
| After rise in prices | 685 | 186 | 871 | 650 | 264 | 221 |

These two families, therefore, rise from the second to the first group.

But while the effect has been beneficial in these two cases, a very large number of families are driven from the second group into the insolvent class. One family rises from the third to the second group, because nearly eighty per cent of its income is derived from land and also because of large interest commitments. It is curious to find the possession of large debts being a source of economic strength under the changed conditions!

The effect of the rise in prices on the position of the six families now belonging to group II. is shown in the following table:—

| Rise in Prices | Income from Land | Labour | Total Income | Necessary Expen-diture | Interest | Excess of Incom |
|---|---|---|---|---|---|---|
| | Rs. | Rs. | Rs. | Rs. | Rs. | Rs. |
| Before | 1,122 | 1,080 | 2,202 | 1,674 | 235 | 528 |
| After | 1,683 | 1,080 | 2,763 | 2,303 | 235 | 460 |

Group III. (or the insolvents) becomes very much larger as a result of this rise in prices. If one family leaves this group twenty-one descend to it, and among these 21 families a total excess of income in the year of Rs. 1,028 becomes a deficit of Rs. 901. The effect of the rise on the condition of the 66 families which originally belonged to this group is equally disastrous, and an original deficit of Rs. 5,914 per annum now becomes Rs. 10,207, being an increase of 72.5 per cent.

Taking the village as a whole, an original deficit of Rs. 3,682 in all the families in the village taken together, has increased in consequence of the rise in prices to Rs. 10,093, or by 174 per cent.

So much for the effect of a rise in prices without a rise in wages. It may enrich a few, but it is disastrous for the village as a whole. But if wages rise also to the same extent as prices, what will be the economic result? The following table shows the figures:—

| Group | No. of families | No. of persons | Income Land | from Labour | Total Income | Necessary Expen-diture | Interest |
|---|---|---|---|---|---|---|---|
| | | | Rs. | Rs. | Rs. | Rs. | Rs. |
| I. | 10 | 30 | 3,233 | 627 | 3,860 | 2,536 | 318 |
| II. | 27 | 140 | 2,541 | 9,974 | 12,515 | 10,086 | 508 |
| III. | 66 | 340 | 6,735 | 9,109 | 15,854 | 23,871 | 1,689 |
| Total | 103 | 510 | 12,509 | 19,710 | 32,229 | 36,493 | 2,515 |

Taken group by group we have in summary:—

1. In group I. (solvent from land only) there was an excess of Rs. 1,324 per annum on ten families, or Rs. 132.4 per family per annum.

2. In group II. (solvent from land and labour) there was an excess of Rs. 2,429 per annum on 27 families, or Rs. 90 per family per annum.

3. In group III. (insolvents) there was a deficit of Rs. 8,017 per annum on 66 families, or Rs. 121.5 per family per annum.

On the whole, in the 103 families for which we have data, there was a deficit of Rs. 4,264, or Rs. 41.4 per family per annum.

The position now reached is, of course, as would be expected, not unlike that originally existing. There were 36 families in a sound position originally; now there are 37. Two rise from the second to the first group on account of increased income from land, and the stationary expenditure on interest; and one family rises from group III. to group II. for the same reasons.

The ten families now in group I. have improved their position. Formerly they had an excess of income of Rs. 750; now this excess is Rs. 1,324. There has, in fact, been a rise of 76 per cent. That is to say the excess has increased in greater proportion than the rise in prices and wages. The twenty-seven families in Group II. have also improved their economic position. Formerly they had an excess of income of Rs. 1586; now this excess is Rs. 2,429. There has, in fact, been a rise in the nett profit per annum of 53 per cent, or again, in slightly higher proportion than the rise in prices and wages. The nett loss of the sixty-six families now insolvent (group III.) according to our standard has increased from Rs. 5,914 to Rs. 8.017, or by nearly 36 per cent. This is very large, but the increase in loss is not quite in the same proportion as the rise in prices.

Generally, we may draw the following conclusions with regard to the effect of a rise in prices and wages to the extent indicated on the economic condition of the population of the village of Pimpla Soudagar.

I. If prices rise 50 per cent without a rise in wages we find as follows:—

(a) The people who depend on land alone, *which they work with their own labour,* and who have sufficient land to maintain them in a sound position before the rise in prices are much better off.

(*b*) Where there is a combined dependence on land (worked by themselves) and on labour, and where there was a sound economic position before the rise in prices, the final position depends solely on the proportion between the income from self-worked land and that from labour.

(*c*) The man who benefits most is he who works his land by labourers, in other words, the non-cultivating proprietor.

(*d*) The general effect on the village population is disastrous, and the annual deficit among the families belonging to the village, added together, increases enormously.

II. If prices rise 50 per cent, accompanied by a corresponding rise in wages, we find as follows:—

(*a*) The people who depend on land alone, which they cultivate with their own labour, and who had sufficient land to maintain them in a sound position before the rise in prices, are again much better off.

(*b*) Where there is a combined dependence on land (worked by themselves) and on labour, and where a sound economic position existed before the rise in prices, the position is still improved, and the improvement is slightly greater than the rise in prices.

(*c*) The non-cultivating proprietor is not appreciably affected, except in so far as he has large debts.

(*d*) The general effect on the village population is to lower their economic position. The annual deficit of the families in the village taken together, has been increased by 16 per cent.

III. With a rise of prices, whether wages increase or no, two general results seem to ensue:—

(*a*) The gulf between the solvent and the insolvent classes tends to widen. The vast majority of the people who were solvent, especially if they were solvent from land, seem to become more solvent: the insolvent previously become in most cases more insolvent.

(*b*) As interest has not risen with rise in prices, those who had large debts incurred previously have suffered less in proportion than the others.

So much for the results obtained by a study of the families in Pimpla Soudagar. But that village has been described as not typical because a considerable proportion of the population obtain their livelihood by working at Poona, or in Kirkee, which is close by. It was, hence, of importance to investigate a place, otherwise of a similar character, but far from any large labour centre. The village of Jategaon Budruk was chosen, being 25 miles from Poona, and having no large non-agricultural demand for labour anywhere near. Here there was a population of 732 in 147 families, and the results of our enquiries were treated in exactly the same way as already described for Pimpla Soudagar.

The economic position of the village families, divided into the same groups as before, was as follows in 1917, that is to say, before the recent rapid rise in prices began:—

| Group | No. of families | No. of persons | Income from Land | Income from Labour | Total Income | Necessary Expen-diture | Interest |
|---|---|---|---|---|---|---|---|
| | | | Rs. | Rs. | Rs. | Rs. | Rs. |
| I. | 10 | 43 | 3,149 | 1.507 | 4,656 | 2,338 | 210 |
| II. | 12 | 25 | 250 | 1.213 | 1.463 | 1,245 | 66 |
| III. | 125 | 664 | 12.408 | 5.826 | 18,234 | 35,393 | 6.479 |
| Total | 147 | 732 | 15,807 | 8,546 | 24.353 | 38,976 | 6,755 |

Taken group by group we have, in summary:—

1. In Group I. (solvent from land only) there was an excess of Rs. 2,318 per annum on ten families, or Rs. 231.8 per family per annum.

2. In Group II. (solvent from land and labour) there was an excess of Rs. 218 per annum on twelve families, or Rs. 18.1 per family per annum.

3. In Group III. (insolvent) there was a deficit of Rs. 17,164 per annum on 125 families, or Rs. 137.3 per family per annum.

On the whole, on the 147 families for which we have data, there was a deficit of Rs. 14,628 or Rs. 99.5 per family per annum.

It is at once obvious that this is a village where the original difference between the more wealthy classes of the village and the poorer inhabitants is much greater than in the case previously studied, and also that the proportion of insolvent families is much greater also.

Let us see what would be the effect of a rise in the cost of commodities by 50 per cent, without rise in wages or in interest. This is shown in the following table:—

| Group | No. of families | No. of persons | Income from Land | Income from Labour | Total Income | Necessary Expen-diture | Interest |
|---|---|---|---|---|---|---|---|
| | | | Rs. | Rs. | Rs. | Rs. | Rs. |
| I. | 12 | 58 | 6,088 | 1.507 | 7.595 | 4,652 | 402 |
| II. | 3 | 4 | 141 | 227 | 368 | 343 | 24 |
| III. | 132 | 670 | 17,474 | 6,809 | 24,283 | 49,952 | 6.329 |
| Total | 147 | 732 | 23,703 | 8,543 | 32,246 | 54,947 | 6,755 |

Taken group by group, we have, in summary:—

1. In group I. (solvent from land only) there was an excess of Rs. 2,943 per annum on twelve families, or Rs. 245.2 per family per annum.

2. In group II. (solvent from land and labour) there was an excess of Rs. 25 per annum or three families, or Rs. 8.3 per family per annum.

3. In group III. (insolvent) there was a deficit of Rs. 25.669 on 132 families, or Rs. 194.4 per family per annum.

On the whole, on the 147 families for which we have data, there was a deficit of Rs. 22,701, or Rs. 154.4 per family per annum.

The position revealed by these figures is extraordinarily similar to that shown in the village previously stated. There were 22 families in a sound position originally, now there are only fifteen. Two families rise from third to the first group on account of the increased income from land and the stationary expenditure on interest, which forms proportionately a very large part of the total expenditure. A very large proportion (75 per cent) of those in group II. fall into the insolvent class: this would perhaps be expected, as they depend for their solvency largely on the income from labour.

The twelve families now in group I. have improved their position. Formerly they had an excess of income of Rs. 2,277: now this excess is Rs. 2,943. There has, in fact, been a rise of 29 per cent, which though less proportionately than the actual rise in prices, is substantial. The three families remaining in group II. have actually lost economically by the rise in prices, and barely maintain their solvency. The net loss of the remainder (Group III.), or the insolvents according to our standard has increased by 51 per cent, a proportion practically the same as that of the rise in prices.

Taking the village as a whole, the effect of a rise in prices of 50 per cent without a rise in wages has been to convert an original deficit of Rs. 14,628 into a deficit of Rs. 22,701, or an increase of 55 per cent. The proportion is not so large in the case of this village as in that previously studied, on account of the larger proportion of the income due to the self-working of the land, and also on account of the very much greater indebtedness of the village.

So much for the effect of a rise in prices without a rise in wages. If wages rise also to the same extent as prices, the economic result is shown by the following table:—

| Group | No. of families | No. of persons | Income from | | Total Income | Necessary Expenditure | Interest |
|---|---|---|---|---|---|---|---|
| | | | Land | Labour | | | |
| | | | Rs. | Rs. | Rs. | Rs. | Rs. |
| I. | 12 | 58 | 6,088 | 2,260 | 8,348 | 4,704 | 402 |
| II. | 13 | .29 | 373 | 2,117 | 2,490 | 2,127 | 90 |
| III. | 122 | 645 | 17,242 | 8,420 | 25,662 | 48,231 | 6,263 |
| Total | 147 | 732 | 23,703 | 12,797 | 36,500 | 55,062 | 6,755 |

Taken group by group, we have, in summary:—

1. In Group I. (solvent from land only) there was an excess of Rs. 3,644 per annum on twelve families, or Rs. 303.7 per family per annum.
2. In Group II. (solvent from land and labour) there was an excess of Rs. 363 per annum on 13 families, or Rs. 27.9 per family per annum.
3. In Group III. (insolvent) there was a deficit of Rs. 22,569 per annum on 122 families, or Rs. 185 per family per annum.

On the whole, on the 147 families for which we have data, there was a deficit of Rs. 18,562, or Rs. 126.3 per family per annum.

The general conclusions are, therefore, almost identical with those reached in the case of the village previously studied. There were 22 families originally in a sound position; now there are 25, or a slight increase. All those previously solvent remain so, and three are added from the insolvent class. The improvement in the position of the latter is due, as before, to the increased income from land, and to the interest being stationary.

The twelve families now in Group I. have improved their position. Formerly they had an excess of income of Rs. 2,277; now this excess is Rs. 3,644. There has been, in fact, a rise of 60 per cent—that is to say, the excess has increased in greater proportion than the rise in prices and wages. The 13 families in Group II. have also slightly improved their economic position. Formerly they had an excess income of Rs. 195; now this excess is Rs. 363. Though the annual profit in these families is small, yet it has risen by 86 per cent, or again, in larger proportion than the rise in prices. The nett loss of the 122 families now insolvent according to our standard (Group III.) has increased from Rs. 17,119 to Rs. 22,569, or by nearly 31 per cent This is again large, but the increase in loss is not quite in the same proportion as the rise in prices.

The results in the two villages under study are, therefore, exceedingly similar, and the results drawn on a previous page for Pimpla Soudagar seem to be perfectly true for Jategaon Budruk. We would particularly call attention to the general conclusions. These show the mitigating effect of a previously high rate of interest or the results of a rise in prices. And more important still they show how a rise in prices will tend to emphasise economic differences, especially where the change in prices does not connote a corresponding change in the rate of interest.

So far we have been dealing with the effect of the rise of prices on the population as a whole: the matter is of special interest, however, *in connection with landholders*, as it is they who would be affected by any rise in assessment which might be proposed on account of such a rise. Let us see how this class would be affected? We have only got figures for Jategaon Budruk, but these are clear.

There are only twelve families in this village who are not landholders, and if the remainder are classified as above, we get the following result.

| | |
|---|---|
| 1. *Nett* deficiency of the landholders per annum before rise in prices | Rs. 14,631 |
| 2. Nett deficiency of the landholders per annum after 50% rise of prices, but no increase in wages | „ 21,937 |

3. Nett deficiency of the landholders per annum, after 50% rise in prices and in wages. „ 18,600

It thus appears that the conclusions already given for the village as a whole are substantially true if the landholders only are considered, and this conclusion seems of great importance.

We are not, at present, prepared to extend the application of the conclusions reached in this chapter, from the typically food producing villages which we have considered, to the places where large crops of cotton, sugar, or similar crops are produced for sale. These cases demand special study, and we feel that it is impossible to say, from general considerations only, how such villages have been affected by the recent rise in prices. We may have the opportunity to make such a study in these cases later on, but, in the meantime, it must be assumed that they have necessarily, as a whole, benefited by a rise in prices such as has occurred.

Leaving such special cases on one side, we may consider that, in general, the evil effects of rise in prices on the general condition of the rural population can, in the Deccan, at any rate, hardly be gainsaid. And if this is clear, the question of the maintenance of low prices by any means may be a matter for much more serious concern than has hitherto been realised.

*Chapter* 10

# INDIA IN THE SIXTEENTH CENTURY*

The way in which we look at present day economic problems in India is, and must be, very largely determined by our opinion as to the progress, or otherwise, of the country within the last three or four hundred years. Is India a richer country than it was at the time of Akbar,—or even in the time of Shivaji? Is it a better country to live in either for the landowners, the town-dwellers or the mass of the rural population? According as we answer these questions, our attitude to almost the whole of contemporary public questions will be determined and it is well squarely to realise that two radically different answers to these questions are current among those who have given attention to the subject.

I must confess that, during the first years of my stay in India, I was utterly bewildered by the absolutely different opinions on what appears to be a simple question of historical fact, put forward by keen, earnest students and competent authorities on this subject. One day I read that the wealth of India was increasing by leaps and bounds and, not only this, but that the general average wealth of the mass of the people was also growing rapidly the next day I was informed, on apparently equally reliable evidence, that the poverty of the vast majority of the people has never been so great as it is now, and that far

*Reprinted with permission from the **Indian Review**, Madras. Vol. XXI (1920), No. 8, pp. 481-482.

from there having been economic progress, the course of general well-being has been backward rather than forward. If there has been,—so the adherents of the latter view held,—any increase in the actual total wealth (and this is doubtful), it is now so concentrated in a few hands, that the mass of the people are steeped in a more hopeless poverty than at any time in their history. A considerable mass of evidence was put forward for each opinion, and an examination of this left me, at any rate, with the conviction that the data now available and accessible are not sufficient to decide the question, that the treatment of such material as exists has nearly always, often unconsciously, been of a partisan character, and that confusion has been widely introduced into the discussion by writers of different periods,—and even by writers of to-day,—using the same terms to mean very different things.

And yet it is of great importance that a clear idea should be obtainable as to the direction of economic change in the centuries during which mediaeval has developed into modern India. There are obviously two or three methods of approaching the question in order to obtain such a clear view. One of these consists of a re-examination of the existing printed data for any period of the past, from a cold-blooded and impartial point of view; a second depends on the accumulation of new material which can be gathered from manuscript records existing in almost every part of India, either in Government record rooms or in private hands; a third will be the result of a close examination of existing conditions in a particular area and, working back from these, the deduction of how the present state of affairs must have arisen. All of these methods will have to be used before we can really come to a conclusion which is satisfactory to ourselves and which will convince others. Nevertheless, we owe a great debt of thanks to anyone who will take the trouble to apply any of these lines of inquiry to the study of the economic conditions of a particular period and their relationship to those of the present day.

Such a debt of gratitude do we owe to Mr. Moreland[1] for the very careful re-examination of existing published accounts of the condition of India at the death of the Emperor Akbar. It does not aspire to gather together new material, but within its scope it is a courageous attempt to give a clear picture of what India was like at the time of the great Mogul Emperor. There are points in the book which I think are open to criticism. I feel that the author has given far too much weight to what may be called the 'travelers' tales' of the sixteenth century; he has quoted as of at least some value, statements which bear on their face their own refutation; he gives undue importance, I think, to the action of the central government in the organisation of life in India in the sixteenth century, and certainly more than they had in the parts of the Mogul Empire with which I am acquainted. But as a collection of information with regard to the India of three hundred years ago, put together from sources not easily accessible, with references given to all authorities quoted, and with a judgment on the data, recorded by one who has almost unrivalled knowledge of the agriculture and Economics of northern India, it will occupy for a long time to come a position of unequalled authority.

But the value of the volume will be, I think, chiefly as a starting point for the collection of further data. The sources on which the author relies are neither sufficient nor satisfactory for the conclusions which he draws. These authorities are essentially of two kinds. First, there are the official reports issued under the names of Abul Fazl and others in the reigns of Akbar and his successors. These are, of course, of supreme importance for the study of the condition of the country at the time. But they are, after all, data recorded and interpreted by an official under an eastern monarch, and it must always be recognised that both in what they include and in what they omit to tell, they reflect this fact. It can be fully recognised that the 'account of the XII Subas' in the Ain-i-Akbari, as Mr.

[1] **India at the Death of Akbar. An Economic Study**, by W. H Moreland, C.S.I., C.I.E London: Macmillan & Co. 1920.

Moreland says, "with all its omissions and imperfections, is still the nearest approach we possess to a systematic survey," but it must be read with a recognition of the bias which the source of the information would suggest.

The use made by Mr. Moreland of his second great source of information is, I feel, one of the heaviest criticisms of the book. This source consists of the many travellers' tales of the sixteenth and seventeenth centuries. The more these are examined, the less value they will appear to have as real interpreters of the economic condition of the country. The travellers of this period chiefly went along certain routes or followed the court of some Indian prince. Their observations were casual, and, in Economics, casual observations are most likely to lead to wrong conclusions. They said little, I am convinced, of the real life of the country, and were humbugged to a very great extent by their informants. They, hence, make what are obviously absurd statements which have nevertheless been constantly repeated as authoritative. The worst example is that of Nunez who says that the peasants paid nine-tenths of the produce of their land to the nobles, and that the latter paid one-half of what they received to the King. Mr. Moreland realises that this, and a very similar statement by de Laet, are absurd and says so (page 98), but there are many quotations scattered about from the European travellers of that epoch which, though not so obviously false, are, I feel certain, equally unreliable. And yet nearly half the authorities quoted by our author are such travellers!

The general conclusions drawn by Mr. Moreland as to the standard of life in India in the sixteenth century are far more cautious than those of most authors. He thinks that there has been a raising of the general plane of living, though not to a very large extent. He considers that the total produce of the land was very much less than it is now, though a good deal of the area actually under cultivation yielded more than at present. His opinion is that in the time of Akbar almost everything pointed to an economic decline in the near future, though some facts (especially the beginning of the foreign trade) pointed to an ultimate brighter economic future. On all these points we want more data, as well as on a multitude of others which rise from the reading of every page. And if these stimulate further inquiry, and lead to vigorous and active investigation in any of the directions I have indicated,—whether they lead to a confirmation of Mr. Moreland's views or not,—the present book will have been one of the most valuable contributions to the history of Indian Economics that have hitherto been made.

*Chapter* 11

# A DECCAN VILLAGE UNDER THE PESHWAS*

The village, in India, is and perhaps always has been the unit of national life. Towns there were, but in former times these have either been the residence of the court of a chief or a prince and his retainers, or only local marts for the exchange of country produce. Wherever we go, and nowhere more so than in the Deccan, we shall find the history of India in its villages, and though this may not be true to the same extent now as formerly, yet even in the present year of grace it is still certain that the heart of India is far away from the great centres of industrial activity, that in the minds of most of its people possession of land counts far more than the mere possession of wealth, that the politics of the village is far more to the vast majority than the politics of the district, the province or the state, and that if we want to understand the history of the country we must seek far more in the obscure unprinted records of village vicissitudes than in the more showy array of documents concerning conquests and Governments, princes and their retainers, or palaces and their inhabitants.

Few historians have, however, taken this view. How would the events usually chronicled have appeared to an intelligent inhabitant of a village? Or, indeed, would they have affected him at all? These questions are rarely

*Reprinted with permission from the **Indian Journal of Economics,** Allahabad Vol. IV (1923), Part 1, pp. 1-17.

asked, but the answers to them are vital to the understanding or to the real appreciation of the past of India. This is probably often recognised, but it is felt that it is impossible to know what the villagers thought of the events happening in the latter half of the eighteenth or the early years of the nineteenth century. And this is doubtlessly largely, though not wholly, true. At any rate in the Deccan where my own enquiries have been made, it is by no means entirely correct to say that the simple annals of the village have left no record. We have, in fact, considerable quantities of material compiled at the time, in the villages themselves, giving in each case an account of the economic condition of the place from year to year. These records are fragmentary; many of them are lost; and many of the odd sheets still available are torn and defaced. But those that remain, fragmentary as they are, but written actually in the villages themselves at that time, furnish a basis for the economic history of the period of more value than all the records of court histories or the despatches of successful generals.

I propose in the present article to take the records of a single village, and with these as a basis, to try and catch a glimpse of the economic life of Deccan villages during the last one hundred and twenty years of Maratha rule, chiefly during the time when that rule was in the hands of the Peshwas. The village selected, located about twenty-three miles to the north east of Poona, is one of the many hundreds for which records exist. One of the crying needs of Maratha history in our time is the working of these records, giving, as they do, the most vivid and authentic view that now exists of the vast majority of the people of Western India.

The village whose records I wish to analyse is Jategaon Budruk in the Poona district. It lies on the banks of a small river or torrent which is often dry for several months in the year. It stands in the area of very uncertain rainfall, where drought and, apart from irrigation, famines are frequent. It lies also several miles from any main road, though that to which it is nearest is the great Mogul high road from Poona to Delhi and North India. There is, even to-day, no regular road to the village itself, the so-called road being largely a cart track across fields.

Our first glimpse of the village whose general characters we have just indicated is in 1698. The army of Aurungzebe had just retired, and exactly one hundred and twenty years of Maratha rule were to follow. The record discloses a village apparently half derelict with only seventeen per cent of its land under cultivation, and only nine acres under irrigation. Much of the remaining area is not even owned, and all there seems to be cultivated is a small patch of the best land in the midst of a wide desolation.

Two facts immediately, however, emerge. There is no sign of any overlord, zemindar, or authority intervening between the owners of land in the village, with their headman or *patil,* and the government in Poona, and the land tax charged by the latter is paid *in money* direct. No sign of payment in kind or by the division of the crop exists at this period. This is important, and shows a relatively advanced type of economic relationship between the government and the villages. The second fact that appears is that only the land actually cultivated is charged with land assessment, and this seems to have been the case right through the period of Maratha rule. Nearly half the cultivated area is held either without any payment or with reduced payment to the Government for services rendered in the village (*inam*) by the *patil* or headman and by the *mahars* or village servants respectively. This latter class, although belonging to the untouchable classes, formed, then as now, one of the most important sections of the village community.

Apparently at this early time there was only one class of land ownership, probably that which was termed the *mirasi* tenure, which included full occupancy right of the land, and which later on became the basis of the general system of landholding in the British period. Some few cultivating landholders had garden or irrigated land, but the majority had then, as they have still, to carry on the annual gamble in rain of which agriculture in the eastern Bombay Deccan usually consists.

Twenty-six years later, in 1724, when the regime of the Peshwas had become fairly well established, and apparently a greater condition of order prevailed in the Deccan, the position of the holders of land in this village became more secure, and the derelict area began to be taken up. Most of this was brought into cultivation under a *kaula,* or written assurance from the paramount power of the security of the cultivator's possession, for a long period, provided the land was cultivated and the assessment paid. It is curious to read in the records of this year the names of the Maratha families which still form the landowners and cultivators of the village,—a striking example of the permanence of the life of a Deccan hamlet.

Accompanying the obviously increasing security of life and property, there is a very marked increase in the demands made on the village by the central Government, in the form of a raising of the land tax on the cultivated area. The total Government revenue had, in fact, already increased, partly by the increase of cultivation and partly by the raising of the assessment, from Rs. 300 to Rs. 526 and by 1727 it had gone further up to Rs. 666. In this last year we get the first indication that an assessment for revenue was not always paid, for no less than seven per cent of the total amount was not recovered. For the first time also, part of the revenue so obtained was diverted to others than the Government itself, Rs. 50 being sent to the *subha* or district officer, and Rs. 100 to the *mokashi.* The latter was probably a grant of old standing for military purposes.

As time goes by, the Government revenue at this period tends to increase. Whether we have struck a time of more than average certainty of crop, or whether the rule of the Poona Peshwas was a great improvement on what had gone before, in 1730 the revenue collected amounted to Rs. 1173, or more than four times the amount collected in 1697. It was not, so far as we can make out, due so much to an increased rate of assessment, though this was partly the cause, but to an extension of cultivation. During the whole period of the Peshwas' rule, the land tax of the village, as we have already said, was only charged on the land actually under cultivation in any particular year. This was a most important provision, and we are not sure how far it was universal. It applied, certainly, however, in the present case, and it may be worth while at once to give figures for a number of selected areas, which prove it.

1698—Area cultivated.

| | | |
|---|---|---|
| (a) Dry or **jirayat** land | ... | 318 acres at $15\frac{3}{4}$ annas per acre. |
| (b) Garden land. | ... | 9 acres at Rs. 4-5-2 per acre. |

1796—Area cultivated

| | | | | | |
|---|---|---|---|---|---|
| (a) **Sosti** or fully assessed land | ... | 865 acres giving | Rs. | 848 | 12 |
| (b) **Upri** or land annually rented. | ... | 288 acres giving | Rs. | 183 | 0 |
| (c) **Kauli** land. | ... | 297 acres giving | Rs. | 69 | 6 |
| (d) Garden land. | ... | 9 acres giving | Rs. | 56 | 0 |
| | | | Rs. | 1 157 | 2 |

396 acres were noted as uncultivated and hence not taxed.

1804—Area cultivated.

| | | | | |
|---|---|---|---|---|
| (a) **Sosti land** | ... | 720 acres giving | Rs. | 800 |
| (b) **Upri** land | ... | 804 acres giving | Rs. | 139 |
| (c) Garden land | ... | 9 acres giving | Rs. | 56 |
| | | | Rs. | 995 |

540 acres were noted as uncultivated and hence not taxed.

If the principle of charging land assessment on the area actually under cultivation only was general under the Peshwas, it would certainly have two effects. It would make the revenue exceedingly variable, which it really is, and it would also tend to reduce cultivation when either physical or political conditions were unfavourable to the success of the crops. It would, on the other hand, make possible a rate of assessment on the land actually cultivated considerably higher, relatively to the value of the crop produced, than can be imposed in the recent British assessments.

The extraordinary variations in the gross, as well as in the net revenue so produced must have been a great cause of anxiety to the Peshwas' accountants, or to those to whom the revenues of a village were assigned. The present village furnishes some figures which well illustrate the point. In 1768 the *tankha* or standard revenue expected from the whole village was declared to be Rs. 1061 per annum. At the same time, the *kamal* or the highest income obtainable from the village when it was cultivated to its full capacity was estimated at Rs. 2100 per annum.

The gross revenue actually collected at different dates was as follows:—

1697—Rs. 275; 1698—Rs. 301; 1724—Rs. 526; 1727—Rs. 620; 1730—Rs. 1173; 1767—Rs. 650; 1770—Rs. 1937; 1785—Rs. 553; 1790—Rs. 493; 1796—Rs. 1311; 1803—Rs. 995; 1808—Rs. 766; 1817—Rs. 639.

The variations are extreme. Even if we omit the first two years quoted, a variation from Rs. 493 to Rs. 1937 is very great indeed. It is not due, as might possibly be supposed, in more than a minor measure to an arbitrary alteration of the assessment, but to the principle of only charging on land which the cultivators found it worth while to put under crop, a principle which appears to be acknowledged throughout. From a cultivator's point of view this was a most valuable provision, though it is obvious that it must have been the despair of the Peshwas' financiers.

The variations of the assessment per unit area cultivated, though considerable, were, as already noted, not so great as would at first sight be supposed. Calculated to our present units of area, the actual rate appears to have been as follows, for dry, *i.e.* unirrigated land on the average: 1698—Rs. 0-15-0; 1770—Rs. 1-5-9; 1796—0-15-9; 1803—Rs. 1-1-9; 1808—Rs. 0-13-0; 1817—Rs. 0-10-7. It may be of interest to compare these rates with those of the regular settlements under British rule, though it must be remembered in these latter cases that all the land in private ownership in the village is now expected to pay every year, apart from the suspensions and remissions in years of famine granted in recent years. The rates under British rule are as follows, but they include those for garden land which very slightly raises the figure and makes it not quite comparable with those given above: 1844-45 settlement, Rs. 0-9-9 per acre; 1874 to 1916-17, Rs. 0-10-0 per acre; 1916-17 to date, Rs. 0-12-0 per acre. The expected revenue from the village under the last of these settlements is Rs. 1581, or about fifty per cent more, in actual money, than the *tankha* or standard revenue under the Peshwas.

The old figures should, if we want to compare them with the rates now prevailing, be expressed in terms of the amount of produce, that is to say, the grain to which they would be equivalent, and if this is done the real character of the assessment is exposed. The rates, calculated as a proportion of the crop growth on the land, appear generally under the Peshwas to be high. In proving this statement we are faced by the difficulty of obtaining records of prices for the years of which village records exist, and also by the difficulty of determining the actual character of the land under cultivation in the years for which we have other data. We have been supplied with price records in Poona for a number of years by Rao Bahadur Parasnis of Satara, and for other years by Rao Bahadur Joglekar of Poona, and these form a basis for the consideration of the question. As regards the character of the land cultivated it has been presumed in each that the best land in the village was cultivated, and that if a reduction in the area cultivated occurred, it was the poor soils that were omitted. How far this is justifiable may perhaps be questioned, but it seems a reasonable supposition.

We will take two dates, 1770 and 1803, for which we have definitely the Poona prices for *bajri*, the chief grain produced in the village, and compare the rate of land assessment per acre, in seers of *bajri* with that in force in British times between 1888 and 1917 and since 1917. To these I have added 1808 where the assessment can be compared on the basis of the prices of 1806. The price of *bajri* in 1770 was 16 seers per rupee and in 1803 was 6 seers to the rupee, while in 1806 it was 10 seers to the rupee. The second year quoted was obviously a year of famine. The average price of *bajri* between 1888 and 1917 was slightly over 13 seers per rupee, and since 1917 about $10\frac{3}{4}$ seers per rupee. The results of the comparison are as follows:—

| | Assessment per acre in rupees. | Assessment per acre in **bajri.** | Assessment of similar land 1888-1917 in **bajri.** | Assessment of similar land 1917 to date in **bajri.** |
|---|---|---|---|---|
| 1770 | Rs. 1- 5-9 | $21\frac{3}{4}$ seers | $11\frac{1}{4}$ seers | $10\frac{1}{2}$ seers. |
| 1803 | Rs. 1- 1-9 | $6\frac{3}{4}$ seers | 12 seers | $11\frac{1}{4}$ seers. |
| 1808* | Rs. 0-13-4 | $12\frac{3}{4}$ seers | $12\frac{1}{4}$ seers | $11\frac{1}{2}$ seers. |

* Calculating on prices of 1806.

It is unfortunate that we do not have village records and prices for the same years in more cases, but in all the years for which we have figures between 1770 and 1791 the prices were as low or lower than at the former date, and so the figures for that year do not represent the grain assessment as heavier than it really was, but rather the reverse. 1803 was, as already remarked, definitely a famine year, and in 1806, the country was apparently recovering from several years of scarcity.

The fairest comparison is, therefore, with 1770, and from this date to 1791 the rate calculated in grain was definitely heavy as compared with that in recent times. It could, in fact, be paid, because it was only charged on land actually cultivated.

The saving of the situation in Maratha times was evidently the fact that the land tax was only levied on cultivated land. With this allowed, the very much higher charge might be (though we are not sure how far they were) far less burdensome to the owner of land than the lower rates of assessment of modern days in the very uncertain climate of the eastern Deccan. It is likely that the system discouraged cultivation. It would be anathema to the disciples of Henry George. The revenue was provokingly varying. But the villages prospered so far as we can judge, and in the first year for which records exist after the advent of British rule, it gave a land revenue of Rs 1900 as against Rs. 300 at the commencement of really effective Maratha rule.

The land revenue was by far the most important village tax under the Peshwas, as it is now, but it was by no means the only one, and the records in hand indicate how close was the knowledge of the central authorities in small matters. There was a regular tax, for instance, on the non-agricultural population of the village (*adansara*). In 1770, for example, the village carpenter was charged Rs. 9, the priest Rs. 5, the *dhobi* or washerman Rs. 3, the potter Rs. 5, the barber Rs. 7, the fisherman Rs. 4, and the shoemakers taken together Rs. 10. The relative importance of these village functionaries is to a certain extent indicated by these figures. But these regular village inhabitants were not the only ones taxed by the central Government. A wandering tribe of gipses (*berad*) squatted on the village land; immediately they pay a tax, albeit a small one, to Poona. The fact that the figures appear in the account of a remote village controverts the idea of any slackness in the supervision of revenue in the palmy days of the Peshwas' rule.

The revenue, in the earliest days of our records, appears to have been paid directly into the Government treasury, with few if any deductions. In 1727, as we have already noted, we first hear of a payment to the *mokashi* and also to the *subha* or head of the district, before the amount reaches the recipients. But as time goes on, a number of officials are able to obtain a share before it reaches its destination. Thus in 1770, a very prosperous year, no less than four people had their fees as follows: the *mokashi*—Rs. 125; the *deshmukh*—Rs. 25; the *deshpande*—Rs. 40; and the *sar patil*—Rs. 24. The latter three were simply supervising revenue officers. These amounts were, however, extremely variable. In some years they were not paid at all, as in 1796. Such revenue officials were constantly increasing in number under the Peshwas' rule, but their pay was dependent on the revenue being secured; if this was not the case, they apparently went without.

In one essential particular the village system of accounts differed completely from that now in vogue. *Then* the whole of the village expenses, even including such minor items as those for lights and stationery, were incurred by the village officers, and the cost deducted from the money sent to the treasury; *now* all the money received is sent, and the budget for village expenses is quite independently supplied from the district or subdivisional headquarters. Under the old system it was easy to find the proportion of the income to the expenditure, now it is practically impossible except after a long search in an entirely different place. The system formerly in vogue makes the old accounts more interesting. We can see recorded in these figures, the arrival of the Peshwa on tour, or of some other high and mighty official. We can see the provision for his food, the arrangements for his tents, the fees paid to his *havildar*, his flag bearer, his elephant drivers, his court priest (*bhat*) and many of his attendants. We can see the forcing of labour to prepare for the great man's arrival, and the recovery afterwards of the amount from the court at Poona.

In these accounts we see contributions to temples, to festivals, even the *mohurrum*, to the cost of a goat killed in the village gate on Dasara day, to charity without explanation, to groups of wandering mendicants. All are considered legitimate charges against the Government revenues, and all appear frequently in these truth-telling accounts. Several times in the later and more degenerate days of the Peshwas' rule, the signs of the passage of raiding and often rebellious forces appear, when apparently they have to be bought off. The Peshwa's army itself imposes contributions. But the contribution or the bribe to leave the village is always charged against the Peshwa's revenue for the year.

Thus, in 1796 we read an item "*Khandani* taken by Daulatrao Shinde's army, Rs. 904-8-0;" in 1803 an item "to . . . . . . Deshpande, who had paid the horsemen of Vinchorkar Rs. 100;" in 1817 an item "Expenses of the Peshwa's army which had come with Baba Saheb from Talegaon—Rs. 200" and so on.

This last series of items takes us a long way from the present peaceful Deccan, and reminds us of the necessity of another item in 1785:—"For the repair of a wall round the village—Rs. 13." For a wall was evidently needed, and the prejudice which still continues in the Deccan, in contrast to the habits in many other parts of India, in favour of living in a walled village rather than out in the fields which are being cultivated, would seem to date from disturbed times such as these.

But there seems no sign in our records of the village being devastated. Much, if not all, of the money given to raiders is recovered from the Government by being deducted from the revenue payable. A force appears. The villagers retire within the wall. The village officers wait trembling before the leaders and offer presents. They are not enough. Large demands are made, and ultimately paid. But the amount really falls on the Peshwa's revenue and it is doubtful how far, in a village essentially consisting of unirrigated land and of *kharif* crops, the passage of a raiding force materially affected the prosperity of the cultivators of the place. There would be little love or hatred of the Government or of its opponents, and what the village lost was passed on as a charge, and an acknowledged charge, on the Government in power.

Another feature to which we should like to draw attention is the high status and value of the position of village *patil* in the time of the Peshwas. More than this, it was not only a duty hereditary in certain families to which the holding of certain village lands was attached, but was also a saleable property of very valuable character. The *patilki* of the present village was held by two families, and these agreed to sell half their right and the property attached to it to an outside family of Marathas for no less a sum than Rs. 3000. This, which at present would be worth considerably more than this amount, seems a great deal for the position with half the land attached to the office, in a very obscure village. It involved possession of only thirty six acres of land, and as a financial investment in a country, and at a time of high interest for money, it would be hard to conceive of anything more unsound. But the honour attached to land holding, and particularly to being the *patil* of a village together with, it is possible, the opportunity for other and less legitimate forms of gain, was such that the amount was paid. We fear that the market value, or in other words, the status of a village *patil*, has fallen much in British times. Some will welcome the fall, others look upon it as almost a disaster, as it indicates the reduction in importance of the village community,—the most stable thing in British India.

There is another feature of the transaction just mentioned which is worth recording. The sale of the village headship had to be registered in the Peshwa's office in Poona. Such registration was paid for by fees. These went first to the Government itself, and secondly to the officials and clerks through whose hands it passed. Out of Rs. 3000 received by the selling party no less than one third (Rs. 1035) went in fees. Of this amount Rs. 800 were to the Government, Rs. 135 to the department of the *phadnis* (or accountant), and Rs. 100 to the department of the *chitnis* (or secretary). The system is well illustrated by these figures, and it is curious to find that Nana Phadnavis, the great revenue officer and administrator of the Peshwas, appears in the records as personally receiving, by virtue of his office, a fee of Rs. 100 for a transaction like that under discussion.

The history of this village well illustrates the truly Indian method of rewarding services by grants of land or of rights in land, and also the system of assigning the revenue of special places to certain sections of the administration, for their support. One man, whose name indicates him to have been a priest, had been at some time previous to 1745 given thirty six acres of land at the village of Shirole, about thirty miles away from Jategaon. When he appeared to take possession of it, it had been appropriated by someone else. He complained, and in 1745 the requisite amount of land, without any payment of revenue, was granted in the village under consideration, to continue to his descendants for ever. Apparently he had great difficulty in getting hold of the new grant, for in 1752 another order appears from headquarters reciting the fact that another man had retained poossession of land. The latter was directed, doubtless much to his chagrin, not only to hand over the land but also to pay four or five years' back rent to the grantee, whose descendents enjoy possession of the land to this day. I should have liked to be present when the court priest armed with his new order appeared to claim his property.

A reminiscence of a similar grant, but this time of the whole village, is contained in the records of 1768. The village had by some means, of which the records have disappeared, come into the possession of a Mohammadan of Chakan, another village some distance off. Probably it was a grant of the revenue of the ordinary type. But the grantee proved a rebel, and the property was confiscated. Instead of the recovered revenue being devoted to the general purposes of the government the question immediately arose after the confiscation as to what department should receive the village revenues. In this case they were assigned to the support of a force of cavalry belonging to the Peshwa's government, and from that date to the latter part of the reign of Bajirao II they were paid to this body. It must have been a sorry sort of dependence for a military force,—being so utterly uncertain. Hence before long a very serious complaint was made. The village lands were not properly cultivated, the crops were not properly reaped, hence the revenue was far below expectations. The result is curious. An enquiry was made, and the revenue which might be expected for each of the following five years was laid down. A rapid rise was anticipated as a result of pressure on the people to cultivate as much as possible, and it was suggested (we fear the officer had his tongue in his cheek) that a revenue of Rs. 950 should be permanently increased to Rs. 1700. The report and estimates were the purest eyewash. The man who wrote the report and made the estimates must have known that the revenue would never be obtained except under very special circumstances. And in 1785 the same complaint as before was made. The revenue has sunk to Rs. 552. Again it is screwed up to over Rs. 1000, but five years later it is again down below Rs. 500.

We suppose that it was this almost intolerable uncertainty which finally caused the introduction of the revenue farming system under Bajirao II. This system at any rate secured the government against the extreme variations of revenue, but at the risk either of rousing the people into general opposition to the ruling power or else of making them into abject slaves. In this case the adoption of the farming system came before 1808 though we do not know the exact year from the village records. At any rate after 1808 there is no longer any mention of the Government cavalry, and in addition, one of the common incidents of revenue farming in the Deccan appears. A note is made, in fact, that over four hundred acres remained uncultivated on account of riots by one of the village *patils* and his people, who drove out the owners of his land. In 1812 the revenue collection (*kamavisa*) was definitely farmed out to the Peshwa's notorious favourite, Trimbakji Dhengle, one of the worst influences at the court of the Peshwa in its last days. In this case, all that the revenue farmer could do did not bring the expected revenue. The actual money obtained from the cultivators went down in 1817 to Rs. 792, and of this collection only Rs. 227 reached the collector, the rest being taken away by armies of the Peshwa himself, by raiding parties of *gosavis*, and by the largely increased village expenses.

It might be thought that in these troubles, during which the sun of the old Poona government was finally setting, the village was largely and permanently impoverished. This has been, at any rate, the view of most of those who have written on the subject. I doubt whether it was really so, however, to the extent suspected. What happened is that the cultivators simply ceased to cultivate more than a small proportion of their usual acreage and ceased to pay revenue for so large an area. Immediately on the return of peace and security, the revenue went up again and in 1823 it is at a higher point than it had reached in any but the very best years of the Peshwas' rule.

Such are the outlines of the story told in these records. There are many other facts revealed which I am not able at present to discuss. The existence of famine appears at times very clearly, and on those occasions there is both remission and suspension of revenue even on cultivated land just as in the most modern system. There are losses of crop from locusts, as in quite recent years, and from other causes. And so on. But the most marked impressions which these records have left upon me are two. The first is the way in which, by limiting the assessment of revenue to land actually cultivated, the Peshwas' government in its palmy days made a very high assessment possible without hardship. But it did this at the cost of the certainty of its own revenue. In evil days, the income of the central authority must have gone down so badly that all its departments must have been badly starved, and this fact probably led finally to the introduction of the farming system by the short-sighted ministers of Bajirao II.

The second and by far the most important impression given by these records is that of the stability of the *ryotwari* village system. I have already referred to this matter, but it seems vital to the understanding of the history of the Deccan. Did it, in fact, matter much to the vast majority of the people whether the rule was that of the old Mohammedan kingdoms, of Aurungzebe, of the Maratha kings, or of the Peshwas, or even, in the early days, of the British? The village was a self sufficient unit; its *patil* a more important man in reality than any officer of the government. A time of disorder meant the reduction of cultivation, the temporary wandering of some of the inhabitants to more settled regions, and the consequent reduction in the government revenue from the village. Drought, leading to famine, meant the same thing. Hardship there was, but hardship is the common lot of man in the eastern Deccan. The assessment charged on the village varied a little, and was generally high, and the variations seem to have been quite arbitrary. But any attempt at a large increase, as in 1768, failed; any attempt to get more by farming the revenue also failed. The latter led to riots, as well as to loss of revenue; the former led to reduction of cultivation, and hence again to loss of revenue. Given peace and order, with good seasons, the income of the government went up to a fairly definite maximum; if there was disorder or bad seasons, cultivation, revenue and, we think, population also automatically adjusted themselves. Truly the village was an "*imperium in imperio*" against which the forces of all central governments spent themselves without disturbing its solidity. Here is the one essential institution of the region,—the village.

Since the time of the Peshwas things have changed, and the village, though still probably the most solidly founded institution in the Deccan, does not occupy the position it did before, and perhaps it is still losing ground as a factor in national life. But that is another story.[1]

[1] I am indebted to my friend and colleague, Mr. N. V. Kanitkar, for copying and translating into English and into modern measures, the records summarised in this paper—H.H.M.

*Chapter* 12

# FOREWORD TO RESURVEY OF A DECCAN VILLAGE*

It was about the end of 1913 that a group of my colleagues at the Poona Agricultural College got together with the object of making a detailed study of a Deccan village. All were deeply interested in rural questions but all felt that if we were to understand the actual condition of the rural population and their way of life a more detailed study of village problems than were at that time available, ought to be made. To carry out such a study on a wide scale was impossible. Nobody could do it. And all the group in question were busy men who could at the most only give part time to such a study, in spite of its recognised importance. Yet it was determined to choose a village accessible enough to Poona for part time workers to visit it frequently and yet isolated enough, not to be seriously affected by its nearness to a large city.

The village of Pimple Saudagar was chosen as a suitable one for the purpose. It was accessible by bicycle or bullock cart, but not easily otherwise. Its relationship with Poona appeared to be slight. It was an *inam* village which rather tended to emphasise its isolation. For two years, more or less, the study continued. Intimate relations with all the village people of every type were established. Crops, wells, irrigation, sub-division and fragmentation of land, were all studied in detail, and the

* Reprinted with permission from P. D. Diskalkar, **Resurvey of a Deccan Village: Pimple Saudagar**, Bombay, **Indian Society of Agricultural Economics**, 1960, pp. v-vii.

intimate contact with the people enabled us to get a fairly complete account of such matters as debts and the general economic position of every part of the population.

The results of these inquiries were published in 1917 under the title "Land and Labour in a Deccan Village" and this gave an insight into the actual conditions prevailing at that time in a Deccan village such as had not been available before. It showed that the economy of the village was unsound, that a large part of the population were living below their own standard of life, though this standard was very low indeed, and the general conclusion was that in so far as the village in question was in any way representative, the economic outlook was black.

What has happened since that time? After forty years my friend Mr. P. D. Diskalkar has undertaken a similar study to that made so many years ago, in the same village, with the idea of seeing how far the many changes, political, social, and economic, have affected the condition of the people, and the present publication gives the results which he has obtained. I have found his description of the changes which have taken place and of the present conditions very absorbing and of very great value. One of the most interesting parts of the study is that relating to the 112 families which was taken up for special study. One point which has interested me much is the apparently complete breakdown of the *Baluta* system which has for centuries been a feature of village life. This of course means that the village is no longer the self sufficient unit, with everybody the servant of community that was formerly the case. I have been very much struck by the analysis of the social and religious structure of the village, a line which I was not able to touch in the original survey. The village has become less isolated as a result of the construction of good roads both to Poona and Kirkee. It has become one of the sources of milk-supply to Poona and to this extent there has been intensification of the agriculture of the village. Many of the men now go to work at Kirkee and elsewhere and to this extent are less dependent on a tricky agriculture.

The main impression that one gets as to the general economy of the village is the decline of agriculture as shown by the abandonment of well irrigation, in spite of the dam across the river and the possibility of great extension of irrigation. There were ten *pakka* wells actually in use in 1917, four were only in use in 1933 and only three in 1952. Not only is this the case but a very large number of the wells used in 1917 are now dilapidated. The account of the lift irrigation scheme and the history of the village is specially interesting. All of this is quite new.

The legend given is very amusing and the memorial stone is the indication of the fact that something of the kind described did take place. The way, in which an account of fragmentation of areas of cultivation is dealt with, is admirable. The chapter on the changes of ownership of land opens up a new field which was untouched in the former survey and I was particularly struck by the increase in the sale price of land in the years since 1917.

The data given as to the changes in the crops grown is very interesting. The two new herbs *Dhorgunj* and *Kapalphod* not included in my list of 1917, seem of some importance.

One of the most interesting points brought out, is the development of a dairy industry in the village for the supply of milk to the Poona market. I have been specially interested in this and in the effort reported to prevent adulteration of the milk.

The increase in the value of cattle from Rs. 10,000 to Rs. 77,000 since the original survey is remarkable especially for the buffaloes from Rs. 1600 to Rs. 38,700. The whole consideration of the cattle is new and is very important, and I congratulate the author on it.

The co-operative movement seems hardly to have touched this village.

Socially, I cannot see from Mr. Diskalkar's account that there has been much change, but others may see a good deal more than I can in the figures he gives and the comments he makes. On the whole, I may suggest that there are points in Mr. Diskalkar's account which are encouraging and others which are definitely disquieting. The summarising account of the work occurring in the concluding chapter is masterly. I would recommend the very earnest study of what the author has found out with a view to see what are the likely changes that may occur in the foreseeable future.

HAROLD H. MANN AND NITYA GOPAL MUKERJI, ABOUT 1905.

# PART III

# TOWN STUDIES IN INDIA

## *Chapter* 13

## CALCUTTA DRINK SHOPS (1903-04)*

The investigation of the grogshops of Calcutta, undertaken by the Drink Problem Committee last year, was commenced with Ward No. 13, lying, as is well known, midway between the European and Indian parts of the city, and bounded by Chowringhee and Wellesley Street on the west and east respectively, by Dharamtala on the north and by Kyd Street and Ripon Street on the south. The results were published in the daily press and separately during the spring of the present year.

The experience gained in that ward has been of immense service in trying to come to closer quarters with the drink traffic as it exists in the more purely Indian parts of the city. One of the entirely native wards has, in fact, formed the theatre of our operations during the past few months. This is that known as Ward No. 6, and comprises the district controlled from the Jorasanko *thana*,—this thana being situated just on the edge of the ward in Chitpore Road.

*General description of the Ward*.—This ward, as we have said, is a purely native one. It is bounded on the south by Machua Bazar Street, on the north by Beadon Street, and by Chitpore Road and Cornwallis Street on the west and east respectively. It embraces a population

* Reprinted from **The Drink Problem in Calcutta**, No. III **Report on Ward No. 6.** The Drink Problem Investigation Committee. Calcutta. 1904, pp. 1-12.

almost as densely packed together as in any part of Calcutta. Within the boundaries mentioned there is an area of 262 acres, and on this there were in 1901 no less than 52,988 people, or 202 persons per acre. This number of people lived in 7,044 houses, nearly 4,000 of which were only kutcha built. There is thus an average of between seven and eight people per house.

It is, moreover, a district whose population has been increasing by leaps and bounds. From 33,000 in 1881, the number reached nearly 42,000 in 1891, and, as has been said, nearly 53,000 in 1901! Out of this population less than 600 were anything but Hindu or Mahommedan, the remainder being divided between these two in the proportion of five to one.

The ward is essentially one of artisans and small shopkeepers, with the addition of a large number (at the south end) who gain their livelihood as personal servants in other parts of the town. Shopkeepers and those dependant on them form nearly a sixth of the total population; artisans working as builders, on the manufacture of metals and precious stones, of glass, of wood and cane articles, of leather and other articles, with their families number more than a quarter of the whole inhabitants. The south of the ward is a typical low class and *busti* neighbourhood. In some parts near Machua Bazar, one may stand and hardly see a *pucca* house for quite a long distance. This part is largely inhabited by Mahommedans, and is one of the centres for *gariwallas* in Calcutta. The neighbourhood of Chitpore Road is, in a large part of the ward, quite a disreputable centre. Brothels and all the connections of such houses abound in this part, and there are no less than 2,700 people who get their living by immorality, chiefly concentrated in the south-west of the ward. As one proceeds north and east from this part the character gradually changes. The streets get better: there is a larger proportion of better houses: the population becomes more and more largely composed of men (and their families) who are in Calcutta offices, until by the time Beadon Square, Beadon Street, Simla Street, &c., are reached, one would hardly conceive that it was the same ward as contained Machua Bazar and its squalid surroundings.

Drink selling itself forms the livelihood of about 83 people. Toddy Sellers are represented by sixteen Hindus. Wine and spirit selling supports 67 people—58 Hindus and 9 Mahommedans. Of these, 46 Hindus (including one Hindu woman!) and four Mahommedans (including a Mahommedan woman!!) are said to be actively engaged in the sale of liquor.

The whole character of the ward may now be briefly summarised. It consists of a low class disreputable neighbourhood in the south and west, and a much more respectable residential quarter in the north and east. The inhabitants are almost exclusively Indians, and hence Hindus and Mahommedans the latter being chiefly in the Machua Bazar and Chitpore Road neighbourhoods. The population is very dense—202 people per acre, and is principally composed of artisans and shopkeepers, with a certain proportion of city clerks and also of personal servants whose work carries them into other quarters of Calcutta.

*The Shops described.*—The drink traffic itself, so far as concerns us in the present investigation, is carried on by means of eight grogshops. Of these three lie in Chitpore Road, two in Cornwallis Street, two in Machua Bazar and one in Manicktolla Street. Every one of these has been watched for at least four hours during the daytime, and note taken of all who entered; and similarly for at least three hours or thereabouts during the evening, though often for a much longer period. The method of observation was intended to be the same as that described in our previous report for Ward No. 13. Owing to the situation of the ward, however, we were not able to rely to the same extent on voluntary observations, and two paid men were employed for a considerable period in connection with these inquiries. In the former report we did not distinguish between Hindu and Mahommedan drinkers, but owing to the arrangements just noted and others, this has generally been possible in the present case.

No attempt was made on the part of the grogshop keepers to interfere with the workers, though they were watched with great interest by those in charge of the shops. In one or two cases, to be noted below, objection was taken to inquiries about the nature of the premises, side-doors, and so on, and in one case (196, Cornwallis Street) the insistence of the workers to see a side-door which was being used at the time by customers nearly led to an assault.

The grogshops, as we have said, are eight in number and are situated as follows. The *monthly* license fee is given in each case.

| | | |
|---|---|---|
| 1. Chandi Charan Shaw, 136, Machua Bazar Street | Rs. | 620 |
| 2. Gosto Behari Nayak, 158, Machua Bazar Street | ,, | 325 |
| 3. Mrigendra Lall Shaw, 14, Upper Chitpore Road | ,, | 460 |
| 4. Surbeshur Shaw & Co., 76, Upper Chitpore Road | ,, | 405 |
| 5. Wupendra Nath Shaw, 89, Upper Chitpore Road | ,, | 550 |
| 6. Naran Chandra Shaw, 196, Cornwallis Street<br>7. Naran Chandra Shaw, 206, Cornwallis Street | ,, | 1,005 |
| 8. Ram Kumar Shaw, 134, Manicktolla Street | ,, | 555 |
| Total | Rs. | 3,920 |

Thus the Government receive at present no less than Rs. 3,920 per month for the right to sell liquor in these eight shops, or an average of Rs. 490 per shop per month.

We will now describe each shop in detail, and give the results of our observations, referring to the appendix for the actual tabulated figures.

[1] This shop and another in Banstolla Street pay Rs. 920 per month. We have taken half for each shop.

(1) 136, *Machua Bazar Street.*—This shop lies immediately facing the end of Halliday Street, a wide road running straight down from Harrison Road. It is, however, right in the centre of what is almost a purely *busti* neighbourhood, and is a one-storey brick building with a narrow passage on each side, not leading, so far as we can make out, to any side entrance. It has four entrances, part of the front of the building being taken up, as is often the case, with *pan* stalls, the presence of which causes part of the interior to be a secluded corner where visitors can sit and chat. The shop has a small verandah in front, which every evening is a mart for the retail sale of grass fodder, several grass sellers coming here for the purpose.

During the daytime business was slack, as with all the shops, as compared with the evening. The average total number of visitors per hour was only 72 as against 96 after 6 p.m. In the morning (before 1 p.m.) business is a little brisker than in the afternoon. The principal day visitors were Mahommedan men, who form in fact 45 per cent. of the total who entered. The low caste Hindus were principally "Beharis" and similar non-Bengali speaking people. In the evening the Mahommedans did not form such a large proportion of the visitors, but still were more than 34 per cent. In a disreputable neighbourhood like this it is not to be expected that many Hindu gentlemen (babus) would visit the shop, and as a matter of fact the proportion is lower than at any other shop save one in the ward. It never the less reached 10¼ per cent. of the total visitors in the day time and 7 per cent. in the evening.

During the day no less than five Mahommedan and three Hindu women were actually in this shop drinking in public during the 8½ hours occupied by these observations; at night, however, it may be noted, only two women were served in five hours. As regards children the evidence shows no less than eleven Mahommedan, five low caste Hindus and two apparently high caste Hindu boys under 16 entering and drinking in the shop, and ten Mahommedans with three low caste Hindu boys carrying away liquor from the place during 8½ hours watching in the daytime; here, again the proportion was less in the evening when, in five hours, only 3 boys entered apparently for the purpose of drinking and five for carrying away.

The bulk of the trade was drinking on the spot both in the day and the evening, the offsale only forming about 1/6 to 1/7 of the total business. One man dressed in European clothes visited the shop on April 21 and drank country liquor.

One of the problems which we have had in mind in watching these shops, is that of the nature of the drinking which goes on;—is it purely caused by being in company or is it independent of this. This answer can only be given by ascertaining the proportionate number of on-drinkers who enter the shop in groups. During the day these reached 92 out of a total of 115, or 80 per cent; in the evening the figure was reduced to 34 per cent. When the proportion is as high as the former of these, there is strong reason for supposing that being in company has something to do with the taking of drink at all.

The general character of the drinking in the evening is thus described by one of our observers:— "The price of the drinks mostly consumed was six pice each, and men seemed to do the drinking without any sociability. A man would come in, drink, pay, and go out. If two came in together and talked together each paid for his own. There were always a number sitting inside smoking, but they did not sit drinking. When a man wanted to drink he left the others, drank, paid and sat down again to smoke."

This was one of the shops where drunkenness was seen. In this case it was a Mahommedan coming out of the shop. He however made no fuss and quietly rolled away.

(2) 158 *Machua Bazar Street.*—A shop paying a license fee of Rs. 325 per month is a small shop situated in absolutely the lowest quarter of the ward. Small though the shop appears, it does a very large trade. The premises are, however, almost the most undesirable we have seen,—dirty, miserable, small, with inside somewhat like a stable and dirty in the extreme. A *pan* stall, together with one selling

(in July) gingerbeer and mangoes, take up part of the front, and allow a sort of snuggery to be produced inside. The shop is only one storey high, but the licensee lives behind. There is a side-door to the shop, in a narrow lane leading to the licensee's quarters and to the *busti*. The surroundings are of the ordinary *busti* character, and the small verandah of the shop is used by children and others as a favourite lounge.

During the day this shop does a brisk business, standing second in the ward for the number of visitors per hour. That it lies in the disreputable quarter of the ward is shown by the fact that in four hours no less than eight Mahommedan women, who all apparently belonged to the immoral class, were served with drink. It is, however, a matter of satisfaction that no children were received here during the day, and only two during our observations in the evening. The proportion of women in the evening was even greater than during the day, there being over three served per hour while we were watching: in this matter this is the worst shop in the ward.

The proportion of better class Bengalis was the lowest in the ward; none being served in the daytime and only 4 per cent. in the evening. On the other hand, the number of Mahommedan men was highest, these amounting to 34 per cent. in the day and in the evening to nearly 41 per cent. During the day three West Indians visited the shop: these have been put down as Europeans in default of better classification, but there really were no European visitors during the periods of our observation. The proportion of customers who purchased for home consumption was about 20 per cent. in the day and sank to 14 per cent. at night.

As regards drinking in company, 45 per cent. of the men on-drinkers entered in groups of two or more during the day and 20 per cent. only at night. The larger amount of what we may call "company drinking" in the daytime will be noticed at each shop in the district.

(3) 14, *Upper Chitpore Road.*—The license held at this place is at present settled with another in Banstollah

Street, the two together paying Rs. 920 per month. In our calculations we have taken it that half of this is paid for each shop and thus the present licence is put down as worth Rs. 460 per month. The shop is situated at the corner of a lane at the base of a large and well-built house, and seems to be a lock-up shop, with licensee living elsewhere. Part of the shop front is occupied by a *pan* stall and part by a tobacco stall. The part thus shaded is used as a lounge for customers. There are entrances into the shop both from the main Chitpore Road and from the side lane. As both are in public thoroughfares we do not think great objection can be taken to this construction. The appearance strikes us as squalid though not worse than the average, especially as this shop is situated in what may be called the lower class end of the ward.

On the day on which this shop was watched, the morning was wet, the afternoon dry, and the result was evident in the small number of visitors in the early part of the day. After 5-30 P.M. the business became much brisker than before, evidently owing to people calling as they returned from work. The bulk of the trade was *on*-drinking country liquor, the proportion who carried liquor away being less than 1/9 of the day customers and just over 1/6 in the evening. Only one woman drank on the spot during all our observations, but two carried away liquor in bottle, a very different state of affairs from what occurred at the last shop considered. Four Hindu boys and one nine year old Hindu girl were served in five hours in the daytime and six in the evening during the same period, three of the latter carrying away the liquor in bottle.

As regards the proportion of classes among the customers the Mahommedans here show a distinct falling off as compared with the previous shops. They only amount to 25 per cent. of the day visitors and just over 12 per cent. in the evening. The "Hindu Gentlemen" are however proportionately much greater in number, reaching nearly 11 per cent. in the daytime, while they form nearly 23 per cent. of the customers at night. No Europeans whatever visited the shop while these notes were being taken.

So far as "company drinking" is concerned, over 40 per cent. of the *on*-drinkers entered in the morning in groups of two or more; in the afternoon nearly 72 per cent.; while as always the proportion was very much lower in the evening when it did not exceed 33 per cent.

(4) 76, *Upper Chitpore Road.*—Is a one storey building at the corner of Ranamusee Ghosa's Lane and Chitpore Road. The principal entrance is in the side street. As usual part of both frontages is taken up with lean-to stalls for sweets, for *pan* selling, and for fortune telling. The part of the shop, however, thus formed into a snuggery is very small. We did notice a side entrance into this shop, and it is apparently a lock-up establishment. The monthly licence fee amounts to Rs. 405.

Observations were made for four hours on the afternoon of July 6 and also on the evenings of August 8, 9 and 12, 1904. At this point we are getting, as the figures show, quite out of the Mahommedan quarter, the number of such customers sinking to a much lower level than hitherto,—only 10 per cent. in the day and under 7 per cent. at night. But the number of higher class Hindus gets greater, the proportion reaching over 13 per cent. during the day and over 19 per cent. at night. A good many of these are casual visitors who drink because they are passing along the main road, but we are, nevertheless, approaching here the residential quarters of the ward.

During the daytime the volume of trade was not very different from the shops already considered,—about 24 customers per hour being registered: in the evening the number of customers is higher than in any shop, except one, in the ward, over 101 per hour. There were no women *on*-drinkers at this place and only one bottle customer—a low caste Hindu woman. Three Hindu boys visited the shop during the day (in four hours) and one in the evening, the latter for *off* consumption. In these respects —women and children—this shop is better than those previously described. But, on the other hand, in the service of already drunken men it cannot be exonerated. On

August 8 between 8 and 9 P.M. there was a drunken row outside the shop, culminating in a fight.

Over 56 per cent. of the day *on* customers entered in company; in the evening this number was reduced to 41 per cent.

(5) 89, *Upper Chitpore Road.*—This shop is situated distinctly in the more respectable end of the ward. It is a house of two storeys, the upper one apparently occupied by the licensee. It is the worst constructed house in the ward, for it has a side-entrance so far down a lane and away from the shop, that no casual observer would connect it with the shop at all, and yet it is used. From this the shop is entered behind the counter. There is no snuggery obvious from the street. The establishment pays Rs. 550 per month for the privilege of selling liquor.

Though apparently small, the shop is visited by as many people as any shop in the ward—24 per hour during the day and up to 104 per hour at night. Very few Mahommedans enter here (under 4 per cent. at night, none in the day), but the number of Hindus of the better class rises in the daytime to over 26 per cent. and in the evening to 20 per cent. of the customers. In the daytime on one occasion no women or children entered the shop; on another no less than 3 women and 4 boys appeared in three hours. At night during our observation, only one woman bought a bottle of liquor to take away while two low caste Hindu boys 13 to 14 years old, actually drank country liquor on the spot.

On one day a series of observations were made to find out what was the busy time for the shop, and the notes kept separate for every hour from 11 A.M. to 4 P.M. From 11 to 12 the shop was busy, then there was a lull till 2 P.M., then ensued another busy hour till 3 o'clock, when the trade almost disappeared till after 4 P.M. At this shop over 59 per cent of the customers entered in company during the day and only just over 28 per cent. in the evening.

Speaking generally it may be said that much more foreign liquor is consumed at this shop than at those previously considered. Most of the customers were lower caste

Bengalis. One of these on the afternoon of July 7 got drunk here and became boisterous, and was taken away by the Police. Hindu clerks often called on their return from offices and during the afternoon nearly all such customers came from the direction of the business quarters of Calcutta.

(6) 196, *Cornwallis Street.*—This shop, whose license together with that of the next in order is sold in one lot for Rs. 1,005 per month, is a one storey building with an upper room at one corner. It has four front doors, a very objectionable side-door entered from a narrow passage on one side, where customers are served after hours. It was at this shop, and in an endeavour to look at this side-door that our observer was all but assaulted on March 17. There are two stalls in front of the shop, and corresponding snuggeries inside it.

All our observers agree that this is a "gentleman's shop," in fact, a still larger proportion of the visitors can be so classed than even at 89, Chitpore Road,—just considered. One worker noted for instance: "The drinkers are mostly the better class babus." The proportion of them among the day customers runs up to 31 per cent. and to 29½ per cent. at night. The Mahommedans served here are almost a negligible quantity. On July 8 three Europeans visited the shop. Two of these came in a *ghari* and drank foreign liquor; the third was a regular loafing drinker. On August 10 one European came in also and, besides drinking on the spot, carried away a pint with him.

In the daytime two women were served in 5½ hours; in the evening none. Children were absent from the day's list of customers, except for one boy, who carried liquor away in bottle from the shop. At this place (when note was taken of the fact) 18 out of 53 men *on*-drinkers entered in company or 34 per cent. In the evening the proportion fell to 28 per cent. The evening observations were taken at this shop, in March and in August, and it is curious to note the effect of the hotter weather on the trade, for while at the earlier, cooler period the customers only numbered 54 per hour in August 62 were served in the same time.

(7) 206, *Cornwallis Street.*—Is a small one storey shop, surrounded by *kutcha* houses, though across the road on every side are brick buildings. There is an entrance into the backyard from Baranosee Ghose's Street, but we have no evidence that this is used by customers entering the shop. There is, moreover, a side entrance up a narrow passage on the other side of the building. Four doors open from the main street into the shop, and there are in addition two stalls in front of the building, thus forming a snuggery in the shop. This is the only grogshop in the ward in which we saw a *punkah* in use.

The premises were under observation, during the day, on July 4 and 9 and in the evening on August 4, 5 and 6, 1904. The results show that while we are here out of the Mahommedan quarter, yet we have not reached such a well-to-do centre as at the last shop described. The proportion of higher class Hindu visitors amounted in the daytime to just over 10 per cent. and in the evening to nearly 29 per cent. Two women were served and eight children entered the shop in eight hours during the day. Two boys, however, counted as children, were probably a little over our standard of 14 years old—they were actually served with drink on the premises. In the evening one woman drank at the shop, and four women carried away liquor in three hours, but no children were served at all. The on-drinkers were nearly seven times as numerous as the off-customers in the day observations, and nearly four times as numerous as those in the evening. Company drinking was, as usual, more frequent in the daytime, but not much more so, for those who entered in groups numbered 31½ per cent. of the total men on-drinkers in the day and 27 per cent. in the evening.

There is one European, wearing the Almshouse uniform, who seemed to haunt this shop during our proceedings. He came down Sukea's Street and regularly took away a pint of liquor. Another day still another man, wearing European clothes, came to the shop and after drinking lay down and slept for two hours in the establishment.

(8) 134, *Manicktolla Street.*—This is the last shop in the ward, and in construction is one of the worst from the point of view of easy supervision. It lies near the north end of the ward and pays Rs. 555 per month, a larger fee than any other house, save No. 1 at 136 Machua Bazar. It is a two storey brick building, the upper storey being used as a residence by someone, though not apparently by the licensee. The main shop appears at first sight to have six doors, the two on the eastern side, however, are not direct entrances but lead to some living premises behind. Inside these doors and quite shielded from observation when they are shut, is a side entrance to the shop. Seeing that many people live behind and can thus get access to the shop at any hour without police or other authorities being any the wiser, we consider this construction as bad and defective. The shop was under observation for two days (July 11 and August 23) and for four evenings (April 15 and 26 and August 10 and 11). The class of customers differs little from that at the other shops at this end of the ward. There were no Europeans and very few Mahommedans. The proportion of the better class Bengalis was smaller than at the Cornwalis Street shop, being 11 per cent. in the day and over 37 per cent. at night. Only two women were served with bottled spirit during nine hours day watching; at night one woman only (in six hours) bought for off consumption. As regards children two boys were served, they came with a crowd of men about 5 P.M. in the evening; 3 boys entered the shop, apparently in order to drink on the premises, while 4 carried away bottles of liquor. The amount of "company drinking" was about normal, 18½ per cent of the men on-drinkers entering in company in the daytime and 39 per cent. in the evening.

At this shop the slackest time for business seemed to be between 12 noon and 1 P.M. and it remained fairly constant from 1 o'clock onward, until the busier evening period.

On the afternoon of July 11 a curious incident occurred. About 5 P.M. twelve municipal sweepers and two boys armed with sticks and two boys came and sat in a row outside and drank country liquor. After a few minutes another party of sweepers appeared in the distance, but seeing the warlike attitude of the first twelve, they drew off and the expected fight did not come off.

We have now described each shop in order, and its customers, so far as is possible, in a short report. There remain two or three matters in which the shops can best be dealt with together. Of these the first is the question of the *time* of greatest drinking. At nearly every shop notes were taken for the three separate evening hours, and there remains no doubt that from 8 to 9 P.M. considerably more drinking is carried on than at any other time in the day. While the average number of visitors to all the shops runs up to 108 during the last hour, it is only 95 between 7 and 8 and 74[1] for the hour previous to that. The difference is as a rule least in the lowest class shops, such as those in Machua Bazar, while in Nos. 4 (76, Upper Chitpore Road), 5 (89, Upper Chitpore Road), 6 (196, Cornwallis Street) and 8 (134, Manicktolla Street) it is very great.

*Infringements of the law.*—We now come to a point of considerable importance. The legal hour for closing licensed premises in Calcutta is 9 P.M. Report has often been made to us that this rule is a dead letter, and in consequence special attention has been given to the point. As a result we must say that we find there is a regular and constant after hours' sale going on at the shops, and we propose now to give notes on the matter with regard to several, giving dates and times.

No. 1. August 5.—The doors were closed at 9 P.M. and all locked except one. This remained open at least till 9-15 P.M. (when observer left) and quite a number of people entered the shop.

No. 2 August 6.—The side door was kept open till 9-15 P.M. at least, not being closed when the observer left.

[1] Average of six shops only.

No. 3. August 8.—The side-door remained open after 9, and a man was stationed at the door holding it, and letting people in and out till 9-10 P.M. when on noticing the observer the house was suddenly closed.

No. 4. August 9.—At 8-30 P.M. a policeman went into the shop and spoke to the licensee and at 8-55 P.M. the same policeman was energetic in driving all the customers out of the shop. The shopkeeper then locked up by exactly 9 P.M.

No. 6. March 17.—Six people went into the shop after 9 P.M. by the side-door. The house was closed very hurriedly after visit paid to the side-door by the observer.

August 10.—At 9-20 P.M., one of the doors was open and people going in and out. At 9-30 P.M. people were still entering and leaving the shop, buying liquor as usual.

August 11.—At 9-20 P.M., the side door was only shoved to and 2 or 3 minutes after about 8 people, evidently drinkers, came out together.

No. 7. August 4.—The shop closed at 9 P.M., but visitors did not leave till 9-10 P.M.

August 17.—At 9-10 P.M. one door was only shoved to, and a man was sitting outside admitting visitors. The house was not locked up entirely till 9-30 P.M.

No. 8 August 11.—At 9-5 P.M. the door was open as usual and customers were going in and out.

August 17.—At 9-20 P.M. the side-door was open, and the place fully lighted. Shortly after five babus came out together.

It thus appears that for every shop in the ward, except two, there is distinct evidence of regular breach of the law with respect to hours of closing. The dates given above were not chosen for any particular purpose and the notes were made in the ordinary course of our observations. In several cases, moreover, *parawallas* were in the locality where the breach of the law was being committed, but apparently failed to check this illicit selling. We consider this state of affairs a very serious one.

*General Conclusions.*—Two more notes remain to be made before we attempt to draw general conclusions from the results given. The first of these concerns the character of the liquor drunk. Only two attempts were made to determine this, and both during the day, one at 136, Machua Bazar (a low class shop) and one at 206, Cornwallis Street (a higher class shop). In both cases nearly all the drinking on the premises was of country liquor, though the gentlemen customers more frequently called for foreign spirit than any one else. Thus at the former shop 4 customers out of 78 had foreign spirit, three of whom were classed as gentlemen; at the latter three out of 38 were so served, none of whom were classed as gentlemen. For off consumption foreign spirit is more used. At 206, Cornwallis Street two men out of six off customers carried away so-called English liquor; at 136, Machua Bazar five out of 31 were similarly served. Practically nothing but spirit is consumed at these shops, no wine or beer, and the title Wine and Spirit Merchant is a misnomer when the business done is considered.

The last note refers to the prevalence of social drinking, that is, drinking by a group who enter the shop together for this purpose. During the daytime this is very general. In one case, as has been noted, 80 per cent. of the men who went into the shop to drink, entered in company. In the evening social drinking is not so frequent, a less proportion being universal in our observations though even then the proportion usually amounts to nearly 30 per cent. But it should be remembered that in these cases, we have not usually social drinking of the European kind, where one man pays for the drinks of a group. Here, on the other hand, each man nearly always pays for his own liquor, however large the group may be, and though snuggeries exist at most of the shops the customers do not use them to stay and chat in except under special circumstances. Generally it is drink, pay, and go.

Such are our observations. It must be admitted that they throw a flood of light on the drink traffic in the native quarters of the city, and confirm the anxious feeling which one gets with regard to the spread of the drink traffic and the change in the character of the drinking habits of the people. That Mahommedans should drink as they do: that even Mahommedan *women* should consume spirit in public, and this not as a rare case but regularly and in large numbers, is a matter of extreme seriousness. Still more serious is the fact that Hindu gentlemen, at least 75 per cent. of whom belong to the higher castes, are found in large numbers as visitors to the grogshops actually being over 14 per cent. of the customers in the daytime and practically 20 per cent. in the evening. We look upon this fact as the most serious and menacing that we have established, not so much because of the evil already done, but because it is a most complete proof of a change in habit and in sentiment, which may lead to the most dire results, not for them only but for the whole people of Bengal.

*What is to be done.*—What then can be done? So far as a general policy on the part of temperance workers is concerned, the future is by no means clear, but the matter demands careful and immediate consideration. On the other hand, there are some matters which while not touching the general and most important question of the spread of drinking habits among Indians, yet demand immediate attention on the part of the legislative or administrative authorities. These are as follows:-

1. A better construction of houses be insisted on. At present all the shops but two have most objectionable side-doors, and in three cases an impression was left in our minds that they were expressly constructed to defy supervision. We consider that at the next licensing session in 1905, it should be insisted that a license will be null and void unless the shop is made a lock up place, that side-doors be all permanently closed, and that the shops be so altered to admit of easy and complete supervision from the street.

2. Prohibition of serving liquor to women and children. As regards the former the customers just reach 1 per cent. of the total. This is small we know; most of the women served appear to belong to the immoral class, but

it is on account of this very fact of the smallness of the number that we would insist on service of women being prohibited. It would cause little or no hardship at present, and would be a most valuable safeguard for the future. We are glad to say that the number of children is considerably smaller than we expected, and very much less than in Ward No. 13. The number in fact only amounts to 0.7 per cent. of the total visitors. These are, however, practically all under 14 years old, the age up to which the new Bengal Excise Bill, as amended, we are given to understand, prohibits service. We do not think any hardship would be incurred and some considerable benefit obtained if the age were raised to 16 years.

3. Three cases of drunkenness were noticed, two of them resulting in a row. We think that the law on this matter should be levelled up to the English standard, and the grogshop keeper be made responsible for seeing that no person whom he serves is drunk and be moreover liable for keeping disorderly premises if a drunken row occurs outside his premises.

4. We think that very much greater activity should be shown in insisting that the law as to closing at 9 P.M. should be obeyed. Unless they suspect that they are watched the Police themselves are not energetic in getting the shop closed to time, and we feel that the results of the energetic action of the Commissioner of Police a year ago are already tending to pass away. The question as to whether the hours cannot and should not be shortened is one which demands close attention.

5. The license fee paid has apparently little connection with the amount of business done. Thus 136 Machua Bazar pays nearly twice as much (Rs. 620 per month) as his neighbour at 158 (Rs. 325), and according to our figures does very little more business. And so on throughout the ward. It is a matter worthy of close consideration whether the present sale of licenses by auction is defensible either from a revenue or from a moral point of view and whether better men would not be secured for the shops were the fee

fixed for each shop at an amount depending not on the length of the licensee's pocket, but on the business done at the place. We only throw out this idea as a suggestion, for the present system of high license even were this change made, is open to very serious objection, an objection not diminished among our workers by what they have seen during these investigations.

*The gravity of the situation.*—In conclusion, we would once more urge that the most serious aspect of the question is the evidence that drinking alcoholic liquor is indulged in by those in all ranks of Society, and while it is certainly a grave matter to find that Mahommedans, whose religion so specifically forbids drinking, enter grogshops in large numbers, it is even more serious that so large a proportion of the drinkers should belong to the higher Hindu castes, whose religion and social code is so strongly on the side of total abstinence. However serious may be the problem of raising revenue, the question cannot be only considered from this aspect. It is recognised that the Indian races have not the drink resisting stamina of the northern nations, and one cannot contemplate the future of the peoples inhabiting Bengal without great misgiving if the drinking habits, now shown to be confined to no class or race or religion, spread among the mass of the community of which they have at present only touched the fringe.

*On behalf of the*
*Drink Problem*
*Investigation Committee.*

HAROLD H. MANN.

Since this report was drafted and a summary of it published, it is pleasing to report that the authorities have taken steps to close some of the objectionable side entrances referred to.

## RECORDS OF INDIVIDUAL GROGSHOPS.

### Day Observations.

| Date 1904. | ON DRINKERS. Indians. Mahommedans. | | | Low Caste Hindus. | | | Hindu Gentlemen. | European Men. | OFF CUSTOMERS. Indians. Mahommedans. | | | Low Caste Hindus. | | | Hindu Gentlemen. | European Men. | Total. |
|---|---|---|---|---|---|---|---|---|---|---|---|---|---|---|---|---|---|
| | M. | W. | C. | M. | W. | C. | | | M. | W. | C. | M. | W. | C. | | | |
| **136. Machua Bazar Street.** | | | | | | | | | | | | | | | | | |
| April 21st (11¼ A.M.—4 P.M.) | 44 | 2 | 4 | 19 | 3 | — | 17* | 1 | 12 | — | 9 | 7 | 1 | 2 | — | — | 121 |
| July 4th (11¾ A.M.—1¼ P.M.) | 21 | — | 3 | 13 | — | 5 | 3 | — | 2 | — | — | 2 | — | 1 | — | — | 56 |
| July 4th (3¼ P.M.—5½ P.M.) | 25 | 3 | 4 | 19 | — | — | 7 | — | 2 | — | 1 | 3 | — | — | — | — | 64 |
| Total | 90 | 5 | 11 | 51 | 3 | 5 | 27 | 1 | 16 | — | 10 | 12 | 1 | 3 | — | — | 235 |
| **Number Per Hour** | 10.6 | .6 | 1.3 | 6.0 | .3 | .6 | 3.2 | .1 | 1.9 | — | 1.2 | 1.4 | .1 | .3 | — | — | 27.6 |

*Two of these were boys.

| | | | | | | | | | | | | | | | | | |
|---|---|---|---|---|---|---|---|---|---|---|---|---|---|---|---|---|---|
| **158, Machua Bazar Street.** | | | | | | | | | | | | | | | | | |
| July 15th (1 P.M.—5 P.M.) | 28 | 8 | — | 41 | — | — | — | 3 | 6 | — | — | 14 | — | — | — | — | 100 |
| **Number Per Hour** | 7 | 2 | — | 10.2 | — | — | — | .7 | 1.5 | — | — | 3.5 | — | — | — | — | 25.0 |
| **14, Upper Chitpore Road.** | | | | | | | | | | | | | | | | | |
| July 5th (11 A.M.—1-10 P. M. | 6 | — | — | 12 | — | 1 | 6 | — | — | — | — | 6 | — | — | — | — | 31 |
| July 5th (4 P.M.—6¾ P.M.) | 21 | — | — | 44 | — | 4 | 6 | — | — | 1 | — | 3 | 1 | 1 | — | — | 81 |
| Total | 27 | — | — | 56 | — | 5 | 12 | — | — | 1 | — | 9 | 1 | 1 | — | — | 112 |
| **Number Per Hour** | 5.5 | — | — | 11.4 | — | 1.0 | 2.4 | — | — | .2 | — | 1.8 | .2 | .2 | — | — | 22.7 |
| **76. Upper Chitpore Road.** | | | | | | | | | | | | | | | | | |
| July 6th (1½ P.M..—5 P.M.) | 7 | — | — | 59 | — | 3 | 5 | — | 3 | — | — | 11 | — | — | 8 | — | 96 |
| **Number Per Hour** | 1.7 | — | — | 14.8 | — | .7 | 1.3 | — | .7 | — | — | 2.8 | — | — | 2.0 | — | 24.0 |
| **196, Cornwallis Street.** | | | | | | | | | | | | | | | | | |
| July 8th (1 P.M..—6½ P.M.) | 3 | — | — | 32 | 1 | — | 15 | 3 | — | — | — | 6 | 1 | — | 6 | — | 67 |
| **Number Per Hour** | .5 | — | — | 5.8 | .2 | — | 2.7 | .5 | — | — | — | 1.1 | .2 | — | 1.1 | — | 12.1 |

DAY OBSERVATIONS—(Contd.)

| Date 1904. | ON DRINKERS. Indians. Mahommedans. M. | W. | C. | Low Caste Hindus. M. | W. | C. | Hindu Gentlemen. | European Men. | OFF CUSTOMERS. Indians. Mahom-medans. M. | W. | C. | Low Caste Hindus. M. | W. | C. | Hindu Gentlemen. | European Men. | Total. |
|---|---|---|---|---|---|---|---|---|---|---|---|---|---|---|---|---|---|
| **89, Upper Chitpore Road** | | | | | | | | | | | | | | | | | |
| **July 7th** (3¾ P.M.—6½ P.M.) | — | — | — | 35 | 1 | 1(?) | 25 | — | — | — | — | 23 | 2 | 3 | 16 | — | 106 |
| **August 22nd** (11½ A.M.—12 Noon) | — | — | — | 12 | — | — | 1 | — | — | — | — | 7 | — | — | 1 | — | 21 |
| Do. (12 Noon—1 P.M.) | — | — | — | 7 | — | — | 1 | — | — | — | — | 3 | — | — | 2 | — | 13 |
| Do. (1 P.M.—2 P.M.) | — | — | — | 7 | — | — | — | — | — | — | — | 2 | — | — | 1 | — | 10 |
| Do. (2 P.M.—3 P.M.) | — | — | — | 18 | — | — | — | — | — | — | — | 8 | — | — | 1 | — | 27 |
| Do. (3 P.M.—4 P.M.) | — | — | — | 4 | — | — | 1 | — | — | — | — | 8 | — | — | 1 | — | 14 |
| Total | — | — | — | 83 | 1 | 1 | 28 | — | — | — | — | 51 | 2 | 3 | 22 | — | 191 |
| **Number Per Hour** | — | — | — | 10.4 | 1 | .1 | 3,5 | — | — | — | — | 6.4 | .2 | .4 | 2,8 | — | 23.9 |

**206, Cornwallis Street.**

| | | | | | | | | | | | | | | | | | |
|---|---|---|---|---|---|---|---|---|---|---|---|---|---|---|---|---|---|
| July 4th (12 Noon—3¼ P.M.) | 3 | — | — | 31 | — | 3 | 4 | — | 1 | — | 1(?) | 4 | — | — | — | — | 47 |
| July 9th (12 Noon—5 P.M.) | 5 | 1 | 2(?) | 27 | 1 | 2 | 6 | 1 | — | — | — | 6 | — | — | — | 1 | 52 |
| Total | 8 | 1 | 2 | 58 | 1 | 5 | 10 | 1 | 1 | — | 1 | 10 | — | — | — | 1 | 99 |
| **Number Per Hour** | 1.0 | .1 | .3 | 7.1 | .1 | .6 | 1.2 | .1 | .1 | — | .1 | 1.2 | — | — | — | .1 | 12.0 |

**134, Manicktolla Street.**

| | | | | | | | | | | | | | | | | | |
|---|---|---|---|---|---|---|---|---|---|---|---|---|---|---|---|---|---|
| July 11th (2 P.M.—6 P.M.) | 8 | — | — | 23 | — | 2 | 7 | — | — | — | — | 7 | 2 | — | 4 | — | 53 |
| August 23rd (11¼ A.M.—4 P.M. | — | — | — | 37 | — | — | — | — | — | — | — | 16 | — | — | — | — | 53 |
| Total | 8 | — | — | 60 | — | 2 | 7 | — | — | — | — | 23 | 2 | — | 4 | — | 106 |
| **Number Per Hour** | .9 | — | — | 6.9 | — | .2 | .8 | — | — | — | — | 2.6 | .2 | — | .5 | — | 12.1 |

DAY OBSERVATIONS.—(**Contd.**)

**Summary—Number of visitors per hour.**

| | | | |
|---|---|---|---|
| 1. | 136, | Machua Bazar ... | 27.6 |
| 2. | 158, | Machua Bazar ... | 25.0 |
| 3. | 14, | Upper Chitpore Road ... | 22.7 |
| 4. | 76, | Upper Chitpore Road ... | 24.0 |
| 5. | 89, | Upper Chitpore Road ... | 23.9 |
| | | Carried over | 123.2 |
| | | Brought forward | 123.2 |
| 6. | 196, | Cornwallis Street ... | 12.1 |
| 7. | 206, | Cornwallis Street ... | 12.0 |
| 8. | 134, | Manicktolla Street ... | 12.1 |
| | | Total | 159.4 |
| | | **Average per hour** | 20.0 |

## RECORDS OF INDIVIDUAL GROGSHOPS.

EVENING OBSERVATIONS

| Date 1904. | ON DRINKERS. Indians. Mahommedans. M. | W. | C. | Low Caste Hindus. M. | W. | C. | Hindu Gentlemen. | European Men. | OFF CUSTOMERS. Indians. Mahom-medans. M. | W. | C. | Low Caste Hindus. M. | W. | C. | Hindu Gentlemen. | European Men. | Total |
|---|---|---|---|---|---|---|---|---|---|---|---|---|---|---|---|---|---|
| **136 Machua Bazar Street.** | | | | | | | | | | | | | | | | | |
| **February 23rd** (6¾—7¾) | 21 | — | — | 56 | 2 | — | 15 | — | — | — | — | 17 | — | 4 | 1 | — | 116 |
| **February 27th** (6½—7½) | 46 | — | — | 44 | — | — | — | — | — | — | — | 3 | — | — | 1 | — | 94 |
| **August 4th** (6.55—8.) | 34 | — | — | 23 | — | 2(?) | 3 | — | — | — | — | 11 | — | 1 | 4 | — | 78 |
| **August 5th** (8.10—9.5) | 18 | — | — | 60 | — | — | 5 | — | 3 | — | — | 6 | — | — | 3 | — | 95 |
| **August 6th** (5.40—6.35) | 22 | — | — | 39 | — | 1 | 11 | — | 10 | — | — | 9 | — | — | — | — | 92 |
| Total | 141 | — | — | 222 | 2 | 3 | 34 | — | 13 | — | — | 46 | — | 5 | 9 | — | 475 |
| **Number Per Hour** | 28.7 | — | — | 45.1 | 5 | .6 | 6.9 | — | 2.6 | — | — | 9.4 | — | 1.0 | 1.8 | — | 96.6 |

EVENING OBSERVATIONS—(**Contd.**)

| Date 1904. | ON DRINKERS. Indians. Mahommedans. M. | W. | C | Low Caste Hindus. M. | W. | C. | Hindu Gentlemen. | European Men. | OFF CUSTOMERS. Indians. Mahommedans. M. | W. | C. | Low Caste Hindus M. | W. | C. | Hindu Gentlemen. | European Men. | Total. |
|---|---|---|---|---|---|---|---|---|---|---|---|---|---|---|---|---|---|
| | | | | | | **158, Machua Bazar Street.** | | | | | | | | | | | |
| July 5th (7—7½) | 10 | — | — | 16 | — | 2(?) | 1 | — | — | — | — | — | — | — | — | — | 29 |
| August 4th (5¾—6¾) | 34 | 4 | — | 42 | — | — | 4 | — | 3 | 1 | — | 4 | — | — | — | — | 92 |
| August 5th (7—8.) | 29 | 3 | — | 35 | — | — | 7 | — | — | — | — | 8 | — | — | — | — | 82 |
| August 6th (8—9¼.) | 36 | 5 | — | 37 | — | — | — | — | 13 | — | — | 13 | — | — | — | — | 104 |
| Total | 109 | 12 | — | 130 | — | 2 | 12 | — | 16 | 1 | — | 25 | — | — | — | — | 307 |
| **Number Per Hour** | 29.1 | 3.2 | -- | 34.6 | | .5 | 3.2 | — | 4.3 | .3 | — | 6.7 | — | — | — | — | 81.9 |

| | | | | | | | | | | | | | | | | | |
|---|---|---|---|---|---|---|---|---|---|---|---|---|---|---|---|---|---|
| **14, Upper Chitpore Road.** | | | | | | | | | | | | | | | | | |
| August 8th (8¼—9.10.) | 11 | — | — | 58 | — | — | 28 | — | 4 | — | — | 6 | — | — | 3 | — | 110 |
| August 9th (5½—6½.) | 7 | — | — | 32 | — | — | 4 | — | 2 | — | — | 7 | — | — | 2 | — | 54 |
| August 12th (6.50—7.45.) | 7 | — | — | 32 | — | — | 11 | — | — | — | — | 16 | — | — | 6 | — | 72 |
| Total | 25 | — | — | 122 | — | — | 43 | — | 6 | — | — | 29 | — | — | 11 | — | 236 |
| **Number Per Hour** | 8.8 | — | — | 43.1 | — | — | 15.2 | — | 2.1 | — | — | 10.2 | — | — | 3.9 | — | 83.3 |
| **76, Upper Chitpore Road.** | | | | | | | | | | | | | | | | | |
| August 8th (6.50—8) | 8 | — | — | 61 | — | — | 22 | — | — | — | — | 32 | — | 1 | 4 | — | 128 |
| August 9th (7.50—9) | 7 | — | — | 80 | — | — | 21 | — | — | — | — | 27 | 1 | — | 3 | — | 139 |
| August 12th (5.40—6.40) | 8 | — | — | 31 | — | — | 10 | — | — | — | — | 16 | — | — | 5 | — | 70 |
| Total | 23 | — | — | 172 | — | — | 53 | — | — | — | — | 75 | 1 | 1 | 12 | — | 337 |
| **Number Per Hour** | 6.9 | — | — | 51.7 | — | — | 15.9 | — | — | — | — | 22.5 | .3 | .3 | 3.6 | — | 101.2 |

EVENING OBSERVATIONS.—(**Contd.**)

| Date 1904. | **ON DRINKERS.** Indians. Mahommedans. | | | Low Caste Hindus. | | | Hindu Gentlemen. | European Men. | **OFF CUSTOMERS.** Indians. Mahom-medans. | | | Low Caste Hindus. | | | Hindu Gentlemen. | European Men. | Total. |
|---|---|---|---|---|---|---|---|---|---|---|---|---|---|---|---|---|---|
| | M. | W. | C. | M. | W. | C. | | | M. | W. | C. | M. | W. | C. | | | |
| | | | | | | **89, Upper Chitpore Road.** | | | | | | | | | | | |
| August 8th (5¾—6.40) | — | — | — | 34 | — | — | 8 | — | — | — | — | 10 | 1 | — | — | — | 53 |
| August 9th (6¾—7.40) | 4 | — | — | 61 | — | 2 | 18 | — | — | — | — | 20 | — | — | 3 | — | 108 |
| August 12th (8—9) | 5 | — | — | 62 | — | — | 21 | — | 2 | — | — | 35 | — | — | 9 | — | 134 |
| Total | 9 | — | — | 157 | — | 2 | 47 | — | 2 | — | — | 65 | 1 | — | 12 | — | 295 |
| **Number Per Hour** | 3.1 | — | — | 55.4 | — | .7 | 16.6 | — | .7 | — | — | 22.9 | .4 | — | 4.2 | — | 104.0 |
| | | | | | | **196, Cornwallis Street.** | | | | | | | | | | | |
| August 10th (5¾—7.50) | 4 | — | — | 34 | — | — | 19 | 1 | — | — | — | 7 | — | 1 | 8 | — | 74 |
| August 11th (8.5—9) | 7 | — | — | 58 | — | — | 15 | — | — | — | — | 18 | — | — | 13 | — | 111 |
| Total | 11 | — | — | 92 | — | — | 34 | 1 | — | — | — | 25 | — | 1 | 21 | — | 185 |
| **Number Per Hour** | 3.7 | — | — | 30.7 | — | — | 11.3 | .3 | — | — | — | 8.3 | — | .3 | 7.0 | — | 61.6 |

**206, Cornwallis Street.**

| | | | | | | | | | | | | | | | | | |
|---|---|---|---|---|---|---|---|---|---|---|---|---|---|---|---|---|---|
| August 4th (8½—9) | 5 | — | — | 36 | — | — | 20 | — | — | — | — | 10 | 1 | — | 1 | — | 73 |
| August 5th (5-40—6¾) | 11 | — | — | 30 | — | — | 18 | — | — | — | — | 15 | 1 | — | 7 | 1 | 83 |
| August 6th (6-50—7¾) | 5 | — | — | 38 | 1 | — | 36 | — | — | — | — | 13 | 2 | — | 6 | — | 101 |
| Total | 21 | — | — | 104 | 1 | — | 74 | — | — | — | — | 38 | 4 | — | 14 | 1 | .257 |
| **Number Per Hour** | 7.6 | — | — | 37.8 | .4 | — | 26.9 | — | — | — | — | 13.8 | 1.5 | — | 5.1 | .4 | 93.4 |

**134, Manicktolla Street.**

| | | | | | | | | | | | | | | | | | |
|---|---|---|---|---|---|---|---|---|---|---|---|---|---|---|---|---|---|
| August 10th (8—9) | 5 | — | — | 57 | — | — | 35 | — | — | — | — | 17 | — | 4 | 12 | — | 124 |
| August 11th (6—8) | 5 | — | — | 41 | — | — | 28 | — | — | — | — | 10 | 1 | — | 5 | — | 90 |
| Total | 10 | — | — | 92 | — | — | 63 | — | — | — | — | 27 | 1 | 4 | 17 | — | 214 |
| **Number Per Hour** | 3,3 | — | — | 30 7 | — | — | 21.0 | — | — | — | — | 9.0 | .3 | 1.3 | 5.7 | — | 71.3 |

EVENING OBSERVATIONS.—(**Contd.**)

| Date 1904. | ON DRINKERS. Indians. Mahommedans. | | | Low Caste Hindus. | | | Hindu Gentlemen. | European Men. | OFF CUSTOMERS. Indians. Mahom-medans. | | | Low Caste Hindus. | | | Hindu Gentlemen. | European Men. | Total. |
|---|---|---|---|---|---|---|---|---|---|---|---|---|---|---|---|---|---|
| | M. | W. | C. | M. | W. | C. | | | M. | W. | C. | M. | W. | C. | | | |
| *Grogshop.* | | | | | | | *Summary.* | | | | | | | | | | |
| 1. | 28.7 | — | — | 45.1 | .5 | .6 | 6.9 | — | 2.6 | — | — | 9.4 | — | 1.0 | 1.8 | — | 96.6 |
| 2. | 29.1 | 3.2 | — | 34.6 | — | .5 | 3.2 | — | 4.3 | .3 | — | 6.7 | — | — | — | — | 81.9 |
| 3. | 8.8 | — | — | 43.1 | — | — | 15.2 | — | 2.1 | — | — | 10.2 | — | — | 3.9 | — | 83.3 |
| 4. | 6.9 | — | — | 51.7 | — | — | 15.9 | — | — | — | — | 22.5 | .3 | .3 | 3.6 | — | 101.2 |
| 5. | 3.1 | — | — | 55.4 | — | .7 | 16.6 | — | .7 | — | — | 22.9 | .4 | — | 4.2 | — | 104.0 |
| 6. | 3.7 | — | — | 30.7 | — | — | 11.3 | .3 | — | — | — | 8.3 | — | .3 | 7.0 | — | 61.6 |
| 7. | 7.6 | — | — | 37.8 | .4 | | 26.9 | — | — | — | — | 13.8 | 1.5 | — | 5.1 | .4 | 93.4 |
| 8. | 3.3 | — | — | 30.7 | — | — | 21.0 | — | — | — | — | 9.0 | .3 | 1.3 | 5.7 | — | 71.3 |
| Total | 91.2 | 3.2 | — | 329.1 | .9 | 1.8 | 117.0 | .3 | 9.7 | .3 | — | 102.8 | 2.5 | 2.9 | 31.2 | .4 | 693.3 |
| **Number Per Hour** | 11.4 | .4 | — | 41.1 | .1 | .2 | 14.6 | .1 | 1.2 | .1 | — | 12.8 | .3 | .4 | 3.9 | .1 | 86.7 |

EVENING OBSERVATIONS.—(Contd.)

(*Observations less complete than the former.*)

| Date 1904. | ON DRINKERS. Indians. M. | W. | C. | European Men. | OFF CUSTOMERS. Indians. M. | W. | C. | European Men. | Total. |
|---|---|---|---|---|---|---|---|---|---|
| **14, Upper Chitpore Road.** | | | | | | | | | |
| February 29th (6.5—7.5) | 57 | 1 | — | — | 11 | — | 1 | — | 70 |
| March 9th (7.5—8.5) | 54 | — | 3 | — | 15 | — | 2 | — | 74 |
| Total | 111 | 1 | 3 | — | 26 | — | 3 | — | 144 |
| **Number Per Hour** | 55.5 | .5 | 1.5 | — | 13 | — | 1.5 | — | 72 |
| **196, Cornwallis Street.** | | | | | | | | | |
| March 9th (6.40—7.40) | 50 | — | 3 | — | 6 | — | — | — | 59 |
| March 12th (8—9) | 50 | 4 | 3 | 1 | 11 | — | — | — | 69 |
| March 14th (6—6.55) | 33 | — | — | — | 1 | — | — | — | 34 |
| March 17th (8½—9¼) | 36 | — | — | — | — | — | — | — | 36 |
| Total | 169 | 4 | 6 | 1 | 18 | — | — | — | 199 |
| **Number Per Hour** | 46.1 | 1.1 | 1.6 | .3 | 4.9 | — | — | — | 51.0 |

EVENING OBSERVATIONS.—(**Contd.**)

| Date 1904. | ON DRINKERS. | | | | OFF CUSTOMERS. | | | | |
|---|---|---|---|---|---|---|---|---|---|
| | Indians. | | | European Men. | Indians | | | European Men. | Total. |
| | M. | W. | C. | | M. | W. | C. | | |
| | | | | **134, Manicktolla Street.** | | | | | |
| April 15th (6.35—7.35) | 40 | — | — | — | 4 | — | — | — | 44 |
| April 26th (8—9) | 50 | — | 3 | — | 8 | — | — | — | 61 |
| Total | 90 | — | 3 | — | 12 | — | — | — | 105 |
| **Number Per Hour** | 45.0 | — | 1.5 | — | 6.0 | — | — | — | 52.5 |

*Chapter* 14

# THE UNTOUCHABLE CLASSES OF AN INDIAN CITY*

In the organisation of society among the Hindu population of India, there is perhaps no feature more interesting than the existence of a large proportion of the people contact with whom means ceremonial impurity to the other Hindus, and who yet remain, as they have done for many generations, a necessary part of the scheme of society. The existence of such a class is, of course, not peculiar to India, but its origin elsewhere can generally be traced with fair clearness, and its continuance is a matter of comparatively short duration. In the present instance, however, its origin is absolutely unknown, and the theories which have been put forward by various observers obstinately refuse to fit all the facts; while the class is so essential a feature of Indian society that its abolition would mean a social reconstruction of the remaining people of an exceedingly fundamental character.

It is not my intention, however, to discuss the whole question of the low untouchable castes of India. For that I have no special qualifications, and without these it would be merely unprofitable and possibly misleading. Suffice it to say, before I turn to the subject of this paper, that little beyond superficial observation seems to have been made and published about more than a very small proportion of the classes to which I refer. It is even impossible to know from the census figures exactly the number of the untouchable castes, because the precise degree of

* Reprinted with permission from the **Sociological Review**, London, Vol. V (1912), No. 1, pp. 42-55.

untouchableness is unknown. Some—and among them my friend Mr. V. R. Shinde of Bombay, who has made a very special study of the question—place their numbers at one-sixth of the total Hindu population, the largest proportion, by far, being in Madras and Southern India generally. I myself think that the number is much smaller than this, but at any rate they form a far larger proportion of the population than is generally conceived.

My own inquiries relate to the Deccan only, and chiefly to Poona City, and it is about the untouchable classes of the latter that I wish more particularly to speak here. But inasmuch as the castes which may be considered untouchable are the same in Poona and the rest of Maharashtra, and inasmuch as Poona City is, like most Indian cities, little more than an overgrown village and maintains its village organisation in many respects, it will be wise first to give a more general idea of the classes with which I wish to deal as they occur in the Deccan villages.

The castes recognised as untouchable in the Deccan are five in number. These are not allowed to live in the village itself, but have special *wadas* or quarters outside the village proper. If there is a wall, as there often is, then they live outside the wall. Of these castes, representatives of two are found in almost every village and have regular functions in the village organisation; the other three may or may not exist, according to the special circumstances of the particular place. The five castes, placed in the order of respect, are as follows, to give them their Marathi names:—

(1) Chambhars or leather-workers.
(2) Dhors or tanners.
(3) Mahars.
(4) Mangs.
(5) Bhangis or scavengers.

Three of them are essentially trade castes and remain as such, the trades with which they are connected being considered impure. These are the leather-workers, the tanners, and the scavengers. The first two only occur where their trades are centred; the last is not required under ordinary village conditions. But the other two groups—the Mahars and the Mangs—are much larger and have a status in practically every village.

The caste of Mahars is by far the largest. Its members, though impure, have definite duties in the villages. They are always the village messengers, the village guides to strangers, often the village watchmen. They erect the tents of strangers, and they have the duty of removing and disposing of dead cattle and other animals. For these duties, which they perform without special remuneration, they have land in the village area which belongs to the various families belonging to the caste and the village. This *watan* land remains with the members of the families wherever they may be: so long as some member of the family is there to perform the duties in his turn, so long the right is maintained. The fact of being a *watandar* Mahar, that is to say a Mahar whose family has landed rights and duties in a village, is a source of pride, and binds a man and his family to his original village by a very firm tie. He may sell his *watan* rights, but his status among his fellow caste-men is much lowered by this, and he becomes homeless and landless, looked down upon wherever he goes. In the villages many Mahars take up ordinary labouring work, or various trades, in addition to doing their duties and working on their land; but so far as villages are concerned by far the greatest number remain in their old condition, depending on the produce of their *watan* land, doing their village duties, labouring for local cultivators during the busy seasons, varied in some cases by a few months in a town at such work as can there be obtained. Education is almost unknown.

The other caste which is commonly found in every village in the Deccan is that of the Mangs. Their position and status are much more difficult to define. They are not attached to a village in the same way as a Mahar. As a class they have no land, and they have no stated duties. And yet, there are some attached to almost every village.

"The Mang has two chief occupations,—he is a ropemaker and he is a musician. In return for services rendered to the cultivating castes he receives from them what is called *baluti,* that is to say, a small portion of the harvest of each field gleaned. It is the duty of the Mangs on their part to supply on demand traces for the cultivator's plough and bullock cart, cords for binding the sheaves and for lowering the bucket into the well, slings for driving the birds from the corn, a whip for his team, a muzzle for his bullock's mouth. Such ropes and cords are spun from the fibre of the *agava* plant. This is the Mang's secular occupation. As a musician he is brought into relation with the religious worship of the village. In some parts it is his duty to beat the drum daily before the chief village temple: everywhere his services are required in connection with marriage ceremonies."[1]

Such is the condition of the two chief castes as found in the villages of the Deccan: of the others we may speak simply as we have met them in Poona City.

Poona, the city wherein the studies of which a slight sketch is recorded in the present paper have been made, is, as all know, the capital of the Bombay Deccan in British India. The chief city of the Peshwas during all the palmy period of the Maratha empire, it was captured by the British in 1818 and its fall practically meant the disappearance of the independent Maratha power. Though it owes its importance partly to its position as a natural trade centre, it would only have risen to importance very slowly if the Peshwas had not made it their capital. It was a court city from the beginning, and it has not forgotten the fact to this day. In our own time Poona as a whole has increased but slowly. Its most rapid increase was between 1881 and 1891, but the ravages of famine and especially of plague in the years following 1896 have given it a set-back from which it has not yet fully recovered. Its population since 1872 has been as follows:—

[1] Rev. N. Macnicol, **Times of India**, April 22nd, 1910.

| | | |
|---|---|---|
| 1872 | | 118,886 |
| 1881 | | 129,751 |
| 1891 | | 161,390 |
| 1901 | | 153,320 |
| 1911 | (provisional) | 158,000 |

This population is that of three separate sections, however. Poona, in fact, now consists of the 'City,' the old capital of the Peshwas; the 'Cantonment,' a purely British development, originally laid out as a camp, though now containing much other population, and still entirely under military control; and the 'Suburban Area' with a municipality of its own, but which is really an outgrowth under modern conditions of the old city. The relative population of these portions is indicated by the following figures from the 1901 census. Those for 1911 are not yet available.

| | |
|---|---|
| Poona City | 111,381 |
| Poona Cantonment | 32,777 |
| Poona Suburban area | 9,162 |

The present paper deals entirely with the first of these three sections. It forms a unit in itself. Its conditions are fairly stable and representative. It was as large as I felt myself able to study. On the other hand, cantonment conditions are so special as to represent nothing but themselves, and the suburban area consists of real suburbs without unity and practically without organisation. Hence I have entirely limited my view to Poona City in all that follows.

Before turning to my special subject perhaps a few more figures would not be out of place, in order that an attempt may be made to give an exact idea of the city with which we have to deal. Of the total population (I speak from the census of 1901, but the figures are substantially accurate) of 111,381 persons, 98,431 are Hindus including Jains, and including also the untouchable classes. There are 11,332 Mohammedans, and the remainder includes 952 Christians, 483 Jews, and 168 Parsis. The number of males is slightly in excess of that of females, but the excess amounts only to 1800.

The city itself lies in a somewhat crowded area on the banks of the Mutha river. It is not attractive to a visitor, being like most other old cities rather a maze of narrow streets, though these have been much improved in recent years. It is unattractive because the backs of the houses usually face the public thoroughfares, gardens lying beyond the houses. If a bird's-eye view of the city is obtained it looks like a forest of trees, so numerous are the gardens on an area which to a visitor at first sight seems like a mass of narrow and crooked lanes. In such a city there are, as far as I can make out, for the published census figures do not help us, about 7000 to 8000 members of the classes which are considered untouchable. They live segregated from the rest and, being segregated, are largely unknown to the general public. As an illustration of this I will quote a description of them from the Bombay Gazetteer (1881) the standard and usually very reliable authority for almost every corner of the Bombay Presidency. It devotes a dozen lines to the untouchable castes of Poona City, and they read as follows:—

"The depressed classes include Chambhars, Dhors, Mangs, and Mahars. They live in dirty huts outside of the town. They are idle, dishonest, given to drinking, thieving and telling lies. Both men and women are of loose morals and husbands and wives are changed at will. Of Mahars, some are in the native army, some are domestic servants to Europeans, some are day labourers, and some are sweepers. Labourers and scavengers begging for remains of dishes served at dinner and for a morsel of food will remain crying at doors for hours together. Chambhars make shoes, Dhors tan hides, and Mangs make ropes and brooms. They live in abject poverty and have scarcely any bedding beyond a blanket. They go almost naked and have no metal pots in their houses. Their women work as day labourers and do house work. They cannot read and write and seldom send their boys to the schools which Government have opened for them.'

At least half of this description is undoubtedly incorrect now, and libellous on what is certainly a lowly and despised section of the population but not one worthy of these remarks. Thus, they do *not*, in large proportion, live in dirty huts outside the town; some do, but these are only a portion. They are as a class *not* idle, *not* dishonest, *not* more given to drinking than some other classes, and *not* specially thieves. They are *not* as a class of loose morals, but how this charge comes to be made can be explained comparatively easily. They are *not* in the abject poverty described, and so on. Some things may have changed since 1881, but much of this description was never applicable.

My own studies form the commencement of an inquiry, suggested to me by Professor Geddes some time ago, into the social conditions of an Indian city. Of these little is on record, and it seemed at once to me that a good beginning could be made in connection with the untouchable classes. The method of work adopted was for myself or my workers to proceed from house to house, make friends with the people, and get them to give particulars with regard to their households, their occupations, their sources of income, where they came from, and the like. As workers and assistants in my own personal investigations, I have had the exceeding good fortune to be able to employ none but members of the untouchable castes themselves, and have so obtained information which is probably more trustworthy than I could otherwise have got. By this means I have obtained more or less complete data within the scope I laid down for myself, for 1400 households, or probably at least between 80 and 90 per cent. of the total. I want it to be understood from the beginning that the object at the back of my mind was essentially economic,—a desire to ascertain the present economic condition of well-defined groups of people living under city conditions in British India. My full figures are not as yet in form for publication, and I can only at present submit a sketch of some of my principal results, and indicate in some way the evidence on which they are based.

The largest of the five castes with which I intend at present to deal is that of the Mahars. These number probably between 2000 and 2500 in Poona City: my figures refer to 2066 persons. They may be divided into two classes. The first of these are the *watandar* Mahars of Poona, the descendants of those who were formerly attached to the three villages out of which Poona has grown. They have their land and *wadas*: though they are segregated yet the land is theirs; the houses, more or less well built, are their own, and this is reflected in their habits and conditions in other ways. In contrast with these, the natural inhabitants of the place, are the large crowds of people who have been attracted to Poona for work and who live under much more squalid and unsatisfactory conditions.

The household arrangements for the latter are very peculiar. The owner of the land on which they have settled does, as a rule, nothing for them except put a latrine in the field. Beyond this, the tenant builds his hut and it remains his. He pays a ground rent of say eight annas per month for the space, but the hut is passed on from tenant to tenant. Such a hut is generally made of mud, and roofed with corrugated iron or old kerosene-oil tins, and it is generally about twelve or thirteen feet by six or eight feet over all, and is only sometimes built on a plinth a few inches high. Inside it is usually divided into two small rooms by a partition which goes partially across the room, made also of mud, and sometimes there is a space in the roof used for storage or for sleeping. There is often a very small verandah, included in the above measurements, on which it may be a goat is kept.

At least three-quarters of the Mahars in Poona City are of this class, drawn to the city by the chance of work, going backward and forward to their villages constantly, and being in every sense strangers and pilgrims. Including both the classes I have named, the population of 2066 with which I am dealing was contained in 555 families. In this number no less than 942 people (45.6 per cent. of the Mahar population) were earning money in one way or another. Of these 574 were men, 297 were women, and 71 were children. The terms men, women, and children are difficult to define, and too much stress must not be laid on the figures for each of these classes. My aim was to include a boy as such until he was about 18 years old, and a girl as such until she went to live with her husband at the age of 15 or 16, but the figures must not be too much relied on in this direction. They are employed as follows:—

| | Men | Women | Children | Total |
|---|---|---|---|---|
| Poona City Municipality (as street sweepers, coolies, etc.) | 53 | 86 | 5 | 144 |
| Factories | 44 | 10 | 15 | 69 |
| Railway | 134 | 5 | 6 | 145 |
| Domestic servants | 91 | 34 | 13 | 138 |
| Coolies | 164 | 137 | 25 | 326 |
| Miscellaneous | 94 | 17 | 9 | 120 |

The miscellaneous occupations are very miscellaneous. They include skilled artisans like masons, sellers of firewood, ward boys in hospitals, and beggars.

The joint-family system touches the Mahars very little. As a rule a husband, when his wife takes up her abode with him, sets up a separate establishment. There are exceptions, but this is a general rule. It makes the economic position a little more easy to consider than would otherwise be the case. At the same time, among the really resident Mahars of Poona it is most common, and apparently most desired, that a young couple should have their parents, or one of them, living with them. One of the older people can then do the housework, attend to the children, and so on, while both of the younger pair can go out to work and contribute to the family income. The average earnings per month, whether as wages in the ordinary sense or as casual earnings, worked out as follows:—

| | |
|---|---|
| Man | Rs. 9.9 per month.[1] |
| Woman | Rs. 4.1 per month. |
| Child | Rs. 9.1 per month. |

[1] The exchange value of the rupee is fixed at one and fourpence; Rs. 15 to £.

The high figure per child is accounted for by the very large proportion of big boys and young men included who can earn men's wages in the factories,[1] on the railway, and in other occupations. The earnings per family amounted to Rs. 13.1 for a family of 3.7 persons. In one hundred families, there were found one hundred men, fifty women, and twelve children, who contributed to the family income.

So much for data concerning the Mahars. Let us now turn to the next great caste, that of Mangs. These are, in my experience, the most squalid of all the groups with which I am at present dealing. They have no status in the community, and have hence far less self-respect than can usually be found among the Mahars. The greater part of them live under conditions similar to those already described for the more unfavourable section of the Mahars, though I know some Mang *wadas* which are well kept. Their houses are not dissimilar in construction to those of the Mahars, and do not need special description. Unlike the Mahars, the Mangs have a very general caste occupation, namely, the making of baskets, brooms, and ropes; and this is work which the women do regularly in their spare time, and so enter into the category of earners although they do not leave home. The number of families of Mangs dealt with in my figures was 526, comprising 1981 persons. The number of these definitely assisting in maintaining the households was 992, or 50 per cent. of the Mang population—a considerably greater proportion than among the Mahars. This is, of course, due to the prevalence of home occupations, as above indicated. Of these earners, 531 were classed as men, 351 as women, and 102 as children. They were employed as follows:—

| | Men | Women | Children | Total |
|---|---|---|---|---|
| Poona City Municipality (as street sweepers, coolies, etc.) | 154 | 104 | 21 | 279 |
| Factories | 5 | — | 1 | 6 |
| Railway | 4 | — | — | 4 |
| Domestic servants | 105 | 8 | 6 | 119 |
| Coolies | 200 | 120 | 41 | 361 |
| Miscellaneous | 92 | 122 | 17 | 231 |

[1] There are two important factories in or near Poona. One is a silk and cotton mill near the city: the other the Government ammunition &c. factory at Kirkee, five miles away.

In this case the 'miscellaneous' includes the men, women, and children to whom broom and basket making is a principal occupation, and, among the Mangs a large number of beggars. The average earnings per month, including all these casual forms of income, worked out according to the people's own statements as follows:—

| | |
|---|---|
| Man | Rs. 9.1 per month. |
| Woman | Rs. 3.8 per month. |
| Child | Rs. 4.5 per month. |

Here the children, not being usually taken on the railway or in the mill, have much lower average earnings than was the case with the Mahars. The earnings per family, however, amounted to Rs. 12.6 per month for a family of 3.8 persons. To obtain this there assisted, in one hundred families, one hundred men, seventy women, and twenty children.

The third case with which I have to deal is that of the Chambhars, or leather-workers. It occupies the highest social position of these untouchable groups, and on the whole is also in a better economic position. Fewer women take a share in the family earnings, and on the whole the children do not work so young. Many, if not most, of the caste work as their own masters and make boots and shoes. The number of families with which I am dealing is 196 containing 798 individuals, of whom 271 shared in earning the family income, or only 34.2 per cent. These consisted of 238 men, 25 women, and 10 children: thus showing how completely the women have disappeared as an earning force. The people were employed as follows:—

| | Men | Women | Children | Total |
|---|---|---|---|---|
| Bootmaking, Leather-working, etc. | 217 | 15 | 14 | 246 |
| Factories | 3 | — | 1 | 4 |
| Railway | 1 | — | — | 1 |
| Domestic servants | 4 | — | — | 4 |
| Coolies | 4 | 8 | 4 | 16 |

The caste is essentially, therefore, a trading group, almost the whole being engaged in their recognised occupation. The average earnings per month among the Chambhars is a good deal more indefinite than among those previously dealt with, —as so large a proportion are their own employers. The figures given, however, are according to the people's own estimates, and I do not think they are very far from the truth. They appeared in the average as follows:—

| | |
|---|---|
| Man | Rs. 10.5 per month. |
| Woman | Rs. 4.5 per month. |
| Child | Rs. 4.3 per month. |

The income per family worked out at Rs. 13.5 per family of 4.1 persons. There was more trace of the joint-family system here, for one hundred families contained 120 men, ten women, and five children who assisted in producing the family income.

I shall say nothing of the Dhors or tanners because they are represented in Poona by under a dozen families, and hence no useful general information can be drawn from their condition. Tanning in Poona is most largely done by Mohammedans.

The last social group with which I have to deal is that of the Bhangis, or scavengers and removers of night-soil. They are the lowest in the social scale, and are looked down upon by all the others. On the other hand, they are not by any means the lowest in comfort, in housing, or in independence. Their houses are usually well-built and clean, and the people are far more permanent residents of Poona than are some of the previous classes.

Of these my records deal with one hundred families, containing a population of 373 souls. A larger proportion among them are wage-earners than in any previous case, simply because as a rule men, women, and children as soon as they are old enough all engage, morning and evening, in the recognised work of the caste. It is remarkable, too, that in this work the earnings of men and women are the same, and those of a working child almost as great. This is, perhaps, what would have been expected, for in their work the employer pays for having the work done, and rarely looks to see who does it. All the wage-earners (241 in 100 families) with whom my records deal were engaged in the work of the caste, 80 per cent. on behalf of the public authority, and the remainder for private parties. Their average earnings worked out as follows:—

| | | | |
|---|---|---|---|
| Man | ... | ... | Rs. 7.7 per month. |
| Woman | ... | ... | Rs. 7.7 per month. |
| Child | ... | ... | Rs. 5.6 per month. |

while the income per family reached the large total of Rs. 18.2 per month. The number of wage-earners per family is high (2.4) as has already been noticed and in 100 families there were 110 men, 110 women, and 27 children engaged in gaining the family income. Such are a few of the most obvious figures resulting from my inquiry. Let us now see what they mean on a cursory examination.

The first point which must strike any one who considers them is the very small families which are dealt with. In every case the actual number of persons per family is below four, except in the case of the Chambhars when it just exceeds that figure. The actual figures are as follows:—

| | | | | | |
|---|---|---|---|---|---|
| Mahars. | ... | ... | ... | ... | 3.7 |
| Mangs. | ... | ... | ... | ... | 3.8 |
| Chambhars. | ... | ... | ... | ... | 4.1 |
| Bhangis. | ... | ... | ... | ... | 3.7 |

These figures are for the family which lives together as an economic unit, occupying one house. Taken in this sense it shows a somewhat astonishing, and certainly serious, state of affairs. For it is obvious that when the average family numbers below four, the population must be diminishing. I have devoted a good deal of inquiry to find out the causes of this smallness of the family among the untouchable classes in Poona, but I can only here give the conclusions which I have reached. It is not due, in my opinion, except incidentally, to the fact that many of the people are wanderers, for it applies to a purely town caste like the Bhangis as well as to those who are constantly coming and going. Two factors seem to be dominant: one, the poor character of the housing, sanitary, and water arrangements; the other, a very large infant mortality. The existence of the latter is a deduction, and I have not got any direct evidence of it. The fact of small families, of diminishing population, seems certain, and forms a very difficult and yet insistent social problem in connection with the labouring classes of our Indian cities. The second point concerns the family income. This according to my figures stands as follows:—

| | Number in family | Family Income | Average per member |
|---|---|---|---|
| | | Rs. | Rs. |
| Mahars | 3.7 | 13.1 | 3.5 |
| Mangs | 3.8 | 12.6 | 3.3 |
| Chambhars | 4.1 | 13.5 | 3.3 |
| Bhangis | 3.7 | 18.2 | 4.9 |

The income per family of course, from some points of view, appears small, and is small really; but it appears to represent a very considerable advance on the amount paid a very few years ago. In most cases (except the Chambhars), and most markedly among the Bhangis, it involves the regular wage-earning of more than one member of the family. And the question at once arises as to how far it is adequate for the needs of the people under Indian conditions, and with the habits of the classes to which the people belong. The question thus posed can only be answered by a consideration of the family budgets of a number of the classes to which we are referring: finding out from them how the money is spent, how far it is adequate according to the people's own idea, and finally how far it is adequate according to a theoretical standard worked out elsewhere. With some difficulty I have collected, for Mahars alone, a number of family budgets more or less complete, and more or less reliable. I cannot enter into the full discussion of these in the present paper, but I may summarise one of them which may be considered typical.

The family in question consisted of seven persons,—an old man and woman, a young man and woman, and three children of the latter aged respectively 10, 7, and 4 years. The old man was unfit to work and simply spent his time about the house, doing most of the necessary housework and attending to the children. The younger man was the principal wage-earner and was a mason. When working he obtained ten annas per day, but his work was irregular, and was considered on the average as simply extending to thirteen or fourteen days per month. The two women worked as coolies, and obtained on the average about two annas per day, or say Rs. 4 per month each. As their work and their earnings were irregular, this is probably a maximum. The children, of course, did not work,—nor did they go to school. The total income of my typical family, therefore, counting thirty days per month, would not exceed as follows:—

| | Rs. | as. | |
|---|---|---|---|
| Man ... ... ... ... | 9 | 13 | per month. |
| Woman (1) ... ... ... ... | 4 | — | per mohth. |
| Woman (2) ... ... ... ... | 4 | — | per month. |
| | 17 | 13 | per month. |

This is equivalent to an annual income of 213 rupees 12 annas, or say 214 rupees. Against this have to be placed the expenses, which are given to me as follows. The house is their own, and no rent has to be paid.

| | Per month. Rs. | as. | Per annum. Rs. | as. |
|---|---|---|---|---|
| **Food.** | | | | |
| Grain (**bajri** and rice) | 5 | 6 | | |
| Pulse | — | 9 | | |
| Fish and flesh[1] | — | 12 | | |
| Vegetables | — | 3½ | | |
| Spices, chillies, etc. | -- | 10½ | | |
| Salt | — | 2½ | | |
| Oil for cooking | — | 3 | | |
| Tea | — | 2 | | |
| Sugar | — | 10 | | |
| Spirits and Toddy | 1 | 12 | | |
| | 10 | 6½ | 124 | 14 |
| **Fuel.** | | | | |
| Kerosene oil | — | 4 | | |
| Fuel (wood) | — | 14 | | |
| | 1 | 2 | 13 | 8 |
| Tobacco | 1 | — | 12 | — |

[1] In addition, there were usually two meals per months of bullock's flesh received as a **watan** right. consumed by men only.

| | Per annum Rs. | As. |
|---|---|---|
| **Clothes.** | | |
| 2 men | 13 | 6 |
| 2 women | 10 | 8 |
| 3 children | 5 | 14 |
| Holidays | 12 | — |
| Religious rites | 3 | — |
| Interest[1] | 13 | — |
| Miscellaneous | 12 | — |
| Total | 220 | 2 |

This shows, then, a net balance against the family of about Rs. 6, which I understand is saved, if necessary, chiefly on the clothes bill.

To anyone accustomed to a consideration of humble Indian life, there is not much to comment on in this statement of account. The simplicity of the food, in which the price of grain is a dominant factor to an extent almost inconceivable elsewhere; the gradual introduction of tea as a luxury for very special occasions; the large amount of expenditure on sugar which is, nevertheless, I believe, one of the most economical parts of the expenditure; the absence of butter or *ghi* from the diet and its replacement purely by a vegetable oil,—these are all factors which may be regarded as almost universal. The large expenditure on alcoholic liquor is particularly noticeable, but known to exist among these castes where the religious influence does not prohibit, though it discourages, indulgence. The present is a very respectable family, and, among such, only the men drink. A woman drinking, at least in public, is considered a disgrace. The indulgence in liquor is the weakest economic factor in the budget and makes what would otherwise be a fairly satisfactory economic position into one which is certainly not satisfactory.

The remaining items call for little comment. The people considered that the legitimate point for saving where this is necessary was in the provision for clothes. The amount put down is what is regarded as requisite for proper clothing.

[1] On loan of Rs. 50 obtained some years ago.

The interest is at the rate of 26 per cent. on Rs. 50, being an old debt contracted in former years,—possibly (though I do not know this), on the occasion of the younger man's marriage. The general economic position of the family is not satisfactory, but this is caused almost entirely by debt and by drink. Otherwise the position would be far different, and saving would be not only possible but easy. The foregoing may be considered a fairly typical case among almost any of the classes with which this paper deals, but is more characteristic of the Mahars than of any others. It illustrates very fairly the condition in which the vast proportion of the respectable members live.

Here I must stop. The data presented are simply some of the first superficial results of my inquiry, and such as they are deal entirely with the economic aspect of the position of the people with whom I am dealing. Of their social position, of the disabilities under which they suffer, of the lack of education and the difficulty of obtaining it, of the devotion to drink which has done so much to keep these castes in a low position, of the institution of *muralis* or girls dedicated to various gods which has done much to lower the moral reputation of the castes,—of all these and a dozen other problems I cannot now speak, but must reserve my results for a future occasion.

*Chapter* 15

# THE HOUSING OF THE UNTOUCHABLE CLASSES IN AN INDIAN CITY*

Some years ago I published[1] a first account of studies which I had made regarding the economic condition of the untouchable classes of an Indian city. The city chosen was Poona, and the method of study was to make a house to house investigation of the people, how they live, what they work at, and under what conditions they carry on their existence. The results I obtained were remarkable in many directions, but there was one point which stood out above all others. It appeared certain that the conditions of city life among the depressed classes were so unhealthy and so unsuitable that the population could not maintain itself without continuous immigration. If new-comers from outside were not continually arriving in Poona, the labouring classes, so far as they were represented by the Mahars, the Mangs, the Bhangis, and perhaps, though less certainly, the Chambhars, would speedily die out, or at any rate become largely reduced in numbers.

The evidence for this important and yet disheartening conclusion is, of course, wholly indirect. The birth rate in none of our cities, except perhaps in Bombay, is taken with anything approaching accuracy. In Poona, the death rate as

* Reprinted with permission from the **Social Service Quarterly**, Bombay, Vol. I (1916), No. 3, pp. 1-10.

[1] **Sociological Review**, Vol. V. (1912), 42-55. Also **Hindustan Review**, Vol. 25, March and April 1912. (See the preceding chapter for this article).

reported is fairly accurate—but the number of births recorded, as elsewhere, is nothing like complete. It is quite impossible, therefore, to calculate for any class of people, or for any locality, the relationship between births and deaths and so get a direct test of the unhealthiness of a particular quarter of the city, or of city life for a particular section of the people. But the conclusion seems nevertheless certain, and for the following reason.

The first point which must strike any one who considers the data which I presented three years ago is the extremely small families among the depressed population. As an average, the actual number of persons per household is below four, except in the case of the Chambhars, when it just exceeds that figure. The actual numbers are as follows:—

| | Average number of persons per family. |
|---|---|
| Mahars. ... ... ... | 3.7 |
| Mangs. ... ... ... | 3.8 |
| Chambhars. ... ... ... | 4.1 |
| Bhangis. ... ... ... | 3.7 |

These figures are for the family which lives together as an economic unit, occupying one house. Now it is obvious that when the average family numbers less than four, the population must be diminishing.

I devoted a good deal of inquiry to find out the causes of this smallness of the family among the untouchable classes in Poona, but I can give here only the conclusions which I reached. It is not due, in my opinion, except incidentally, to the fact that many of the people are wanderers, for it applies to a purely town caste like the Bhangis as well as to those who are constantly coming and going. Two factors seem to be dominant: one, the poor character of the housing, sanitary, and water arrangements; the other, a very large infant mortality. A consideration of the latter may form the subject of a subsequent paper; the former in so far as it concerns the housing is what I wish to deal with in this article.

The housing conditions among the untouchable classes wherever they are met with in the Deccan, are, as is well-known, deplorable. Under ordinary village conditions they are excluded from the village site, and live outside in an area which is rarely visited by others, and in which the houses are, so far as by far the largest proportion is concerned, of the most crude and primitive type. Usually made of grass, with, in some cases, a sheet of corrugated iron forming a part of the roof or walls, sometimes covered with kerosene oil tins beaten out flat, or with thatch composed of grass or leaves, they form under usual circumstances the picture of neglect.

This condition of things continues, in a large measure, in the city of Poona,—though not by any means entirely. The depressed classes are no longer excluded from the city area. Though the untouchable quarters are usually marked off from the rest, yet these quarters are intimately mixed up with those in which the remainder of the people reside. They are hence not so entirely neglected as under village conditions, and yet a visit to the parts of the city in which they live gives (except as regards the Bhangis) an impression of conditions in which neglect prevails to an almost inconceivable extent. It is seen at its worst in the rainy season, and at that time visits to some of the quarters inhabited by Mahars and Mangs show a state of things which has to be seen to be believed.

The data which I present to-day deal with the housing of considerably over 1000 households. In each of these cases I have ascertained the number of rooms occupied by the family and (by eye estimation only) the area of the accommodation provided. Beyond this, I have found the number of people who own the houses in which they live, and the number who either rent the house or the ground on which it is built. The proportion of *pucca* houses has also been determined, by *pucca* house being meant any building of brick, either burnt or sun-dried, or of more substantial material.

The first caste I will deal with is that of the Mahars. These number probably between 2000 and 2500 in Poona City. I have records referring to over fifteen hundred of them as indicated in my previous paper. These may be divided into two classes. The first are the *watandar* Mahars of Poona, the descendants of those who were formerly attached to the three villages out of which Poona has grown, and those who have come to be associated with them. These have their land and their cottages. Though they are segregated, yet the land and the houses are theirs, the houses are largely *pucca*, the surroundings are *fairly* clean, and the sanitary arrangements, though inadequate, are much better than in many other parts of the city.

The most typical area of this kind is found on a small mound in Mangalwar Peth near the river, which is wholly in the occupation of the Mahars. There is a water supply at the foot of the hill, latrines in an open space at one side, a temple on the top of the mound, and houses crowded together with very narrow lanes on the rest. In this quarter fifty-six families were residing at the time of my investigation; at least a dozen broken-down houses also existed, some of which seemed to have been in ruins for a good many years.

Though the houses were more *pucca* than elsewhere, the rooms were, if anything, smaller, and the accommodation, less. The actual figures resulting from my inquiry were as follows:—

| | | |
|---|---|---|
| Number of households. | 56 | |
| Number of **pucca** houses. | 36 | |
| Number of kachha houses and huts. | 20 | |
| Number owning houses in which they live. | 16 | |
| Number paying rent. | 40 | |
| Average rent. | 10½ | annas per month. |
| Average number of rooms per family. | 1½ | |
| Average area of rooms per family (by eye estimation). | 70½ | square feet. |
| Average area of rooms per individual | 17½ | square feet. |

When one considers that this is in many respects the best part of the city for Mahars, and that here they have a quarter which has apparently belonged to them for centuries, the condition is very bad indeed. In some respects the *pucca* character of the houses would seem at first sight to be a disadvantage, for it means that they are much more closed in than is possible in the huts in a great part of the remainder of the city, and this with the extremely limited space must make the conditions decidedly unhealthy.

The pattern of the houses varies considerably, but that which seems most common in cases where one room or a little more can alone be obtained is as follows. The room is divided across the middle for about two-thirds to three-quarters the distance by a wall of mud or other material. One side of this is the living room, and contains the brass pots the family usually possesses, and the *chula* on which the cooking is done. There is no chimney. The other half of the house forms a sleeping room, and is often, if not usually, covered with a ceiling, about five feet from the floor, composed of slight wooden beams. The space above this is used as a store-room and often as a sleeping place for some members of the family. Outside the house there is in many cases a small enclosed compound a few yards square, containing a few plants, and often a goat.

When we turn from these comparatively developed city conditions of life—however unhealthy they may be—with a more or less permanent population, to the areas which are occupied by the other portions of the depressed classes who are not permanent residents of Poona in the same sense, a very different and much more primitive condition of things is found to exist. I will try and describe two of the areas occupied, which are typical: namely the *Gadi Tal* and the *Bhokarwadi* areas. These, however, possess many features in common.

In both of these areas, as in most others, the population is a shifting one. It consists chiefly of those who have come to Poona for work, and who, though they may have been there for years, are still passing between their villages and Poona occasionally, and who look upon themselves as pilgrims and strangers in the city. The conditions under which they occupy their houses are somewhat as follows. When they come to the city, they settle on a piece of spare land, with the consent of the owner, and pay him rent for the piece of land on which they build their hut. The owner of the land, as far as I can find, takes no responsibility for housing or for anything, except in so far as he is compelled by the municipal authorities to provide sanitary arrangements. The houses or huts are built by the people themselves, subject to the condition of paying the landlord Re. 1/- or Rs. 1/8/-, per annum. Once built, they are passed from hand to hand, and transferred from person to person. The result is that a colony of Mahars in Poona City in such an area as I am describing is a collection of huts and houses of the most various kinds, generally clinging close to the edge of a road, with some latrines provided, under municipal pressure, by the owner of the land,—and with nothing else. Sometimes there is pipe water in the immediate neighbourhood which the people can use,—more often not. If not, then they have to go often to a very considerable distance to get it. The waste water from cooking or washing, generally runs away in a *kachha* drain, often half blocked, and hence the home of countless insects of the fly and mosquito type.

*Gadi Tal* is a typical area of this kind, and occupies part of what always was, and still is, a resting place for carts on the edge of the city. A large part of it is now full of wretched *kachha* huts. The immediate edge of the main road has, partly at any rate, been occupied by shopkeepers with slightly better structures than those inside. The whole colony is on black soil, is lowlying, and possesses no made roads except the main road running through the centre. Most of the huts have no real plinths, though they have the soil heaped up somewhat where the house actually stands. A few are better than this, but what I have described is the general pattern. The houses are low,—the doorway about four feet to four feet six inches high. A very small verandah sometimes exists. The walls are five feet high or thereabouts, and the roof necessarily projects far over the walls to keep them even passably dry in the rainy season.

The result of such a state of things is obvious. In the dry part of the year, the conditions are not very bad. There is plenty of air and open space outside the hut,—and the chief cause of unhealthiness, I should suppose, is the absence of adequate sanitary arrangements and the fact that water has to be brought from a distance of over a quarter of a mile. In the rains, the whole place becomes a quagmire. Although the actual hut itself is usually kept dry, the surroundings are impossible to describe. What with the black soil, the torrential rain, the low *kachha* huts, the lack of plinth, the hopeless condition of approaches, one could hardly fancy that one is dealing with a portion of the municipal area of a large and advanced city.

Under such conditions as I have described, in this one area there live eighty-seven families of Mahars, and forty-eight of Mangs, with a population of three hundred and thirteen Mahars, and two hundred and fourteen Mangs. The families of Mahars are small, slightly lower than the average for the city among the same class of people. All the families, save six, hold their houses under the conditions I have described. All houses, save five, are *kachha* dwellings containing one room each, sometimes however, with partial divisions. The area per family, by estimation, was 100 square feet and hence contained twenty-eight square feet per individual.

Such is one of my two typical areas. The other is *Bhokarwadi* and here the conditions are very similar in essence. The land, however, in this case belongs to the Municipality, in part at any rate, the houses are not quite so primitive, and cling more closely to the roadside. Fairly adequate latrines are provided and a good tap water supply. The waste water flows away (or often does not flow!) in a *kachha* drain through adjoining municipal land till it meets a street drain at some distance. One of the features in this area is that the inhabitants have in some cases closed in what was originally intended for a verandah with a low roof, and left an entrance so low that one has to creep into it in order to enter the hut.

The conditions are distinctly higher, however, than those last described. The huts and houses do not lie, in the rains, in the middle of a field which is an area of mud, and a water supply is available, but the size of the families remains the same, being about 3.6 per household.

I have examined in this way all the areas in which Mahars live in Poona City, and the general result for the whole comes out as follows, on an average. I give these figures, not because I have much faith in averages in matters of this kind, where there is often such a large gap between the best and the worst—but to show how near the whole Mahar population comes to the condition I have described in these typical areas I have taken.

| | |
|---|---|
| Number of households dealt with. | 539 |
| Number of **pucca** houses. | 223 |
| Number of **kachha** houses and huts. | 316 |
| Percentage of **pucca** houses. | 41 per cent. |
| Number said to own houses or huts in which they live. | 252 |
| Number paying rent either for house or for ground on which it stands. | 287 |
| Percentage owning house or hut. | 47 per cent. |
| Average number of rooms per family. | 1.3 |
| Average area of room per family (by eye estimation). | 86 square feet. |
| Average area of rooms per individual. | 24 square feet. |

So much for the Mahars. Let me now turn to the next great caste, that of the Mangs. These live on the whole in greater squalor than the Mahars, or any of the other castes among the untouchable classes. They have not the status in the community that the Mahars possess, and none among them can be considered, in the same sense as the Mahars, permanent residents of Poona. A small proportion of the total, however, own their houses and in a few areas where they have settled permanently, they own a large proportion of the houses which they occupy. The conditions as to housing do not widely differ from those already described for the Mahars, but the houses are usually smaller, and, if anything, more primitive.

Taking the Mang community as a whole, the following statement shows the average housing conditions. I give the figures with the reserve already stated.

| | | |
|---|---|---|
| Number of households dealt with. | 477 | |
| Number of **pucca** houses. | 86 | |
| Number of **kachha** houses and huts. | 391 | |
| Percentage of **pucca** houses. | 18 | per cent. |
| Number said to own the house or hut. | 149 | |
| Number paying rent either for house or for ground on which it stands. | 328 | |
| Percentage owning house or hut. | 31 | per cent. |
| Average number of rooms per family. | 1.0 | |
| Average area of rooms per family. (by eye estimation). | 81 | square feet. |
| Average area of rooms per individual. | 21 | square feet. |

I am not able at present to give similar data regarding the Chambhar community, as my records are defective with regard to them, but they enjoy a higher condition of comfort on the whole than either the Mahars or the Mangs. With regard to the Bhangis,—socially the lowest, but in actual material comfort as high as any of these castes,—I have records for 89 families or between eighty and ninety per cent. of their total population.

The Bhangis, or scavengers and removers of nightsoil, are essentially a town caste, being usually unrepresented in the villages. As most of the members of the family generally work as soon as they are old enough to do so, the caste as a whole is comparatively well off. While (as I showed in my paper three years ago) the average income per Mahar family was Rs. 13.1, per Chambhar family Rs. 13.5, and per Mang family only Rs. 12.6 per month,—that of the Bhangis per family was Rs. 18.2 per month.

Partly as a result of this, the conditions of life among the Bhangis, as far as housing is concerned, are much higher. The houses are generally *pucca*, usually with a single room, but this is larger than with others. The quarters are clean and well-built, and there is an appearance of comfort which does not exist elsewhere. The households are, however, still small (3.7 per family), a fact difficult to account for. The figures which I obtained are, in summary, as follows:—

| | |
|---|---|
| Number of households dealt with. | 89 |
| Number of **pucca** houses. | 84 |
| Number of **kachha** houses and huts. | 5 |

| | |
|---|---|
| Percentage of **pucca** houses. | 94 per cent. |
| Number said to own the house or hut in which they live. | 83 |
| Number paying rent either for house or for land on which it stands. | 6 |
| Percentage owning house or hut. | 93 per cent. |
| Average number of rooms per family. | 1.2 |
| Average area of room per family (by eye estimation). | 143 square feet. |
| Average area per individual. | 38 square feet. |

Such are a few of the figures resulting from my enquiry into the housing of the depressed classes in Poona City. They show the condition of crowding which exists and the large proportion of purely *kachha* huts, often of a very primitive type (except among the Bhangis). These facts are well illustrated in the following summary:—

| | Percentages of **pucca** houses. | Percentages owning house or hut. | Average living area per house. |
|---|---|---|---|
| Mahars. | 41 | 47 | 86 square feet. |
| Mangs. | 18 | 31 | 81 square feet. |
| Bhangis. | 94 | 93 | 143 square feet. |

I must own that this inquiry into the conditions of the housing among the depressed classes does not give us any very clear results as to the effect of the character of the housing on the healthiness of the people. The size of families is almost exactly the same in all the three communities above specified, though the percentage of *pucca* houses is so different. Among the *Bhangis* who live almost entirely in *pucca* houses, the family is of exactly the same size as among the Mahars and almost the same as among the Mangs whose housing is of a very *kachha* character. More than this, there does not seem very constant relation between the size of families in a particular quarter and the proportion of *pucca* or *kachha* houses. Taking, among the Mahars for example, the quarters with varying proportion of *pucca* houses, the number of members per family is as follows:—

| Localities in which the proportion of pucca houses is as follows:— | Number of families included. | Average number of individuals per family. |
|---|---|---|
| 71% to 100% | 93 | 3.5 |
| 51% to 70% | 62 | 5.4[1] |
| 31% to 50% | 49 | 3.6 |
| 11% to 30% | 51 | 3.7 |
| Below 11% | 212 | 3.5 |

Except for the quarters containing between 60 and 80 per cent. of *pucca* houses where it varies considerably, the number of members per family varies very little. For the Mangs a similar table would show little, because these so uniformly live in *kachha* huts and so few of them occupy *pucca* houses. And a consideration of all the figures available certainly indicates no relationship between the character of the house and the undoubtedly very large mortality among the community as a whole, as revealed in the small families which actually exist.

At the same time, an examination of the existing conditions makes one ashamed that they can continue to exist under modern city conditions. Some would have many of the worst areas closed, and the people driven elsewhere. This has been, for some areas, repeatedly suggested by the consulting health officers of Poona City. To do this, however, merely means transferring the problem and not solving it, but possibly even making it worse by planting the people elsewhere under even more wretched and unsuitable conditions.

What I feel is wanted is a constructive policy: by which accommodation shall be secured which will at least be sanitary and in which it will be possible for self-respecting people to live. I proposed in the Poona Municipality some time ago the construction of model colony as an experiment. The matter was sanctioned, the scheme was prepared, and land was given for the purpose, and its carrying out only awaits the provision of money. The difficulty is that such

[1] These are practically all in one locality where the supply of water is close and good. All the others are from a considerable variety of areas.

a scheme could not be expected to yield more than four per cent. on the money expended, and hence would hardly attract private capital. But it seems a matter in which the honour of the municipality is concerned. All who wish to see our cities to be really the abodes of happy, healthy people must feel that the housing of the untouchable classes is a matter which demands earnest and early attention. The future of the community as a whole seems involved in its readiness to deal with the problem.

*Chapter* 16

# THE SUPPLY OF MILK TO INDIAN CITIES*

The supply of milk to Indian cities is known to be at present in a very unsatisfactory condition. Whether looked at from the point of view of the healthiness of the milk obtainable, or of the quality of the milk to be had, or of the quantity available,—it is open to the most strenuous criticism. There is *no* control at present whatever in most cities,—except possibly the Presidency towns and one or two others. The Act for the prevention of adulteration (Act II of 1899) in Bombay is a dead letter. And as a result of these things, the condition must be recognised as unsatisfactory from every point of view. This would be inexcusable except for one fact which increases the difficulty of dealing with the problem. The cost of milk is already very high: anything which would increase the price, even though it made the supply a better one, would but increase the difficulty of getting it to those people of the poorer classes who are at present suffering most from the present conditions. At every point, in every proposed method of dealing with the matter, one is met by the difficulty of controlling, of improving the supply without increasing the cost,—for, and I repeat it, if a remedy involved a material increase in the cost, it would be perhaps worse than the disease.

* Reprinted from the **Agricultural Journal of India, Calcutta**, Vol. IX (1914), Part 2, pp. 160-177.

To show exactly where we stand, it would perhaps be best to give a description, obtained from data collected, many of them specially for this paper, of the present supply of a city, namely, Poona. It is typical of many other places: the difficulties which surround the problem here are those which are found elsewhere: and it is probable that a thorough appreciation of the present position may at once lead to a clearer understanding of the methods which are feasible to improve the situation.

Poona City contains, according to the last census, a population of 1,17,256. Its population is a fairly representative one, save that perhaps the so-called higher classes are in slightly greater numbers than usual. There is a limited amount of manufacturing industry, but Poona cannot be called a factory town. It lies in a district which is dry. almost semi-arid, except where irrigation occurs,—but both above and below the city there is a large area watered and irrigated by the Mutha Canal extending from eight to ten miles above the city to fifteen miles below. Irrigation from wells is frequent in the country round,—and that this country is adapted to the production of fodder is seen by the train loads of lucerne which are sent to Bombay every day for feeding horses.

The city like all others is partly supplied from milk animals kept in the city, and partly from milk brought in from outside. I will deal with the latter source of supply first, and am able to give figures obtained from an actual census of the amount brought on two days into the city boundaries. There are fifteen entrances to the town, and the milk is brought from no less than forty-seven villages.

The actual supply brought in, in one day amounted to 5,560 pounds, almost all of which was brought in by hand and, in a few cases, by bullock cart. The train was only used for the milk from three villages,—and the amount brought only amounted to 120 pounds. For the remainder the milk came as follows.—

| | |
|---|---|
| From villages three miles or less from the centre of Poona. | 1,168 pounds or 21 per cent. of the total. |
| From villages from three to six miles from the centre of Poona. | 2,688 pounds or 48.3 per cent. of the total. |
| From villages from six to nine miles from the centre of Poona. | 1,582 pounds or 28.5 per cent. of the total. |
| From villages over nine miles from the centre of of Poona (by train). | 120 pounds or 2.2 per cent. of the total. |

It will be seen, therefore, from how restricted an area Poona City draws its milk supply. It is practically all brought in from the immediate neighbourhood, the railway is not used to any appreciable extent, and the means of bringing it are of the crudest.

As already stated, it is usually brought by hand and generally in brass pots slung over the shoulders of the milkmen, or carried on their heads. The number of people engaged in this work, and for many of them it must mean the loss of at least half a day, is 351, and the average amount brought per man is only sixteen pounds. It is evident, therefore, that this portion of the Poona City supply is brought and distributed in an exceedingly uneconomical manner, and involves a very large waste of labour without any corresponding advantage.

But what of the price and quality of the milk thus brought into a city like Poona? In the first place, by far the largest amount is buffaloes' milk. Out of the animals supplying the milk, in number 956, only 178 were cows—and taking the yield of a cow in the Deccan at five pounds and of a buffalo at ten pounds of milk per day, this would indicate that only 8½ per cent, of the milk brought into the City is provided by cows, and the remainder by buffaloes. This indicates the extent to which even in the Deccan, the buffalo proves itself *the* milking animal of India.

Bought at the entrances to the City, the price varies with the quality. The results of all my previous investigations would indicate that for such cows as occur in the Deccan, all genuine cows' milk contains at least 3½ per cent. of fat and 8½ per cent. of other solid matters ("Solids-not-fat' ), while the corresponding figures for buffaloes' milk are 5 per cent. of fat and 9 per cent, of other solid matter. These figures are, in fact, exceptionally low, and ninety per cent. of actual genuine samples will be far richer than these figures show. In determining the number of adulterated samples, I have allowed all samples to be genuine which, even on this basis, showed less than ten per cent. of water added. On this basis, out of fifty-one samples collected at the entrance to the City, only ten were genuine or less than twenty per cent., the remainder were adulterated with more than ten per cent. of water. The amount of adulteration varied with the price charged here, as well (as we shall see later on) as in samples brought at shops in Poona City. Of samples bought

(1) at 4 seers (8 pounds) per rupee, all were genuine

(2) at 5 seers (10 pounds) per rupee, sixty-four per cent were adulterated with water, and the average amount of water added was *twenty-nine parts* to 100 parts of genuine milk;

(3) at 6 to 7 seers (12 to 14 pounds) per rupee, fifty per cent. were adulterated with water, and the average amount of water added was *forty-seven* parts to 100 parts of genuine milk;

(4) at 8 to 9 seers (16 to 18 pounds) per rupee, ninety-two per cent. of samples were adulterated with water, and the average amount of water added was *seventy-two parts* to 100 parts of genuine milk.

(5) at 10 seers (20 pounds) or more per rupee, all the samples were adulterated with water, and the average amount of water added was *one hundred and seventeen parts* to 100 parts of genuine milk.

Thus the close connection even before the milk reaches the City between the price and the amount of water added is evident,—and the fact of the almost general adulteration of samples now brought from outside seems proved.

We are in fact able to conclude that at present, in so far as Poona is a typical city, that

(1) the milk brought from surrounding villages for City consumption is almost all brought in small quantities and carried by hand,

(2) this milk is adulterated with water in the case of practically eighty per cent. of what enters the City,

(3) the amount of water added varies very closely with the price at which the milk is to be sold,—and there is not even an even chance of getting pure milk if more than six seers (12 pounds) of milk are obtained per rupee, while the cheaper milk contains already more than its own volume of water. This is at the entrance to the City, before the retailers' profits are added and the retailers' adulteration occurs.

In one matter I have been rather pleased with the condition of the milk,—namely, in the amount of dirt it contains. This is not nearly so great as would have been expected, and does not show the addition of quantities of dirty water. I have divided the milks into four classes, and the results of examination of fiftyone samples are as follows, the dirt being determined by Gerber's dirt tester':—

| | |
|---|---|
| Clean, or nearly clean samples | 56.7 per cent. |
| Fairly clean samples | 33.3 per cent. |
| Distinctly dirty samples | 7.8 per cent. |
| Very dirty samples | 1.9 per cent. |

But, after all, the milk brought into Poona City only forms a portion, and not a very large proportion, of the milk consumed there. A much larger quantity is obtained from animals maintained in the City itself, and for information with respect to this, I am indebted to a census of cattle taken by the City Health Department some time ago,—as well as to inquiries by my own assistants.

According to this census there were *2,688* milking animals kept inside the City, of which *1,532* or 57 per cent. were kept for private use only, and *1,136* or 43 per cent. kept for sale of the milk. I do not propose to refer further to the former class, except to note that among these animals kept for private use by far the larger proportion were cows and not buffaloes,—these amounting to 73.5 per cent. of the total number so kept. This probably arises from the fact that a cow is much more easy to attend to than a buffalo, and can usually be tended by a member of the household. A buffalo, on the other hand, requires a servant, as not only is its feeding more troublesome but it has to be taken out every day, and requires far more thorough washing. But it seems nevertheless, from the preference for it by milk-sellers, that the buffalo is found to be a more profitable milking animal.

Among the animals kept for the sale of their milk *(1,156)* 224 or 19.75 per cent. were cows, and 932 or 80.25 per cent. were buffaloes.

The sheds in which they were kept were distributed all over the city, and were mostly very small. Over eighty per cent. of these sheds had less than ten animals, and over fifty-five per cent. had less than five animals. They were situated in all sorts of positions, without any control. Nearly six per cent. were under a dwelling house: over six per cent. were either *in* a dwelling house or on the verandah of a dwelling house: over fifteen per cent. were attached to and continuous with a dwelling house, while sixty-three per cent. were in the compound of a dwelling house. Though the pavement of the milk sheds was good (being made of stone) in thirty-seven per cent. of cases, yet the remainder were very bad. No pavement whatever existed in over thirty-two per cent. of sheds, and in sixteen per cent. it was so bad that it might as well have not been present. So far as these two classes of sheds are concerned, they are abominably insanitary, the urine sinks into the ground and remains there. The sheds, too, are overcrowded. Allowing six feet as the necessary length for an animal to stand, the width allowed in the shed for each animal is only 3⅓ feet in three per cent. of cases, less than five feet in 14 per cent. of sheds, and less than 6½ feet in twenty-three per cent. further of the sheds.

The milk produced by these animals can only be estimated, but taking the same rate of production as has previously been suggested (5 pounds per cow and 10 pounds per buffalo per day) the amount available from animals kept in the City will be.

| | |
|---|---|
| **Private sheds.** | |
| Cows (1,126 at 5 pounds) | 5,630 pounds |
| Buffaloes (406 at 10 pounds) | 4,060 pounds |
| | 9,690 pounds |
| **Sheds for sale.** | |
| Cows (224 at 5 pounds) | 1,120 pounds |
| Buffaloes (932 at 10 pounds) | 9,320 pounds |
| | 10,440 pounds |

This gives a chance of estimating the total daily supply required by a city of the size of Poona, and if the above figures be added to those of the milk brought in from outside, the amount comes to 25,690 pounds; of this 21.25 per cent. is obtained from outside sources, 37.72 per cent. from animals kept by house-holders for their own use, and 40.63 per cent. from animals kept by *gowalas* in the City for the sale of the milk. This gives an idea of the problem which is before anyone who wishes to improve the milk supply of a large Indian city.

We have now to consider the organisation for the sale of milk. There is one regular milk market (Jogeshwari) where all outsiders who bring milk for general sale, are found, and where retailers congregate to purchase this supply. By the time it is exposed for sale here the milk has become dearer, and has also become more adulterated. We purchased only ten samples here, but they were obtained in several days, at prices varying from four to nine seers per rupee.

(1) One sample was bought at 4 seers per rupee and was genuine.

(2) Three samples were bought at 5 seers per rupee: two of these had water added, the average amount *being sixteen per cent. of the original milk.*

(3) Three samples were bought at six and seven seers per rupee: all were adulterated, and on the average the milk as sold consisted at least *to the extent of one-half of added water.*

(4) Three samples were bought at eight and nine seers per rupee: all were adulterated, and on the average the milk as sold contained 135 *parts of water for every* 100 *parts of original milk.*

The chance of adulteration, therefore, and the quantity of water added varies almost exactly with the price,—and even at the wholesale milk bazar in the city, if more than 4 seers (8 lbs.) of milk are obtained per rupee, there is an almost absolute certainty that pure milk will not be obtained.

When we come to the milk as retailed in the shops in Poona City, the state of things is still worse. Here, even if the very high price of four seers per rupee[1] is paid there is no certainty of getting pure milk. Even at this price out of nine samples obtained only six could be passed as pure even with the exceedingly low standard we are setting up. Of the others, thirteen parts of water were added on the average for 100 parts of milk. If the price is lower, the adulteration becomes a certainty. At 6 to 7 seers per rupee, there is, on the average, 47 parts of added water for every 100 parts of genuine milk; and at 8 seers per rupee, there is no less than 92 parts of added water for every 100 parts of the original milk.

I am able to get an idea as to whether the condition of things is getting worse or no by comparing these results with those obtained by one of my assistants (Mr. S. R.

[1] It may interest those acquainted with the milk trade in England to note that 4 seers per rupee is equivalent to nearly $2\frac{1}{8}$d. per pint.

Paranjpye) in Poona city in 1911. He then found the results indicated in the following table:—

| | Price 4 seers per rupee. | Price 5 seers per rupee. | Price 6—7 seers per rupee. | Price 8 seers per rupee. |
|---|---|---|---|---|
| Number of samples taken | 1 | 2 | 5 | 18 |
| Genuine samples | 1 | 0 | 2 | 2 |
| Adulterated samples | 0 | 2 | 3 | 16 |
| Water added for 100 parts of pure milk | | 23 parts | 49 parts | 37 parts |

It would seem evident that while adulteration is no more general than two years ago, yet it is more shameless, and the quantity of water added for milk of the same price is much more than it was previously. In other words, *pure* milk *is* rising in price very rapidly, and the probability of getting pure milk at any particular price is getting less.

I may now summarise my conclusions with regard to the present condition of the milk supply of Indian cities, so far as they are illustrated by experience in Poona, considered as a typical city.

(1) The milk supply is partly (to the extent of about one-fifth) brought from the outside of the city. This amount is drawn from the immediate neighbourhood and is almost all brought in by hand in small quantities.

(2) The remainder is produced inside the city,—partly by animals kept in or in connection with houses for private use only,—and partly by animals kept for the most part in unsatisfactory and insanitary small sheds.

(3) Except for the animals kept for private use, about ninety per cent. of the milk is produced from buffaloes.

(4) The milk as it enters the city from outside is already very largely adulterated with water,—the adulteration varying with the price. No milk

costing less than 4 seers per rupee can be expected to be pure.

(5) The milk as obtainable in the city itself is adulterated to a still greater extent. Again no milk bought at a lower price than 4 seers per rupee is likely to be pure,—while it is practically certain that all milk costing 6 seers per rupee or less will be adulterated.

(6) The practice of adulteration is increasing in shamelessness, the amount of water added is greater and greater,—as the price of pure milk tends to rise.

(7) The dirt in the milk is not so great as would be expected from the insanitary and dirty conditions under which the milk is produced.

Such being the conditions of the milk supply at present in Poona,—and there is no reason to suppose that the conditions would be materially different in other big cities—the question at once arises as to what can be done to deal with the situation both as a piece of sanitary improvement, and as an agricultural problem.

From a sanitary point of view what we want to do is to ensure—

(1) That the cattle—cows or buffaloes—giving milk are healthy. Of this there is no control whatever at present.

(2) That they are kept under sanitary conditions. For the animals inside the city I have shown that this is not the case now, and there is every evidence that the conditions in the villages from which the milk comes are no better. except for the fact that there is plenty of fresh air.

(3) That the milk should be in good condition when delivered to the customers. The milk is delivered only once a day, and hence has to be kept without pasteurisation or other method of preserving, in a hot climate for many hours.

(4) That the milk should not be mixed with water or creamed before sale.

(5) That the arrangements for bringing to market should be such as to prevent the milk becoming dirty, or liable to be infected with objectionable materials.

This represents only one side of the question. From an agricultural point of view we want to find out the most economical method of meeting the large demand for milk from our cities, and placing it at their disposal at the cheapest rate possible. Once more I would state that in my opinion the question of price is vital. Any attempt to impose sanitary measures which will result in a rise in price of pure milk will be a mistake, and the problem is to find a method of so economising in the production of milk, and in the method of its conveyance to market, that it will be possible to impose restrictions in the interests of sanitation without raising the price.

Is this possible? I think it is, judging by the experience of the last few years here, and by the experience of almost all other countries in which the dairy industry is highly developed. I shall not be able, in this short account which follows, to give all the reasons for the positions I put forward, but the evidence is available.

In my mind, three things are radically wrong with the present milk supply, as illustrated by Poona, and which lead to the production of a very bad supply at a very high price:—

(1) The animals which are used for milk supply are subject to defects both in the case of cows and buffaloes. In the case of cows, they are not produced for milk, but are in most cases incidental in the production of a breed of working cattle. In the case of buffaloes, though they are primarily produced for milk purposes, the number of really good milking animals or milking strains is decreasing,—and such animals are not obtainable now at all easily. In other words, the amount of

milk produced by the animals used is very small for the amount of food consumed.

(2) The animals are kept, in large proportion, where they ought not to be kept, namely, inside a big city. They are there, in very expensive quarters under almost necessary insanitary conditions, where all the food has to be carried to them,—where it is in many cases necessary, for economy's sake, to slaughter them when they become dry—simply because of the difficulty hitherto felt of carrying milk in good condition from any considerable distance. This difficulty can be got over, and, if this is true, milk can be delivered in Poona, as in most other large cities from animals kept outside, more cheaply and under purer conditions than by the present arrangements. It seems to me essential to take the cattle out of the cities,—and to arrange to transport milk.

(3) The milk already produced outside is transported to the cities in the crudest, the most expensive and almost the most insanitary manner possible. If the bringing of the milk from villages to the cities can be organised, it can be cheapened, more can be produced by the same people who are now producing it, and better milk will be available in the city.

It seems therefore of vital importance, if the milk supply is to be improved:

(1) that the animals should be improved so as to give a larger amount of milk for the food employed;

(2) that the animals should be in cheap and natural surroundings outside the city instead of inside;

(3) that the transport of milk both from adjacent points by road, and from distant points by railway, should be organised, so that cheaply produced milk should be delivered in the cities in good condition.

With regard to the first of these points, namely, the production and use of a better animal, the difficulties are perhaps greatest, and will take long to solve. For the present, all a dairy owner can do is to take the best breed available, whether of buffaloes or cows, and the best animals among them with the best of male animals for the herd, and ensure that the best is got out of these animals by suitable feeding.

In connection with the development of better animals than are in use at present, my colleague, Mr. J. B. Knight, who has had large experience, has recently expressed himself as follows:—

The development of a profitable dairy animal from any of the types existing, without outside blood, is the work of generations, working on scientific principles, consistently, intelligently and persistently. The agency for effecting this is difficult to point out. The necessity for continuity in this work is so great that it places it beyond the scope of Government Institutions, because of the necessary frequent changes of personality. The only agency which could undertake this work, with a reasonable expectation of accomplishing marked results does not evince any great interest in work of this character. We refer to the hereditary Chiefs and Sardars, who have estates adapted to the work, and who, if they were really in earnest, could carry on this work from father to son, along fixed lines and accomplish something. If the above noted gentlemen could be aroused to form breeders' associations and definite ideals laid down, and authentic records of breeding and production kept, some noticeable improvement might begin to appear in ten or fifteen years.

Till such time as this is done it will be a question of utilising the best of the existing milk breeds and milk animals. That even among existing breeds there are very great differences is shown by the following figures from the records given me by Mr. Knight:—

| Breed of Animals. | Average yield of milk per annum, in lbs. | Average yield of best animals, in lbs. | Maximum yield per annum, in lbs. |
|---|---|---|---|
| Sind Cows | about 2,000 | about 3,500 | 4,700 |
| Kankrej Cows (Gujarat) | about 2,000 | about 3,500 | 4,265 |
| Gir Cows | about 2,000 | | 3,687 |
| Jafferabadi Buffaloes (Poona records) | | 3,500 | 5,438 |
| Delhi Buffaloes | 3,000 | | 6,000 to 8,000 |
| Surti Buffaloes | 2,780 | | 5,034 |

The Deccan breeds of both cows and buffaloes give a much less quantity than this per annum.

The question of getting the largest amount of milk from animals kept is largely connected with suitable and regular feeding. Such feeding is very rarely done at present. There is hardly ever any arrangement for the regular growing of succulent fodder, and for its provision in the form of silage during the drier seasons of the year. Without this the best will never be obtained from the cattle kept,—and the milk supply, if it is to be in the greatest quantity and best quality that the animals can produce, must be obtained from a situation where this regular feeding is possible, where growing and storage of succulent fodder can be made, and where the animals can utilise it best in the production of milk. If we look on the animals, as we ought to do, as simply milk-making machines—the cost of the fodder for each pound of milk produced becomes the vital consideration in the economic success of milk production. At present I am convinced that the production is not economic,—because animals are used giving very little milk, and the feeding is not such as to lead to their giving the maximum amount of milk.

Again, the animals should be in cheap and natural surroundings away from the city. I would do everything to encourage the growth of country dairies or of the keeping of cattle, for milk supply to towns, in villages,—and everything to discourage the keeping of cattle for commercial milk production in the city, at least when a satisfactory supply from outside is arranged for. I think that the provision of municipal sheds for milk cattle, and all such arrangements, while they may be necessary in certain cases at present to ensure sanitary milk at all, are essentially a move in the wrong direction. The object should be to get the cattle removed away from the city entirely, and transport the milk.

For villages or, in fact, for centres of milk production either near or far, the chief question seems to be the organisation of the supply either by large dairies, or by co-operative arrangements for collecting and transporting the milk.

There has been much talk in the last two or three years about the organisation of large dairy companies for the supply of cities. Such large dairies are likely to be successful if located suitably on cheap land with regular water supply,—with good railway (or road, if near enough) connection to the market,—and if good animals are kept. As an illustration of what can be done in this direction, I may note that the Military Dairy in Kirkee can now and does now place milk in Bombay at a cost of 2 annas per pound. This is of course done because it is a large concern with well selected animals with excellent feeding arrangements, including the regular use of silage when green grass is not available, and with pasteurising and cooling arrangements to ensure that no milk goes bad. With the perfecting of pasteurising, there is no need for such a dairy to be near the city supply, provided there is a good railway connection; it can be located wherever, say, within a hundred miles, milk production will be the cheapest and most successful.

Apart from the establishment of large dairies as just suggested, the only method of ensuring a pure milk supply at a cheap rate, is the organisation of village production and transport. It always seems to me,—and it really is,—a great pity that such a large amount of labour should be employed in merely taking milk to market, which could be used in tending more animals and producing more and cheaper milk. The whole is a matter of gaining the confidence of the village cow-keepers, and then their organisation. The winning of their confidence is a thing for which no rules can be laid down, but it is an essential preliminary to the organisation of the supply,—either by a middleman who buys it in the village, or better by a co-operative organisation of the milkmen themselves.

I only know one such co-operative organisation in India at present,—at Benares,—and the account of this (which I owe to Mr. Ewbank, Registrar of Co-operative Societies) illustrates its possibilities and also its difficulties. I will quote the account of this by the Registrar of Co-operative Societies, United Provinces, and then give a summary of its accounts for the last half year of 1912. In this account an "Ahir" is simply a cow owner.

"The Vishweshwar Co-operative Dairy at Benares is an institution started and owned by Ahirs and not by capitalists. To remove at least partially the great complaint about the impure milk supply of the town, the idea of starting a dairy to supply pure and fresh milk was suggested to the Ahirs and some of them took it up. They organised the business with the help of Babu Shiva Mohan Lal, Inspector of Co-operative Societies. A manager was appointed to work under the guidance of a committee of Ahirs who elected Babu Motichand, a leading resident of Benares, as their President.

"The Society was registered in June 1911. The shareholders are almost exclusively the Ahirs and they have paid Rs. 354 in shares so far. They will continue to buy shares from their savings. No other capital is employed if we exclude the cash credit of Rs. 100 which the Benares Bank has sanctioned and the amount borrowed to build sheds. Sales are made for cash and the sale proceeds are enough to meet current expenditure and leave a margin for profit. The manager keeps the accounts. There is also a credit branch for advance of loans to the Ahirs.

"That dairy management purchases milk from the individual shareholders (the Ahirs) at ten seers for a rupee. The feeding of cattle and the milking is conducted in the presence of the manager, the headman of the Ahirs, and some servants. The milk is collected in cans, locked up by the manager and carted to the city in charge of the salesmen. The cans have pipes attached to them, the contents can be taken out, but nothing can be poured in. One horse cart and one *ekka* are at present employed for cartage. They take the milk morning and evening to the five shops which have been opened in the various quarters of the city and the milk is sold there to the general public. The rate is eight seers a rupee. Shops have been taken on rent and one salesman works at each shop. The manager goes out from time to time to see how they work and to attend to complaints. There is a great demand for the milk. The crowd at the shops is sometimes as large as at the third class booking offices in large railway stations.

"The dairy is situated outside Municipal limits, about four miles from the centre of the city. A plot of land was taken in the beginning on lease and is still in occupation. An opposition set in soon after on behalf of vested interests; the other Ahirs and professional dealers left no stone unturned to discredit the organisation and to injure it. They spoke to the owners of the shops to eject the salesmen, and also to the landholder and his agents not to allow any more land to the dairy. Shops had to be changed or rents raised and the servants of the landholders had to be propitiated. On one or two occasions some Ahirs armed with lathis came to threaten the salesmen at the shops. The landholder and his tenants raised difficulties when they were approached to give some more land for extensions. In this matter the aid of the district authorities was sought through the Registrar of Co-operative Societies and they settled reasonable terms with the landholder and the occupancy tenants. One tenant is still objecting. This is about the only matter in which help has been given by the authorities. No help has been given by the Municipality.

"The success of the dairy was soon noticed. People thought there was a good deal of profit and hardly two months were out when a rival shop was opened in the immediate vicinity. In this case some men combined to buy milk from the Ahirs in the city and put them in cans similar to those of the dairy, rented a shop adjoining the dairy shop and brought the milk there for sale. As they had not to traverse any distance they often managed to reach their shop earlier and customers who did not like to wait bought from them. They had also a comparatively small quantity to sell. The rival shop is now on the decline. A capitalist however who had his dairy at Calcutta and who could not make it a success there is shifting to Benares and some of his cattle have already come.

"The management has found from experience that there is not much profit in this business if it is run on modern lines, but that it can pay its way. In Benares specially even well-to-do people are not willing to pay a little extra price for good and pure milk. The dairy has three ponies and it had to pay on account of their feed, treatment, occasional breakdown and hired labour, Rs. 900 during the last year for transport. There were also occasional surpluses of milk which resulted in loss. *Dahi* or butter is generally made of the unsold milk, but they yield much less profit than milk. The surplus is occasional, so it will not pay to engage a permanent establishment to deal with it. An attempt was made to sell the extra milk to sweetmeat-makers at a slightly reduced rate, but they are somehow in opposition and are determined not to give any help.

"It has also been found that the best customers are the common people who buy for cash and give no trouble. Formerly the manager made it a point to supply to public institutions and rich men, but delays, deferred payments, the question of commission to servants, etc., came in the way and he has now definitely given up the practice. The manager has also hitherto failed to solve the question of a house-to-house supply, for Benares is peopled mostly in lanes through which no carts can pass. A sufficient number of customers living in one locality has also not come forward to pay the extra cost of labour.

"The dairy has about 100 head of cattle of which about 30 are dry. There are more buffaloes than cows. The outturn of milk now is nine maunds per day, but the quantity goes down in the summer."

Of course this is only a small concern: there are some groups of villages near Poona, for instance, where a larger society could probably be organised than that of which I have spoken. This Benares Society works, however, on a very narrow margin of profit. Its accounts for July to December 1912 showed a gross profit on the sale of 120,700 pounds of milk, of Rs. 1,595 and the net profit was Rs. 235.

Perhaps this is not a *brilliant* success,—but it sells pure milk at 8 seers per rupee in Benares, a price at which pure milk is *never* obtainable in Poona.

This organisation of milk production away from the cities, and of transport to the cities does not seem beyond possibility in Western India, whether by large dairies, or by the utilisation of village supply. Exactly the mechanism for doing it will have to be worked out. Neither the data nor the means are available for giving details of such a mechanism. But every evidence exists to show that it can be done.

When it is done, then the time will have come to apply strictly sanitary rules with regard to the milk cattle-sheds inside the city, to the quality of milk sold in the streets or in the shops, and generally to bring the milk supply under the sanitary control which is so very essential. Till such organisation has been shown to be workable, without raising the price of milk, then I fear that the supply as at present arranged cannot be interfered with. It is a great pity that this is the case, but anything which tends to raise the present very high price still further would defeat its own object.

In conclusion, I have tried to show the deplorable condition of city milk supply at present in Western India, the difficulties in improving it, and the methods which seem to offer the most hopes of success. The position is by no means hopeless, but it is difficult. The data I have put forward will, however, I hope, lead to a wide discussion of the subject, and to the improvement of one of the very worst conditions in city life in Western India at present.

*Chapter* 17

# THE PROVISION OF MILCH CATTLE STABLES IN BOMBAY*

## I. *Introduction.*

The whole question of the provision made for the housing of the cows and buffaloes necessary for the production of the milk required by the population of Bombay came before the Municipal Corporation of that city in 1915, and a Committee was appointed to consider and report on what action should be taken. I was, at that time, asked to assist the Committee in their work, and Government gave their permission to my doing so by their resolution No. 8200 (R.D.) of July 31st, 1915. I attended a meeting of the Committee in September of the same year, and was then requested to investigate the local conditions in connection with various letters from the Municipal Commissioner placed before the Committee. A long time has elapsed since then, and, in fact, since my local inquiries were completed, but conditions have been so rapidly changing, and the expansion of Bombay and the rise in the value of land has gone on so much faster than was thought probable even in 1915, that it is likely it will be possible to deal with the question from a much more comprehensive point of view than would have been the case at that time.

* From a report made to the Department of Agriculture of the Government of Bombay, November 5, 1919 (Letter to the Secretary No. 13855).

In 1915 the whole question was as to whether the milch cattle stables should be removed to definite sites outside of the town; and four sites had been suggested as possible. When I came to go over these, I was informed that only one of them, near Mahalakshmi Station, was still available. This I saw, in conjunction with the staff of the Corporation's Health Department, and found it to be entirely unsuitable for the purpose in view. It was low-lying and would require a very large amount of making up. It was surrounded by mills and a rapidly increasing population. It was therefore, likely to cost a lot of money before it became ready for use, and moreover it would not, for many years, be likely to be healthy for cattle and would be, in a very short time, likely to be a considerable nuisance to the neighbourhood in which it is placed.

The unsuitability of these sites enables us to consider *de novo*, the problem of the future of the milk supply of Bombay, and to see what ought to be aimed at so far as the means and method of its supply are concerned. If we do this, we can, of course, frame our policy for the future so as to get the maximum supply of good milk, produced under healthy conditions, and with a minimum of nuisance to the citizens.

## *II. The Problem of the Bombay Milk Supply.*

In entering upon an examination of the methods employed in meeting a problem of any complexity, it is essential to have, at the beginning, a clear idea of the character and terms of the problem itself. Let us consider, first, therefore what the milk supply of a big city like Bombay involves, and, as nearly as we can, the advantages and disadvantages of every method by which it can be met. The present system of supply has grown up haphazard. It meets the need, certainly, but has many disadvantages, and in the form which it now takes contains very many undesirable features from the point of view of (1) the character of the milk supplied, (2) the effect on the sanitation and health of the city which it produces, and (3) the result on the permanent supply of good milking animals and hence on the cost of the milk. Perhaps a description of the existing system of supply may be useful as a preface to any criticism of the methods by which that supply is secured.

The supply of milk to the people of Bombay amounts to 0.27 lb. per head per day[1]. This quantity which is about the same as that used in most of the cities of Western India means that for a million people a supply of 270,000 pounds of milk per day will be required. If the population increases beyond this number, the daily requirements will increase in proportion. I do not here discuss whether this supply is sufficient. There is little doubt, however, that if milk could be supplied cheaper the consumption would increase, and that such increase would mean the saving of a lot of infant life.

The present supply, too, is produced to the extent of 83 per cent in the city, while the remaining 17 per cent is brought from outside. In 1916 to supply the quantity produced in the city there were 18,000 buffaloes and 3,000 cows kept in stables in Bombay itself. These numbers must have materially increased since that time.

We have, in fact, three essentially different sources of the milk supply of Bombay. By far the largest quantity, as noted in the previous paragraph, is produced from cattle kept in the milch cattle stables in the city or at any rate in the municipal area. The second source of supply consists of the animals kept in villages within a few miles of the city and within the range of the suburban trains. I visited many of these in both the Great Indian Peninsula and Bombay, Baroda and Central India Railway lines in connection with the present inquiry, and notes on what I saw will be found below. This supply is really similar to the first in that the cattle are kept by *goalis* in cattle stables similar to those in the town, but which not being subject to Municipal inspection are kept in a worse condition. The third source of supply, which has, up to the present, been

[1] Report of Committee to consider measures for the improvement of the milk supply in large cities, Poona, 1916, Page 2.

of a rather spasmodic character consists of milk brought from a considerable distance from Palghar, from Gujarat, from Kirkee, and from a few other places, from animals kept under genuine dairy farm condition. I will proceed to give some account of each of these sources of supply as it exists at present, its disadvantages, and its good points.

The first of these and at present the most important is the supply from buffaloes kept in stables situated in Bombay itself. A few years ago these stables were most insanitary places, in which the buffaloes were kept in filth, and the manure was not regularly removed, and hence the stables were centres of nuisances of various kinds—smells, flies, etc., etc. But this kind of state is being gradually got rid of. I have seen some of them, but the application of the new rules has gradually converted most of the milch cattle stables into as sanitary places as they can be when situated in the middle of a large city and surrounded by buildings. The present rules for milch cattle stables in Bombay are excellent, and I do not think that any serious improvement in the situation can be effected by an improvement in the rules.

But large milch cattle stables located in a big city cannot be really sanitary, nor can they be entirely free from nuisance. I am not saying that animals kept to provide milk for private consumption or a stable containing a very few animals under very careful supervision and strict rules need necessarily be a serious nuisance or a danger to health. But large stables like those in Bombay at present, unless there is a supervision and rules which are quite unpractical, cannot, I think, avoid being a real sanitary danger. The collection of animals housed close together, the accumulation of dung even for a few hours, the smell which cannot be avoided when large numbers of animals are packed into a confined space in a highly populated area, the collection of milk in large quantities in the dusty and possibly disease-infected air of towns, all these tend towards the lowering of the standard of milk produced, lead to the creation of a nuisance in the neighbourhood, and may result in the possible dissemination of disease by means of the flies which inevitably follow cattle stables.

I know that the newer stables built have been located in special areas left for them, and that they are not likely to occupy such congested areas in future as has been the case in the past, but I think it must be conceded that the system of locating large milch cattle stables, housing many thousands of milch cattle among the human population of a big city is not desirable if any other system can be found. Even when such stables are placed in what appears now as an open area outside the town, they are liable, in a city as rapidly growing as Bombay, quickly to find themselves in the heart of a populated area.

There is, however, no evidence at present that the actual milk produced under these conditions is inferior, provided the feeding is all right. It appears to be equally good with that produced under more natural conditions, but the danger to it commences from the time it is drawn.

There is another objection which to me as an agriculturist is very important. The animals kept in the middle of a large town like Bombay are usually slaughtered after their milking period is over. They are rarely covered by a bull and hence rarely come in calf during the time they are in Bombay. When dry they must be taken to cheaper surroundings or disposed of. When disposed of it generally means that they are slaughtered, and very few milch cattle owners in Bombay maintained a farm in cheaper surroundings to which the dry cattle can be taken.

I am aware that this argument of the loss of cattle after their milking period is over, is not an inevitably fatal argument against the present system. If there was a regular breeding industry outside, constantly producing in large numbers animals as good as those kept in the city stables, it might be uneconomical to slaughter the animals after the lactation is over, but that is all. The future milk supply would not be endangered. But there is no such breeding industry on anything like the scale required, and hence the tendency is for good milking animals to become more and more difficult to get and hence more costly, and so the price of milk must necessarily get higher. This is, in itself, a sanitary objection to the present system, for the rise in the price of milk is to an appreciable extent due to the very large rise in the price of good milking animals which are only kept for milking in Bombay for a single year. Then again if there were any organisation for buying up animals whose milking period was over and taking them out of Bombay until again in milk and if this could be done economically, the evil I am at present discussing would be avoided, for it would save the animals from slaughter, would preserve the young stock produced, and so prevent the fearful loss to our best milking stock which occurs every year. But there is at present no such organisation. A few people have done it on a small scale. I know of one or two places which offer to take care of dry milch stock at a fixed sum per month until they came again into milk. But it has never been done on any but a small scale, and the bulk of the cattle in the Bombay milch cattle stables are still slaughtered when they become dry. Against these disadvantages of the system now prevailing, of large milch stables in Bombay, there are certain advantages to be placed, which have hitherto justified its continuance. These are—

(a) The milk is produced on the spot, and can hence be delivered to customers fresh, without any treatment whatever, and without being dependent on any railway company or other body.

This is very important, and its importance must not be minimised. Fresh, unboiled and unpasteurised milk has certain properties which make it superior to milk treated by any process. It may not be much superior but it has some advantages. But, in the plains of India where milk cannot be kept even six hours without much of it going sour, to get such fresh milk, the milch animals must be on the spot, and a railway journey of even a few miles may often mean the wastage of very much milk. Therefore milk produced at any distance is of necessity 'treated' milk whether pasteurised or otherwise prepared.

(b) The milk produced and the animals from which it is produced can be under much more complete inspection in a large city than when they are away from the town. In other words it can be seen at once by the municipal inspectors whether the cattle are healthy, whether the conditions are approved, and in accordance with by-laws whether the vessels used are clean, and whether proper precautions in collection and preservation are taken. Where the conditions in any respect are unsatisfactory the animals or the animal owners can be eliminated.

This is a matter of great importance. As will be seen when I deal with the stables outside Bombay, say at Kurla or similar places, the cattle outside the city itself tend to be kept under much worse conditions, and this is simply because they are able to escape the Bombay inspection and the Bombay bye-laws. But it is not necessarily an insuperable difficulty or an irreplaceable advantage and I shall try and show later how the same inspection may perhaps be obtained without the disadvantages of the stables in Bombay.

(c) Urgent supplies of milk can be obtained of known quality. This has been represented to me as a very strong argument for keeping milch cattle in Bombay, but I do not think it is a very important one. All that is really needed is that milk dealers in Bombay shall have cold stores which enable them to keep large enough quantities of milk in their stores to supply any probable demands.

There are two further economic reasons for considering the production of the greater part of the milk consumed in Bombay undesirable if it can be avoided. Of these the first is the necessity of bringing into Bombay very large quantities of forage which tends to clog the railways leading into the city. When it is considered that each animal used for milk in Bombay will require a minimum of thirty pounds of *kadbi* or grass per day, generally the latter, it means that over six lakhs of pounds of fodder must be brought daily into the city apart from the concentrated food like grain, gram, or oilcake which the animals eat in addition. All this is to produce from two to two and a half lakhs of pounds of milk. It is obvious that if the latter could, equally well, be produced outside and brought into the city the cost of the milk, other things being equal, should be decreased.

The second economic reason against the present dependence on milch cattle contained within the city is the rapidly rising price of the land on which the stables are located, and the demand for it for other purposes, including the housing of human beings. Every year adds to the cogency of this reason, always provided that milk can be equally well produced outside the city and transported into the city for the use of the people.

Summarising the arguments for and against the present arrangement of having large milch cattle stables in Bombay town itself, or indeed inside the municipal area, we have as follows:—

(A) In favour of the present arrangement it may be said that—

(1) the milk can be used fresh, without pasteurisation or other treatment, and the supply of such milk is not dependent on railway companies or other outside bodies.

(2) the conditions of production can be under more complete inspection and control than otherwise

(3) extra and urgent supplies of fresh milk can be obtained at once.

(B) Against the present arrangement it may be said that—

(1) large milch cattle stables among large and growing human population are certain to be a source of smell, flies, and other nuisances to the human population.

(2) the large milch cattle stables of a big town are not healthy for the cattle, though for the pe-

riod of one lactation they may remain in good condition ;

(3) the system of keeping cattle in large stables in the municipal area makes it difficult to bring the animals into calf again. and so tends to lead to slaughter of the animals when they become dry and so to the loss of the best milking stock.

(4) the system now in vogue leads to the choking of the railways with fodder (grass, etc.) which need not otherwise be brought into Bombay. If the milk were brought in instead, it would mean then only about one-third of the weight would have to be carried;

(5) the rapidly rising value of the land where the cattle stables are situated is reflected in the rapidly rising price of milk while the demand for the land for human habitation is great and pressing.

It is, therefore, clear that, provided adequate supplies of fresh milk can be so kept in Bombay as to be always available, and provided the conditions of production can remain under as careful a control as at present, there is almost everything to be said for the production of milk outside the city instead of having, as at present by far the greater part produced inside the town. In saying this, I do not wish to make any reflection on the management of the present milch cattle stables. They have been bad, but all my observations indicate that they have rapidly improved in recent years, and are, on the whole, at present carefully controlled and sanitary places.

It must be understood, too, that I am not recommending a sudden and revolutionary change, or the wholesale immediate transfer of the milk producing cattle into the mofussil. But I think the Municipality should make the supply of the city from outside the goal of its milk policy, and should arrange restrictions, assistance, and encouragement to would-be suppliers with this in view. How best this can be done I will consider later in this report.

*III. The Supply of Milk from the Suburbs.*

Although no less than eighty per cent. of the Bombay milk supply is produced in stables in the city, the remainder is already coming from outside, nearly all in fact from the many suburban stations as far as Virar on the B. B. & C. I. line, and as far as Kalyan on the G. I. P. Railway.

The following figures, though collected in 1915, give an idea of the amount from the different stations. I have not obtained new sets of figures in 1919, but I do not think they would be relatively widely different:—

| | | | | |
|---|---|---|---|---|
| **G. I. P. Railway—** | | | | |
| Kurla | 12,300 | lbs. | milk | per day. |
| Ghatkoper | 720 | ,, | | ,, |
| Bhandup | 3,000 | ,, | | ,, |
| Thana | 720 | ,, | | ,, |
| Dombivli | 2,400 | ,, | | ,, |
| Kalyan | 2,160 | ,, | | ,, |
| **B. B. & C. I. Railway—** | | | | |
| Bandra | 4,800 | ,, | | ,, |
| Santa Cruz | 360 | ,, | | ,, |
| Vile Parle | 900 | ,, | | ,, |
| Andheri | 1,320 | ,, | | ,, |
| Goregaon | 600 | ,, | | ,, |
| Malad | 1,200 | ,, | | ,, |
| Kandivli | 900 | ,, | | ,, |
| Borivli | 3,000 | ,, | | ,, |
| Bhaynder | 3,300 | ,, | | ,, |
| Bassein Road | 300 | ,, | | ,, |
| Nalla-Sopara | 6,000 | ,, | | ,, |
| Virar | 5,700 | ,, | | ,, |

There is a little milk received from beyond these points, but the quantity may be neglected. I will deal with the considerable and growing amount from Palghar later, as it represents a different system of production.

I have, however, made a close examination of the milk production at Kurla and Bhandup on the G. I. P. Railway and near Virar on the B. B. &. C. I. line. It may at once be said that the system followed is practically the same as that in Bombay itself without the advantage of the close and careful sanitary and other control over the milk production which exists in the Bombay stables. With all its disadvantages I would prefer from the point of view of getting good milk, the produce of the Bombay stables to that produced at any of the places I have visited and seen. Stables are nearly always crowded, the water supply is often unsatisfactory, the removal of dung is not nearly as strict as in Bombay and the stables are usually infested with flies and mosquitoes, resulting in damage to the milk and often unhealthiness to the cattle. To show that the mere removal of milch cattle stables to the mofussil by itself is not a remedy for the present evils. I may quote my notes in my visit to the milch cattle at Bhandup.

The sheds, at the time of my visit, had been recently set up and *goalas* from Bombay had been persuaded to take their cattle there by the low charge of Re. 1 per month for each cattle stall. They found however that this advantage was largely outweighed by the disadvantages. They had anticipated that the local price for grass would be much cheaper, and though this was to a certain extent the case, it was not so much as expected, as nearly all the local grass was already contracted for. This, of course, would naturally be the case unless a large enough number of animals were kept there to enable them to make their own independent contracts. Other foods were dearer as they had to be brought from Bombay. The stables were unpaved, and the removal of the drainage water was not complete. A pond of filthy water remained near the stables. Hence one of the chief complaints was that the place was unhealthy both for the *goalas* and for the animals, that flies and mosquitoes abounded and that the loss of cattle had been great. The chief objection appeared to be, however, in connection with railway difficulties as, on account of no central distributing organisation existing in Bombay, men had to take the milk to Bombay themselves and distribute it. One man took one maund of 40 Bombay seers (or 60 lbs. of milk) to

Bombay twice per day. Fifteen seers could be taken free, thereafter the charge, at the time of my visit, was said to be one anna per 15 Bombay seers.

This was hardly an extreme case of the conditions in the suburban milch cattle stables. Similar conditions prevail elsewhere. I have seen, in the rains, milch cattle lying in mixed mud and dung four to six inches deep in such stables. And though in the older centres like Kurla, conditions are better yet it may be said at least that the stables in the outskirts of Bombay are worse than in the town and less under control and that the chances of danger from such milk is greater than in Bombay. In all the centres now under discussion, the cattle are therefore largely kept under conditions not widely dissimilar from those in the Bombay city. That is to say, the animals are kept tied up in stables and the amount of grazing, if there is any, is small. The conditions are almost as unnatural as in Bombay, and are in almost all cases not farm conditions, but city conditions. In summary they have the advantages over the city as follows:

(1) There is generally though not always the chance of a certain amount of grazing, and to this extent, the animals are in a more natural condition;

(2) there is a greater chance of the animals being brought into calf again, and if this is secured there is less reason for the good milking stock to be slaughtered when dry;

(3) there is less need of bringing fodder into Bombay, though much of the other cattle food is actually obtained by purchase from the city;

(4) the cost for rent of stalls is smaller, but this advantage is compensated for by the cost of railway transit, as in the absence of a large distributing agency in Bombay, the milk has to be accompanied by a man in practically every case.

The disadvantages as at present conducted are:—

(1) There is very little, and (in places like Kurla) much slacker control of the condition under

which milch cattle are kept than in Bombay. Hence the milk is more liable to contamination and to be the produce of diseased animals;

(2) there is often, though not always, a much inferior water supply for the cattle;

(3) the railway transit involves the milk being often long in coming to market, though this delay is not perhaps great where the suburban train service is frequent. Inasmuch as there is in no case any arrangement for cooling the milk except the ordinary water supply, the milk is likely to be put on the market, especially in the hot weather, already considerably changed and not in a condition for infant consumption. In fact what I have seen of the present supply from the villages within 20 miles of Bombay has convinced me that from the point of view of a cheap, healthy, milk supply the conditions are worse than in Bombay. Any attempt to drive the milch cattle stables from Bombay into the area outside unless it is accompanied by ownership of the stables or at least effective supervision of the stables by the Bombay authorities will be a step backward and not forward. The present city stables must remain until the Bombay Municipality are prepared to own cattle stables themselves, or to devise a system of effective supervision, or until a totally different organisation of the milk trade comes into existence (of which more below).

Let us see what would be the essentials of a sound system of milk supply from animals kept in stables on the outskirts of Bombay say in Salsette, or at any rate within the reach of suburban trains or of any easy motor-lorry journey. I may say at once that I am not recommending the system as the best and that it would only be applicable near enough to Bombay for the milk to be brought into Bombay without treatment. If the milk has to be pasteurised I fancy it will be found advisable to go much further afield and have genuine dairy farms, etc., equipped with pasteurising plant for treating the milk.

But the essential conditions of what I may call a suburban milk supply are as follows:—

(1) The stables and animals in them must either be owned or be under the supervision of the Bombay Municipality. Owing to the very large increase in the value of land near Bombay, I fancy that suitable land for this purpose could not be obtained within the Kurla-Andheri line. Beyond that there is fair amount of land. The land now covered by the Powai estate would be in many respects ideal for large dairy stables, or even for a dairy farm (see below). But you must have (*a*) high and well-drained land, (*b*) a certain grazing and exercise area which even when the cattle are kept in stables should not be less than a fifth of an acre per animal. This grazing area, under the scheme I am outlining, could almost be neglected as a source of food; (c) cattle sheds built on sanitary lines.

To secure the conditions required as to control or supervision of the stables would be difficult, I understand, in the present state of the law, unless the land and stables were at least owned by the municipality This would involve a very large expenditure. The number of milch cattle now in Bombay is considerably over 20,000: we must therefore calculate on 30,000 Suppose half this number was transferred to the suburban system and suppose the stables would require a capital expenditure of Rs. 200 per head as a minimum including the cost of land; the capital involved would be Rs. 30 lakhs for stables alone. Of course, the process of transference would be gradual and it might be possible to achieve the end by laying out Rs. 5 lakhs per annum. The financing of such an undertaking is however beyond the scope of the present report. If the stables built were full, I may say, that they ought to pay as an investment.

(2) If such stables were owned or even if they were built on land owned by the Bombay Municipality, supervision could be made easy, and would involve

simply an extension of the present inspecting and controlling system. In any such scheme I would strongly urge that not more than six hundred cattle should be located in one set of stables. Each such set should be in its own ground, and separate from any other set, on account of the extreme danger in India of rinderpest and other epizootic disease. With this proviso, however, there would be no difficulty in locating the sets of stables in two or three areas as might be found most convenient.

(3) The transit of the milk to Bombay must be organised either by the municipality itself or by a joint stock company or by a co-operative organisation, on a large scale. This I count essential. Small producers without an organisation for transit must be very near a station and hence must be on land not selected for its suitability for cattle stables, but for other reasons. But if the transit to market on a large scale is provided and its distribution is arranged for then the question of site can be settled almost without reference to the very great nearness to a station. I insist that this must be part of the scheme for milch cattle stables. It is no use starting a scheme for stables except on a large enough scale to justify an organisation of transport as well.

(4) The distribution in Bombay must be organised too. To put milk production even into the suburban area and keep the supervision in hand, means that the producing and the distribution of milk will necessarily be dissociated from one another. This is, of course, what has always happened in the really large cities of the world. London in which thirty years ago the milk was still largely distributed by small producers, is now almost entirely in the hands of the large distributing companies. In Bombay, at intervals, distribution has been undertaken by one or two companies deriving their supplies from outside but the matter has hardly been more than experi-

mental. Whether such distributing companies would arise, whether they would also arrange for the collection of the milk from the *goala* producers in the municipal stables and carry it to Bombay is not yet certain. I believe they would do so, as the business would be a very safe one and the profit could be fairly easily calculated, provided the stables where the milk is produced are concentrated as I have suggested.

(5) The bringing of milk into Bombay would have to be licensed or else I fear the whole scheme would break down. That is to say, no one would be allowed to bring milk into the city unless he had a licence to do so, issued by the municipality. I need not go into this matter; I think the necessity is fairly obvious.

In summary, I think it may be said that the present suburban supply (that is to say, the supply from beyond Dadar and within the reach of suburban train services) is worse than that produced in the city itself from almost every point of view. The essentials of a sound supply from the suburbs based not on daily farming as such (which is unsuitable there) but on the provision of milch cattle stables in rural and not in town areas are as follows:—

(1) the power of supervision and control must be retained by the municipality. This means ownership of the land at least on which the stables are built, and possibly ownership of the stables;

(2) the concentration of the stables in a limited area or number of areas must be secured. I have suggested units for 6υ0 animals each with 100 acres of land. A number of these can, however, be placed in one area, if the area is otherwise suitable. If the land be secured such units could be built one by one;

(3) the stables must be located at least beyond the line from Kurla to Andheri on high, well-drained land, but with security for water supply;

(4) the transit of the milk to the city would have to be organised either by the Municipality, by joint stock company which would also undertake the distribution of the milk, or by a co-operative organisation of the producers who would deliver the milk to a distributing agency in Bombay. The arrangements for transit of milk could also be used for transit of food from Bombay and fodder from producing centres;

(5) the distribution of the milk would have to be organised independently of the producing of the milk. This could probably be done best by a joint stock company or series of joint stock companies who would have pasteurising plants and cold stores;

(6) all people taking milk into Bombay would have to be licensed or in the end, the licence would be withdrawn if the conditions of production were not satisfactory.

*IV. Production of Milk on Dairy Farms.*

While I think the scheme just outlined, if carried in all its sections,—suburban stables, control, transit arrangements, distribution arrangements, licenses,—will be a great advance on the present system, I do not think it represents the last word with regard to the future milk supply of Bombay. To keep cattle either under city conditions or even in suburban stables as I have described above, is neither the cheapest nor the most satisfactory method. It has the advantage of giving fresh milk without treatment but, every other advantage it possesses can be secured much better by the production of milk under farming conditions far way from the city or even from the suburban area, and transporting this to the city after pasteurising and cooling, coupled with cold storage in the town itself. And a system like this, for the development of which Bombay is peculiarly well situated, is what I think the Bombay Municipality should look forward to, and towards which it should work.

There have been a number of moves in this direction in recent years. The Military Dairy Farms Department have long supplied Bombay Hospitals with milk from Kirkee, and it has been placed there at a lower cost than that at which milk could be purchased in Bombay. Messrs. Duncan Stratton & Co. have successfully brought milk from the Charotar in Gujarat. But there is even no need to go so far as the two centres named. Dr. Dubash has developed a dairy farm just beyond Palghar, and is sending pasteurised and cooled milk into Bombay every day. And a number of more such schemes are mooted. I consider, in fact, that at present there is room for the production of almost all the milk required by Bombay, if land which is now almost useless and is only producing very poor crops of grass beyond Palghar on the one side and beyond Kalyan on the other can be made available for dairy farming. Much of such land is at present very poor forest—much more is under grass and produces very little per acre—but wherever there is a water supply, which can be made perennial and used (1) for dairy purposes, (2) for watering cattle, (3) for irrigating at least in the *rabi* season fodder crops for the cattle, there is scope for the development of a regular dairy farming system—such as is in use in other countries for the supply of large cities. Such dairy farms would, properly worked, not only secure a supply of good milk, but would save us from the scandalous waste of our best dairying stock which is now going on. Good animals would then live out their natural life, would produce at least seven or eight calves per animal,—and city milk supply would lead to the improvement of the dairying stock of the country (as has happened at the Government Farm at Surat) instead of to its injury if not to its destruction.

And, given the best of stock, and good management, such dairy farms for city milk supply will pay. Where dairy farms of reasonable size have not paid it has either been due to the use of inferior animals, to the lack of good management, or to epizootic diseases. In Gujarat, even where land is very expensive and assessment very high the keeping of buffaloes for milk is one of the most paying of agricultural businesses. At Surat, the dairy herd at the Government Farm pays well, because the yield per animal has so increased that no animal is to be retained in the herd in future which gives less than an annual yield of 3,000 lbs. of milk, and yields of up to 6,000 lbs. are obtained. The position is such, in fact, that if suitable land can be made available on reasonable terms and if good management can be secured a good many people and firms are ready at present to go into the milk supplying business on a fairly large scale, on the basis of the dairy farming idea as I have described it above.

The development of milk supply on these lines can be encouraged by the Bombay Municipality, and in my mind should be so encouraged. Let us discuss the essentials of such a business.

1. The primary necessity is the provision of land and a good deal of land, at a comparatively cheap rate, at such a distance from Bombay that the milk can be there within say four to six hours as a maximum. Such land must be cheap, well supplied with water, and must in large measure be now growing grass and be fairly open country. I take it that about two-thirds of the fodder required would be supplied by grazing, or from grass land, and about one-third from cultivated crops. If the amount of cultivated crops can be increased (and this is all a matter of the character of the land and the supply of water) then the land will carry a larger amount of stock, but at the rate in question a herd of 600 animals would need an area of 1,000 acres or thereabouts. This would be approximately the case whether the animals were fed on the grass in the stall or grazed on most of the grass land. In making this valuation, I am taking into account the grass producing capacity of the areas suitable for the purpose and conveniently situated near Bombay.

Now, though large joint stock companies can acquire such land, and to this extent develop the milk supply, it will be necessary for the Bombay Municipality to take up themselves considerable areas if they are to make possible the enterprise of smaller men including many of the present *goalas*.

The land suitable for the purpose now under conside ration will be found beyond Palghar on the B. B. & C. I line, in a long stretch to the east of the line which now produces very large supplies of grass, which are shipped from Bulsar and other stations. There is every evidence that water can easily be found by digging or boring on this land, and it is already yielding grass. Much is forest land, but yields little forest produce except grass. Similar areas exist within working distance of the G. I. P. main N. E. line beyond Kalyan, but I do not know these so well. On the G.I.P.-S.E. line there is much less area. I fancy, of the class I am discussing.

To get the land required it will be necessary to approach Government to allow the taking up of a good many areas of forest land for the purpose in view, under such conditions as they cared to impose. I am not aware what would be the attitude of Government to such a proposition, but I do not fancy that the scheme I am outlining could be carried out without their co-operation. It would be necessary, I fancy, for the Bombay Municipality themselves to approach Government in the matter and not leave it to individual companies to deal independently.

Counting at the rate specified above, to establish sound dairy farming yielding, say, 100,000 lbs. of milk per day at the present yield would require 17,000 acres of land. The yield from this area would increase, I hope, with the improvement in the stock, which ought to occur. Such an area need not be together, but must either be in large blocks, or else in blocks near enough to a central pasteurising and cooling station.

2. Given such land with water supply and other necessities, the next requirement would be, for each block of, say 1,000 acres (600 animals), sufficient stables, and other buildings of a dairy farm. An essential part of these will be a dairy building, fitted with appliances for pasteurising and cooling the milk. Including the cattle, the capital cost invol-

ved in such a standard herd of 600 animals and necessary buildings and appliances would be from 2½ to 3 lakhs of rupees. Of this the cattle would cost from Rs. 1,00,000 to Rs. 1,20,000. If the area was taken up by a company or large capitalist, these would naturally find the whole of the capital; if taken up as I hope it would be in part, by *goalas* and cultivators who would use the central arrangements for treating the milk, the provision of the capital would be divided accordingly.

3. The Bombay Municipality would have to make arrangements with the railways for satisfactory transit of milk to Bombay. This is obvious, and I need not dwell upon it.

4. The distribution in Bombay would involve one or more large cold stores in the city, whether these were provided by the Municipality or by private company. There would have to be sufficient to hold at least one whole day's supply, so that in case of interruption of railway connection the city would not be short of milk.

The whole organisation of distribution would be separate always and often independent of production of milk, and this must be recognised. But it would lead me too far to discuss in detail the method of organisation most suitable for different conditions obtaining in Bombay.

It may be well to summarise the necessities if a system of milk production on dairy farms far distant from the city is to be adopted, as I think it should be, as the central feature of the milk supply system of the future. They are:—

(a) the acquirement of sufficient land by the Bombay Municipality to allow the production of, say, 100,000 lbs. of milk per day or more, situated as above described and with adequate water supply;

(b) the establishment on the area thus acquired of units of approximately 1,000 acres capable of maintaining 600 cows and producing at least 6,000 lbs. of milk per day. These units might be established by large milk-producing companies or by groups of small producers (*goalas* and others);

(c) to each of the units just described would be attached arrangements for pasteurising and cooling milk so that it need only be sent to Bombay once per day;

(d) regular milk trains would be arranged for on the railways to carry the milk to the city. The arrangements for these would probably be the duty of the Municipality;

(e) such dairy farms, by whomsoever owned, would be regularly inspected by the Health Department of the Bombay Municipality.

(f) a necessary corollary of such a system would be sufficient number of cold stores in Bombay, to hold and keep at least a day's supply, and a regular organisation for distribution.

The advantages reaped by such a system are obvious. The milk would be produced under healthy natural conditions, by animals living a normal life,—and yet would be obtained under conditions capable of inspection and control. The production, if the management was good, would be cheap, and there would be no reason for the slaughter of the best milking stock in the country,—but it would rather tend to improve. The young calves would not die but would become, if properly managed, better milk-yielders than their mothers. Immediate pasteurisation, followed by cooling and keeping in cold store, would tend to prevent milk ever coming on the market half sour. The other evils of city production would be got rid of. But, on the other hand, the organisation would be a more difficult matter, and a greater responsibility would inevitably rest on the Municipality than at present. There would always be the possibility of interruption of the supply, though this danger is largely got rid of by always having a day's supply in store in Bombay, and by having the dairy farm on two or three different lines which are not likely to be interrupted at the same time.

## V. *Other Sources of Supply.*

There is likely always to be a certain amount of milk produced at an even greater distance than that contemplated in the dairy farm scheme outlined above, and brought to Bombay after pasteurisation or sterilisation and cooling. Such will come particularly from the dairy districts of North Gujarat, and such a supply should be welcomed. It will, as a rule, be brought in by large collecting companies, and the only difficulty will be the control of the quality of the milk, or of the conditions of production. Naturally, such milk should only be brought in under licence, and the licence would be liable to be withdrawn if the quality were not what could be accepted. But the control of such a supply could really only be exercised by frequent examinations of the milk itself and not of the actual conditions of production.

*VI. The Future Milk Supply of Bombay.*

The action to be taken at present in regard to milch cattle stables must be largely determined by what it is desired shall be the nature of the milk supply, say, 20 years hence. If we clearly have this in the mind, then all developments, which must in any case be gradual,—would be made to lead towards the system which it is considered ought to prevail. Hence I am tempted to suggest the type of milk supply which ought to exist say, twenty years hence. For this purpose I am placing the daily needs at 300,000 lbs. of milk: if they are larger, as they probably will be, the principle of the scheme will not be altered. In comparing what I suggest with the present arrangements it should be remembered that now from 80 to 85 per cent. of the total quantity is produced in the city itself, that almost all the remainder is what I have called a 'suburban' supply and that an almost negligible proportion is brought from larger distances after pasteurisation and cooling.

The system which I hope will be in vogue twenty years hence will be one which depends for its supply on four sources, as at present, but will depend on these in a totally different proportion from what we have at present.

1. I hope to see the present supply from animals kept in the city almost done with, though not entirely. There

will always be a few animals kept by private owners and a few small stables. But out of a daily supply of 300,000 lbs. of milk, I hope not more than 10,000 lbs. will be produced in the city itself. This will only be brought about gradually, but as other sources develop, the present stables will be removed to make way for the much more necessary purpose of housing the human population.

2. I hope to see a limited suburban supply, from milch cattle stables with a limited amount of land attached, on selected suitable sites belonging to the Bombay Municipality outside the Municipal area. These sites should not be within the Kurla-Andheri line, and I have indicated one or two suitable sites. The land must belong to the Municipality to ensure inspection, possibly the stables might be in many cases erected by the Municipality itself. Each set of stables should not house more than 600 animals. By this means I should hope that not more than 60,000 lbs. (out of a total of 300,000 lbs) of milk per day would be produced for Bom bay use. It would be advantageous to have such a supply which could be supplied without pasteurisation, but only with cooling.

3. I hope to see the main bulk of the milk needed produced by animals living under farm conditions, at any distance up to six hours run from Bombay by railway. Such farms would grow their own fodder for their cattle,—and would pasteurise and cool all milk before sending to the city. This would involve a system of cold stores in the city which would maintain a day's supply in hand. The organisation of distribution would be quite separate from that of production, even if the same company is doing both. Out of a total supply of 300,000 lbs. per day, I should expect 200,000 lbs. to 220,000 lbs. to be produced in this manner. This would involve between 30,000 and 40,000 acres being devoted to the production of milk for Bombay use.

4. I hope to see a limited amount of milk brought to Bombay from much greater distances, produced by ordinary cultivators, and the produce being collected by large collecting companies or associations, co-operative or otherwise. I

do not expect, however, that this supply will, if the total needs are 300,000 lbs. per day, exceed 30,000 lbs. per day.

Thus if my scheme is sound we should have, twenty years hence, out of a total supply of 300,000 lbs. of milk a day;

(a) 10,000 lbs. produced in the city (3.3 per cent.);

(b) 60,000 lbs. produced in milch cattle stables in the suburban areas (20 per cent.);

(c) 200,000 lbs. to 220,000 lbs. produced on dairy farms from 4 to 6 hours railway journey for Bombay (66.7 to 73.3 per cent.);

(d) up to 30,000 lbs. collected from cultivators in Gujarat or other great distances (10 per cent.).

The change to this can only be brought about gradually. That I recognise. But, if my outlook is correct, it should be faced at once, and action taken on the one hand to prevent its becoming impossible, and on the other to encourage any scheme of milk production which tends in the same direction. I am convinced that, if encouragement is given by the Bombay Municipality, the part marked (c) above and which is the centre of the scheme I recommend, will be taken up almost at once by a number of companies and others who are prepared to launch out largely in the production of an adequate and satisfactory milk supply for Bombay.

**PART IV**

**THE SOCIAL SETTING OF AGRICULTURAL DEVELOPMENT IN INDIA**

*Chapter* 18

# THE LINES OF DEVELOPMENT OF INDIAN AGRICULTURE.*

The second session of the Indian Science Congress is the first at which it has been possible to organise an agricultural section. To some people the very existence of an agricultural section is an anomaly. To these there is no science of agriculture. It is simply an art, a trade, a profession, an industry,—call it what you will,—which involves the application of a number of sciences which are much better dealt with each from its own point of view. We have, however, the best precedent for the devotion of one of our sections to the consideration of agriculture, for in recent years the British Association have given the great industry a special place in its organisation.

We meet then for the first time as a section of the only representative body of Indian science, to consider the applications of science affecting agriculture, and we do so at a very interesting moment. It is only during a very few years that Indian agriculture has been thought worthy of special scientific study. The beginnings of experiment in agriculture in India were almost pathetic. Apparently those responsible for it had hardly any ideas beyond, on the one hand, repeating here the experiments on the value of artificial manures

* Address delivered to the **Indian Science Congress** (Section of Agriculture and Applied Science) on January 14th, 1915, and specially revised for the **Mysore Economic Journal**, Vol. I (1915), pp. 11 18.

which had been carried on at Rothamsted and elsewhere without considering in the least whether manuring at all was the most vital question for Indian agriculture, and on the other hand, introducing foreign varieties of crops and new crops, without much consideration as to the likelihood of their success either as culture or as business.

This stage was maintained very largely until about fifteen years ago, over much of India. The first advance beyond it took place, I think, in Madras and Bombay, where excellent work in many lines has been carried on with a very much broader outlook by Benson and his coadjutors in the former, and by Mollison and his pupils in the latter area. But any truly Indian outlook on the development of agriculture is the product of the years since that time. The departments of agriculture in the Indian provinces date their foundation as living and working bodies from about twelve or thirteen years ago. At that time scientific men were first obtained on a liberal scale, and Government and the public now realise from what has been already done, what tremendous scope Indian agriculture holds for improvement if developed in a scientific spirit.

On the whole we may make it that agriculture in India is of two kinds. On the one hand you have an extensive agriculture conducted without much capital, with primitive implements and methods, and yielding poor results when compared with any western standard. On the other hand you have a number of comparatively small, but highly organised industries, conducted largely by planters, growing special valuable crops with adequate capital, and yielding exceedingly high returns. Of the latter type, the planting industries of tea, coffee, rubber and the like, I have little to say to-day. I spent several of the happiest years of my life as the employee of one of the largest of these planting industries. And from my experience there, I can say that there are probably few agricultural enterprises on the whole conducted with more skill, a wider outlook, a greater wish to apply science as an aid to progress, and a more complete readiness to adopt recommendations than most of the planting industries of India

But my business to-day is with the much larger problem of the extensive agriculture, conducted with a minimum of capital, and with corespondingly unsatisfactory results, which prevails over the greater part of India I often wonder whether it is generally realised how unsatisfactory the results of the average Indian agriculture really are. We have no census of production in India as yet, and really satisfactory figures are not possible. But in a few cases it is possible to give figures which make us realise the position. In wheat production, for instance, the yield per acre is certainly not more than ten bushels per acre, and is probably nearer eight, or one-third of what might be considered as a good crop, more especially in this case where so much of the land devoted to it is irrigated and hence not dependent on a very variable rainfall. In the case of cotton, the figure is equally striking and in this case the area and production is very fairly well-known. From twenty-two million acres, the produce runs over four million bales or about seventy-five pounds of lint per acre, while in America, with an equally uncertain rainfall, the production reaches two hundred pounds per acre. Again if we take an intensive crop grown very largely under irrigation, like sugar-cane, the yield works out at under one ton of raw sugar per acre, as against a world's average of about two tons at least. In this case the average in India is very much lowered by the miserable outturn from Northern India and the crop in Bombay and Madras is fairly up to the world's average. These figures are so striking that a newcomer to the subject is apt to think that improvement is easy, and that the raising of the standard of cultivation towards that reached elsewhere is not very difficult.

This, however, is not the case. In no country perhaps is progress more difficult. One is hindered at every point by unexpected difficulties. The lack of anything more than a minimum of capital has been often considered as the most outstanding of these hindrances, and it is very important, though not the only one. The very great conservatism of the Indian cultivators has often also been mentioned. In this matter I wish to state very emphatically my opinion that Indian cultivators are not more conservative than their situation demands. When a man is working on a minimum of capital, when any excess capital costs probably from twelve per cent upwards for interest, when the money is only turned over once or at most twice in a year, it is the only policy to be extremely conservative. And yet my experience has been that when our Indian cultivators have gained confidence in the person advising them, they are ready to adopt almost anything, and take risks that if I were in their position, I should hesitate to accept.

But though the conservatism is not greater than is really necessary, yet it does form a barrier to progress. And there are other difficulties. In the early days of attempts at agricultural advice many recommendations were made which were considered as absolutely certain to give good results. And they failed, one after the other. Deep ploughing was recommended, and it ruined the land in one famous instance. American cotton was introduced wholesale, and has only maintained itself in one or two small concerns of the cotton-producing areas. Carrot seed was imported in large quantities as a famine crop, and was a dead failure. Artificial manures were recommended largely and have been so little successful that, except in case of intensive crops cultivated by planters, the whole consumption of the country is probably at the most a few hundred tons. In fact, I know no case in the world where improvements, suggested on an empirical basis, have so uniformly failed.

We have realised, in recent years, that this failure is not due to conservatism, and is not entirely due to lack of capital. It has been clearly the result of something beyond this, namely the non-realisation that Indian agriculture is something special, and that owing to differences of climate and conditions we have to work out our own problems here on the spot, and that few if any, improvements can be transplanted bodily from any other country. I want to be specially emphatic on this point

With the exception of certain parts of the United States of America, I doubt whether any country where large agricultural progress has been made has a climate and other physical conditions at all comparable with the greater part of India. China and Japan might perhaps be cited. But very little consideration will show how widely the conditions of life there differ from ours in the greater part of the country, and how difficult it would be to apply their experience to ourselves. And moreover we cannot do so, because we know really so little about their method and work, except in a superficial way. We must then develop very careful study of Indian agriculture, not as a branch of what is being done in Europe or America, but with a view which looks out from the eyes of our Indian cultivators, which sees things from their point of view, surrounded by the difficulties which they know and feel are real. We must devote ourselves to solving the problems which they, the cultivators, place before us, and which they know are the hindrances in their way.

I have tried to realise how things appear to a cultivator to a certain extent in Western India, and perhaps I might place before you some of the results which long contact with the cultivators has made me believe are the vital lines of development in our part of the country They will probably not be the same elsewhere. India is a continent, and the problems of the Deccan are not the problems of the plains of Bengal or of the deltas of Eastern and Southern India. But in every case, and on this I want to insist, progress will most effectively be made by following what may be called the line of least resistance,—working from the cultivators' point of view and trying to meet and satisfy their needs.

Now among the cultivators of Western India one need is dominant. And that is the necessity of water The people realised before anyone else the fact that the supply of water was the limiting factor which determined their crop. They tell me that manuring a dry crop is absurd, unless you can ensure water to bring it to maturity They say, and say with some truth, that it is little use cultivating the soil very well by any method which they know, if the water proves insufficient for the crop they grow. They are ready to spend money, borrow money, to pledge almost all they have if a prospect of a water supply is offered. Many and many are the cultivators I have met who have ruined themselves over an unsuccessful well,—and many are those who would be prepared to lay down almost all they have, if they had an assurance of success in digging one.

The first business of those who are responsible for suggesting lines of progress in Western India is then to meet this need. Either we must show how a water supply can be obtained, which is utilisable, or we must show how it can be done without, and still much better results obtained than at present. The first of these (apart from large irrigation schemes which are extremely welcome when they come, but which, in Western India, progress very slowly) resolves itself into the detection and tapping of underground supplies of water, and the storage by tanks of rainfall. For us, the second is of minor importance, though elsewhere it is of vital interest I know, and the improvement and cheapening of the making of tanks is an object well worth the most careful investigation.

But among us the detection and tapping of underground water is more vital. The urgency of the need of detecting underground water is shown by the existence of crowds of men with supposed powers of water detection. We have a few of such men in England; in the Deccan they abound, and have always abounded. I unearthed a short time ago a Sanskrit work on the detection of underground water, abounding in the most fantastic statements, which has recently been published in a Marathi translation! The difficulty is shown, moreover by the fact that forty per cent at least of wells sunk by our people to as great a depth as their water-lifting power and their means will allow, are failures. I suppose when I speak of detecting underground streams of water, I may be laughed at, but, laughed at or not, the finding of water is the thing which our people are most keen about. and if science has any help to give, here is a place where it can be given with a certainty that it will lead to immense advantage. I have for the last two or three years been experimenting with an automatic waterfinder based, apparently, on the detection of electric currents which accompany underground streams, and the results are promising though often difficult to understand. The instrument is the invention of an empiric, and I do not understand it. It gives results, however, and of one thing I am certain, that if there is anything in the principle of this instrument, and scientific men, physicists, would give attention to the question, a far more effective machine would quickly be evolved.

But supposing that any satisfactory instrument for detecting underground streams of water be not found, it would seem that so far as very large areas are concerned the matter is of such importance for the future development of agriculture of large areas, that I would commend to my geological friends the very careful study of many of our areas in regard to water supply, particularly in the direction of the obtaining of artesian or subartesian water supplies for agricultural purposes. Very little study has been made, apparently, on this point. One recent paper in the publications of the Geological Survey, considers the whole question of artesian water supplies in India, but its perusal serves to impress one with the little that is known of Indian formations as water producers, and their possibilities.

In the absence of such data as only a very careful geological study can give, my colleague Mr. Schutte, the Agricultural Engineer to the Bombay Government, has recently made a number of comparatively shallow borings in the Deccan trap, none exceeding two hundred feet in depth, and has succeeded in obtaining considerable subartesian supplies in a number of most unexpected places. Our practice has been to test a site with the automatic water-finder previously referred to, and then bore at a place indicated by that instrument. The boring is almost all through hard basalt, and in all the places indicated water has been found, under more or less pressure, and is available for us. One case is worthy of quoting, though I might describe ten or a dozen. At one site, in the driest belt in the Bombay Deccan, in a village notorious for failures of well sinking, water was thus struck at fifty-three feet deep, at a pressure of ten feet head of water, and though this was at the driest season of the year and at the end of a famine, fifteen hundred gallons per hour have been pumped from the bore hole ever since. Another case gave a head of water of no less than eighty-three feet, though the supply was less than in the case already quoted. The point I want to make here is this. Over a large part of India agricultural progress depends, and is felt by the people to depend, on the discovery and utilization of underground supplies of water. And one of the most useful works that can be done by science at present for these enormous tracts is to invent or improve the means of detecting such underground supplies, especially those occurring as running streams in fissures, and to study more closely the rock formation of India as water producers. In both of these directions there is a big field for scientific activity.

But in the absence of such supplies, howsoever detected and utilised, the insistent demand of our cultivators for water for their crops can be only met by improvements in the methods of utilising the water which actually falls. In the drier tracts of India, even where the slopes are only slight it is probable that forty per cent of the water which falls is not utilised at present, but runs off the surface carrying soil with it, and doing harm rather than good. This is an inevitable result of heavy falls of rain on the hardened ground of a dry tract unless special precautions are taken. Some precautions are taken by the people in many of the tracts, and large amounts of money are borrowed to level lands, to create basins, to collect silt from hills, and to build embankments to prevent such wash. But these are the elementary methods, and I feel that an immense amount can be done by studies in the soil physics of many of our drier tracts and especially of those with a fairly shallow soil to increase their capacity for retaining water capable of being utilized by agricultural plants. Of course a great deal has been done in America, South Africa and elsewhere in this direction, but the conditions are so widely different, the character and resources of the cultivators involved are so varying that very great modification of the implements and methods used there will be necessary.

Three methods of investigation seem important in this connection. One of these, the study of soil physics, I have already mentioned. It seems very important here, especially as several of our types of soil are peculiar, the methods of increasing their absorbing and retaining capacity for water under our conditions require very careful investigation. A second investigation, that of implements, is one which is very much needed. We are recommending a good number of foreign implements, but they are in many cases, a *pis aller*. They are expensive, the importing charges are heavy, they are designed for other conditions, and few hence are ideal, and while we sold on the farm where I stay no less than a value of Rs. 21,000 last year chiefly of ordinary farm implements, yet I always feel that implements specially designed here could, probably, meet the circumstances of the case better. One of my colleagues, Mr. S.S. Godbole, has recently worked out an improvement of the country plough which considerably increases its efficiency and at no appreciable additional cost. And it is more than possible that implements based on those already in use and which are worked out here may have a great future.

A third method, of which perhaps I have little more doubt, to increase the utility of the water which actually falls in our drier tracts, is the development of drought-resisting varieties of plants. I must say that our work in this direction has not been very productive hitherto. In cotton by a process of selection for the mixture of types, we have been able to get a seed which is immensely in demand in Khandesh, our largest cotton district, but apart from this, our efforts to bring to the notice of the people what seemed to be better types of seed than they had been growing have not succeeded. But what I want is more than this. It is the working out on the spot of a type which is especially develop ed for our dry and severe arid conditions, and for the soil which it suits. The only case of this being done in India, that I know of, is the brilliant success achieved by Howard's wheat in Bihar.

In my contact with the cultivators in Western India, they will ask for nothing until help is given on this water question. But once it is solved, once a canal serves their land, or a successful well is obtained, or in any other way a cultivator feels he is more or less free from the vagaries of the season, then he asks for much else. In my experience he rarely inquires for seed, his demand is rather for superior implements again, or for manures.

With the supply of water comes the chance, the people feel, of real improvement, and the way in which improved implements have been taken up by those who have the water is a real revelation to me. In the Deccan, ten years ago, I believe that the number of genuine ploughs, (that is to say, implements which do more than tear the soil apart like a "cultivator") could be counted on the finger. Now the number probably exceeds forty thousand, and is increasing fast in those tracts where the water difficulty has been met. And it is almost the same with other things. But my previous remarks on the subject of implements are sufficient.

The other demand is for manures, and is one which is much more slowly felt, but which is coming. Here again, it seems essential to proceed from the cultivator's point of view and not from European experience. Artificial manures are probably the last step, and not the first, to improved crops, and the devotion of the people to coarse manures, and to the rich manures containing organic matter is, I believe after special study, fully justified by the condition of high temperature and dry atmosphere under which most of the irrigation crops are cultivated. I am certain that in such districts as I have had to deal with in the last seven years, the crying need of our Indian soils is humus, in other words, for organic matter. This is shown by the great advantage which results from green manuring, an advantage which our people are quick to recognise. In fact I believe that in the extension of the practice of burying green crops lies one of the large improvements which are at our doors, once the demand for water is satisfied. But an immense amount of experiment is still needed in ascertaining the suitable crops, and the best condition, and the supplementary manures which are needed to obtain the best results

I suppose we have in the Poona district a limited area which is more highly manured than any part of India. It is irrigated of course, and grows sugar-cane which is excellent. It is curious to find that even in this tract, the concentrated manures which hold the field are oil-cakes, and to a less extent fish. Many attempts have been made to use artificial manures, but except for a beginning in the utilisation of sulphate of ammonia, those mentioned still hold the field. This is a consequence of what I have already said is the true, but unconscious, conviction that under our conditions organic matter is a necessity even in a concentrated manure.

But the coming of water raises many difficulties as well as introduces new needs. And some of these difficulties the people are always asking. The first difficulty that follows irrigation is very often the development of salt in the land. And a very large proportion of the matters raised by cultivators in such places depend on it. We find that seventy or eighty per cent probably of questions raised regarding citrus orchards, are the result of salt in the land. Now I am not sure we have got to the bottom of the question of the occurrence and rise of salt in our soils, and it would make one feel much more confident of the future of some of our best irrigation tracts could be with certainty prevented at a reasonable cost.

This is only a type of the difficulties which such a cultivator brings. Now appear inquiries regarding all sorts of crop diseases and crop pests, which are felt to be serious, and a student of insects or of fungi who had practical methods worked out here and adapted to our condition would do a great service to the people. Despite all that has been done regarding the relationship between plants and their parasites, I feel we are only at the beginning of the subject. I would like to see those take up the study who could approach it from more detached point of view than some of us have time to give. It would be a glorious thing if one could sit down and not have reports called for at frequent intervals, and be able to study under some of our Indian conditions the progress of some disease in a plant, purely with a desire to get to the bottom of the parasitism, and the results would, I feel, be very important, for they would give the chance of applying principles so worked out to their cases, and possibly lead to wholesale methods of dealing with pests when at present, we can only 'nibble' at the matter.

At present, these are the matters which are continually brought up to me as an agricultural officer by the people, and I feel that in working from their realization of the real problems before them, and trying to answer their needs, one is working in a real scientific spirit, and one that will lead to the real advance of the industry we represent. The prospect may not appear so attractive as many of the by-paths of pure science which are continually opening out. One is tempted to turn aside to the seductive fields which, deserting the agricultural outlook, pure science seems to offer. But whenever I feel like this, I turn to a passage which I once found in the preface to Pasteur's account of his experiments in the diseases of silk, and go on content to serve the realised wants of the humblest cultivator of our country. The passage is as follows,—with that I close:—

"Although," says the great master, "I devoted nearly five consecutive years to the laborious experimental researches which have affected my health, I am glad that I undertook them. The results which I have obtained are perhaps less brilliant than those which I might have anticipated from researches pursued in the field of pure science. But I have the satisfaction of having served my country in endeavouring to the best of my ability to discover a remedy for great misery. It is to the honour of a scientific man that he values discoveries which, at their birth, can only obtain the esteem of his equals far above those which at once conquer the favour of the crowd by the immediate utility of their applications; but in the presence of misfortune it is equally an honour to sacrifice everything in the endeavour to relieve it."

## *Chapter* 19

# ROYAL COMMISSION ON AGRICULTURE IN INDIA: DR. MANN'S WRITTEN EVIDENCE (1927)*

As to the methods of getting improvements introduced into Indian Agriculture, I would refer the Commission to a series of four reports, which I edited, which were published from 1909 to 1912 by the Imperial Department of Agriculture. I do not think that any new methods have been devised since that time, though, of course, with increasing experience, the stress laid on the different methods has altered.

But it must never be forgotten that the essential difference between Agricultural Departments in the East and in the West is that the latter have arisen to meet the spontaneous demands of the cultivators of the soil, while the former have been created by a Government anxious to give all the help it can to its agricultural citizens in India; therefore,—and this certainly applies to Bombay, it is necessary for the Agricultural Department to put forth every effort, first to ascertain the needs of the cultivators and then to demonstrate how they can most effectually be met.

I would, however, urge that on the whole the people with whom we have to deal are not more conservative than they have to be. In fact, I am inclined to think that they are less conservative than an average English farmer. The position of a cultivator, however, faced with a recommended improvement, is that if it costs any money he can only re-

* Reproduced with the kind permission of the Controller of Her Majesty's Stationery Office from Dr. Mann's written testimony, submitted in the volumes of **Evidence**, before the **Royal Commission on Agriculture in India**, Vol. II, Part II, **Bombay** (London, H.M.S.O., 1927), pp. 95-102.

ceive it with very great caution and only after such local demonstration as makes its value completely certain. Hence, all advertisement whether by leaflets, by speeches, by meeting, or by lanterns and cinematographs are of very little effect unless accompanied by actual demonstrations on the spot. This is a definite opinion after a good many years observation. The place of these things is as an accompaniment to demonstration and they will have little effect apart from it.

Now, we have on the whole, been successful in introducing a number of implements, a number of improved and imported seeds, a certain number of artificial manures, certain remedies for plant disease, and a few other things, and in creating an interest in better cattle, in pumps for irrigation, in land improvement to prevent erosion in certain areas, and so on.

Thus, within the past twenty years the iron turnwrest plough has almost crushed out the old Deccan plough in many parts of the Deccan, and the Egyptian plough has almost completely replaced the Sindhi plough in large areas in Sind. In the matter of seeds, the demand for improved seed of cotton in many areas, of tobacco in Northern Gujarat, of rice in certain areas in the North Konkan is far greater than we can supply,—even though a good deal is spent in maintaining and multiplying the seed. Among artificial manures, the success of Sulphate of Ammonia as a more efficient partial substitute for fish and oilcake in the Deccan Canal tracts only really dates from 1919, and the consumption is now nearly 3,000 tons per annum. The treatment of jowar seed for smut has very rapidly increased under intensive propaganda in the last three years during which the area for which treatment has been sold has increased as follows:—

| | |
|---|---|
| 1923-24 | 250,348 |
| 1924-25 | 343,978 |
| 1925-26 | 682,188 |

One of the most successful pieces of work of this kind has been the spraying of grapes on the Deccan for mildew.

The industry was a dying one in 1908 on this account, now it is advancing very rapidly in the favoured areas and nobody would think of growing grapes without using the remedy. There are a few established successes—and I may note a few principles which I think have to be adopted if real success in introducing improvements is to be achieved.

1. The man who approaches a body of cultivators must have previously gained their confidence either by previous successful recommendation or in some other way. An expert, *qua* expert, will get little hearing. When I first worked among the potato growers of the Poona District, they listened but would do nothing: Only when I had been able to check potato moth,—after guaranteeing them against financial loss if they used my method,—was I able to get a hearing with regard to other matters. Now I can get men willing to test anything I recommend.

2. This winning of confidence, as a necessary preliminary to any success in propaganda, makes it necessary that practically all field demonstration should be done on the fields of cultivators. Generally this has been done under a guarantee to make good any loss, but if care is exercised no financial liability has in my experience ever resulted. Government demonstration farms, or even plots managed by the agricultural department are usually very ineffective. I pin my faith to working with the cultivators on their land,—just coming in the one operation that we desire to recommend.

3. The necessary for gaining confidence emphasises the necessity of non-official agency being used for propaganda wherever possible. Until recently local agricultural associations had not been a great success except in a few cases. But we think, in Bombay, we have now got to the stage when Taluka Development Associations as established by Sir Chunilal V. Mehta when Minister of Agriculture, under a scheme which I drew up, promise to become one of the very important factors in agricultural advance here. They have certainly become very popular in several of our

divisions though not universally and are doing very good work.

These bodies are formed for a Taluka containing 100 to 200 villages, and are intended as the agents of co-operative, agricultural and other forms of development. They receive an annual subsidy from Government equal to their other income but limited to Rs. 1,000, on condition that they employ a demonstrator approved by the Agricultural Department. These have undertaken the spreading of the use of improved implements, of improved seed,—the formation of co-operative societies for various purposes, and they also supervise the non-credit co-operative societies in their areas. They are supported by subscriptions from individuals, from Co-operative Societies, and, in the last years or two, by grants from Taluka and District Local Boards. In recent cases, the tendency has been to obtain in the first instance a large capital fund, and make the interest on this a large factor in their income. I was at the inauguration of a Taluka Development Association in Gujarat a few days ago, which started with a capital fund of Rs. 12,000 specially collected for the purpose.

Their value depends on the guarantee of work which is given by the presence of an approved fieldman, and on their non-official character, while the Government subsidy gives a title to stimulate work, and guide it in suitable directions.

4. Propaganda must usually follow a realised need or else improvements must be so introduced as to make a need realised. Many efforts have been failures because while good in themselves, they have not met a real need. I may illustrate by the many efforts to introduce winnowing machines. They are not wanted. The existing method is good, only requires more time,—and there is plenty of time.

5. Improvement of organization and finance may be wanted to make technical improvement possible. This indicates the need for close local study area by area, and even village by village.

6. The improvement must be worth while. That is to say, the increased return must make enough difference to make the risk worth while. In the case of a crop, I generally will not push an improvement unless I am confident it will give an increased nett return of 15 to 20 per cent.

Most of these points are perhaps platitudes, but success in propaganda, in my experience, depends on

(1) Winning confidence.

(2) Keeping official people in the background.

(3) Having an improvement which meets what people want, and giving a large nett return.

(4) Being able to command finance, and to organize financial help if necessary.

One difficulty occurs when the success of propaganda depends on common action by a large number of people as in schemes of land improvement in the dry areas of the Deccan, or as in schemes of fencing of large areas against wild pig in which we have had considerable success in Western Dharwar. In these cases, we do need legislation to provide that when such schemes are agreed to by a large proportion (say 75 per cent) of the owners of land, the remaining landowners can be compelled to come into the scheme.

In Bombay, propaganda is now carried out as a combined effort of the Co-operative and Agricultural Departments. A programme of work is drawn out for each unit of area under the control of one agent of either department, —and the work contained in that programme whether for co-operative development, or for technical agricultural improvement, is looked after by one man in that area. These men have their programmes passed and their reports received by the Divisional Board of Agriculture, composed of four non-official members, the Deputy Director of Agriculture, and the Assistance Registrar of Co-operative Societies, and work directly, as to policy under its control.

I really think that now, in many areas of the Bombay Presidency the idea of agricultural improvement has got into the minds of a considerable proportion of the rural population,—and I anticipate that advance may be much more rapid in the next ten years than in the last. My fear for the future is rather that we shall not be able to meet the demands of the people.

*Fragmentation of Holdings*

The actual condition of things with regard to the fragmentation of holdings, as a result of the present laws and customs of inheritance has so often been described that there is no need to stress the fact. But there are one or two points which perhaps may be made clear.

Apart from a general increase in the value of land the number of holdings, and the number of fragments seems to tend to increase to a definite maximum. I have recently re-investigated, at the request of the Hon. Mr. A. M. K. Dehlavi after nine or ten years, two Deccan villages which were studied in 1915 and in 1917 respectively. The following table shows the number of holdings, in the villages at different periods, and the change in the number of land fragments at an interval of nine or ten years.

| Number of Holdings | Pimpla Soudagar | Jategaon Budruk |
|---|---|---|
| 1771-72 | 24 | — |
| 1785 | — | 42 |
| 1790 | — | 34 |
| 1791-92 | 41 | — |
| 1796 | — | 23 |
| 1811-12 | 48 | — |
| 1817 | — | 36 |
| 1823 | — | 50 |
| 1829-30 | 52 | — |
| 1840-41 | 54 | — |
| 1914-15 | 156 | — |
| 1917 | — | 146 |
| 1926 | 156 | 148 |

Comparing the character of the holdings at Pimpla Soudagar in 1914-15 and at present, we find a tendency for the size of holding to concentrate at about 1 to 5 acres. This means that agriculture on a man's own land is becoming more of a partial occupation, with labour for wages as subsidiary.

The number of each size has been as follows:—

**Pimpla Soudagar.**

Number of Holdings of different sizes.

| | 1914-15 | 1926 |
|---|---|---|
| More than 40 acres | 1 | 3 |
| 30 to 40 acres | 1 | 1 |
| 20 to 30 acres | 9 | 5 |
| 10 to 20 acres | 18 | 20 |
| 5 to 10 acres | 34 | 32 |
| 1 to 5 acres | 71 | 81 |
| Less than 1 acre | 22 | 14 |

Similar figures for Jategaon Budruk are as follows. It must be remembered that two severe famines have meantime occurred.

**Jategaon-Budruk.**

Number of Holdings of different sizes.

| | 1917 | 1926 |
|---|---|---|
| More than 150 acres | 1 | 1 |
| 100 to 150 ,, | 1 | 1 |
| 50 to 100 ,, | 5 | 3 |
| 40 to 50 ,, | 4 | 3 |
| 30 to 40 ,, | 6 | 12 |
| 20 to 30 ,, | 16 | 14 |
| 10 to 20 ,, | 43 | 37 |
| 5 to 10 ,, | 34 | 35 |
| 1 to 5 ,, | 25 | 34 |
| Under 1 acre | 11 | 8 |
| | 146 | 148 |

I think the tendency is probably in the same direction here as at Pimpla Soudagar.

**2. Fragments of Land Separately Owned.**

**Jategaon Budruk.**

Number of Fragments of different sizes.

| | 1917 | 1926 |
|---|---|---|
| Over 30 acres | 3 | 4 |
| 20 to 30 acres | 8 | 8 |
| 10 to 20 ,, | 51 | |
| 5 to 10 ,, | 95 | 101 |
| 3 to 5 ,, | 69 | 68 |
| 2 to 3 ,, | 58 | 73 |
| 1 to 2 ,, | 98 | 111 |
| ¾ to 1 ,, | 34 | 26 |
| ½ to ¾ ,, | 31 | 28 |
| ¼ to ½ ,, | 51 | 61 |
| ⅛ to ¼ ,, | 40 | 27 |
| Under ⅛ acre | 13 | 21 |
| Total Plots in village | 551 | 574 |

The general character of the distribution has not been altered, but there is evidence that some consolidation is tak ing place as well as further sub-division. The number of separately owned plots under one acre in size has gone down from 169 to 163.

In this village I noted in 1917 that the units of cultiva tion (in which from an agricultural point of view we are chiefly interested) were larger than the units of land held, and that a man appears to cultivate a less number of frag ments than he owns. In this connection the following two statements will be found interesting:—

**Number of Holdings and Areas of Cultivation held by one man**

| | 1917 | | 1926 | |
|---|---|---|---|---|
| Acres. | Holdings. | Areas cultivated by one man. | Holdings. | Areas cultivated by one man. |
| Above 100 | 2 | 1 | 2 | 1 |
| 50—100 Acres. | 5 | 5 | 3 | 4 |
| 40— 50 ,, | 4 | 9 | 3 | 2 |
| 30— 40 ,, | 6 | 9 | 12 | 17 |
| 20— 30 ,, | 16 | 21 | 14 | 13 |
| 10— 20 Acres. | 43 | 27 | 37 | 41 |
| 5— 10 ,, | 34 | 24 | 35 | 20 |
| Under 5 Acres. | 36 | 18 | 42 | 22 |
| | 146 | 114 | 148 | 120 |

2. **Number of Fragments held and cultivated by one man.**

The table shows the number of men holding and cultivating various number of fragments.

| | 1917 | | 1926 | |
|---|---|---|---|---|
| | Holdings | Cultivation | Holdings | Cultivation |
| | Number with specified number of fragments | Number with specified number of fragments | Number with specified number of fragments | Number with specified number of fragments |
| 1 Fragment. | 30 | 24 | 28 | 19 |
| 2-5 do | 87 | 59 | 89 | 65 |
| 6-10 do | 26 | 29 | 24 | 26 |
| 11-15 do | — | 2 | 4 | 7 |
| 16-20 do | 2 | — | 3 | 2 |
| 22-25 do | — | — | — | 1 |
| above 25 do | 1 | — | — | — |
| | 146 | 114 | 143 | 120 |

These figures seem to show pretty clearly that the natural process of consolidation has now proceeded, in the last ten years as rapidly as that of further fragmentation. This means of course the sale and consequent recombination of the holdings. As regards cultivation, the units are decidedly larger than those of land held; we have, in fact, in all appearance, reached a stable position in which, unless something happens to increase the value or the productivity of the land, it is not likely that the situation in this respect will get materially worse.

The position is, however, bad enough now,— and it is urgently necessary to deal with it. The Punjab experience has shown that in flat land, with limited variation, it is not an insurmountable problem to deal with the problem by co-operative methods. With us in Bombay, such areas occur to a very limited extent, and in the areas where the problem is most acute,—the irrigated areas of the Deccan, the Konkan, &c.,—the soils differ so widely that restriping is a matter of the extremest difficulty. In spite of this, a Committee of which I am the Chairman has, by the orders of Government, taken in hand the planning of a restripement in the irrigated village of Manjri Budruk. The present position will be shown to the Commission on a map,--and the proposed lines of work explained.

For the purpose of restripement by voluntary agency, I feel only one piece of legislation is needed,—namely to provide that when the owners of 75 per cent of the land agree to a plan, the remainder can be compelled to come into the scheme. For the present, I propose that Government should bear the cost of survey and replanning,—and the remapping of the area including the setting up of boundary marks.

As regards future fragmentation, legislation is under consideration in Bombay to prevent this taking place beyond a certain limit. But as I am not in close touch with the position of this proposed legislation, I cannot say anything about it.

*Marketing.*

(a) Do you consider existing market facilities to be satisfactory? Please specify and criticise the markets to which you refer, and make suggestions for their improvement.

I am not sure I have understood the meaning of this question. Marketing of all sorts of marketable produce in Bombay presents no difficulty at present, in the sense that there is no real trouble in getting a market for it. The arrangements for disposing of any agricultural produce are, in fact, very highly organised, and with those which are common market commodities a man need usually not leave his holding in order to sell, and many products, like fruit, are often taken possession of before reaping and reaped by the purchaser. So that I cannot say that in Bombay there is any lack of facilities for marketing; whether the arrangements are such as to give the biggest return to the producer is a different matter, and will be dealt with under the next heading. But if the question intends to ask whether there is usually any difficulty in selling produce after it is grown, I should say No, as far as the Bombay Presidency is concerned. I do not know of a case where produce is not grown because it cannot be disposed of except in the case of new or improved products, and here there certainly is difficulty.

I need, therefore, only refer to the marketing of such new and improved products and quote the case of certain improved cottons. When an improved cotton has been produced, it has been found generally difficult to get a proper price for it. This has happened when we introduced Broach cotton in Dharwar, improved cotton in Surat and American cotton in Sind. In the first of these cases Broach cotton in Dharwar, in order to secure a fair price for an infant production, we introduced special Government auctions of kapas in Dharwar, which were successful. We did the same with the lint of improved cotton at Surat, but there the problem was not so difficult as Surat is a lint market and not a Kapas market. In Sind the difficulty was temporary and now I think that the American kapas gets its proper price:

(b) Are you satisfied with the existing system of marketing and distribution?

To this I think it is obvious that the answer must be No. Nothing like this is ever perfect. But I would insist that for important products the system is very highly organised, and every link in the chain has its function. The repeated and frequent failure of amateur attempts to replace the present system by something apparently much simpler shows either that the system on the whole works well, or else that there are vested interests in it so powerful as to be able to resist any but very well considered changes backed by very general support of the producers. Probably both alternatives are correct.

But I am not sure that, except in a few cases in the Bombay Presidency, we know very much about the whole course of the marketing of any product. This certainly came to light in connection with the enquiries I recently organised for the I.C.C.C.[1] in the finance of cotton production and marketing in Khandesh and in Upper Gujarat. First a questionnaire was issued to people thought to be likely to know the whole of the process through which kapas passes before it reaches the larger buyer in the markets. Answers were received from a considerable number of people. Then it was decided to organise an extensive enquiry in a number of villages and in all the local markets. The result, in Khandesh at any rate, has been to show that the impression given by the answers to the questionnaire was very largely incorrect,—to show that the reputed dependence of the cotton grower, in regard to the marketing of his produce, on the money lender is a myth, that there is generally a preference to sell his produce in the village to petty traders over taking it to market because of bad market practices. The complaint was not against the recognised market charges, but on account of (1) disputes about the rate after weighment had commenced and (2) arbitrary deductions in the market.

Now I am not aware that any investigation of a really serious character, apart from the above, has been undertaken in any part of the Bombay Presidency, on the course of marketing of a product from the actual grower until it reaches the big merchant and so enters the ordinary stream of commerce. Hence it is not now possible to give an account of the chain of hands through which a product passes, as desired in the Commission's questionnaire. What we do know is that

(1) with almost every product, there is a certain amount of forward sale before crops are reaped. The proportion of the whole to which this applies is quite unknown in any case.

(2) with almost every product there are a considerable number of compulsory deductions in the market, for religious and charitable purposes, for samples and the like.

(3) in many cases, in the same market, the unit weight for buying from the cultivator is larger than the

[1] Imperial Central Cotton Comiittee.

unit weight for selling by the dealer. Thus in the Gul market in Poona, the cultivator-seller has to supply 256 lbs per *palla*: on the other hand, the unit for a purchaser of gul from the middlemen is 240 lbs. per palla. In the same market, potatoes are purchased from the cultivators at 300 lbs. per palla and sold to the public at 280 lbs per palla.

(4) in most products there are, at present, two necessary middlemen in any market, if a cultivator is selling,—the adti or aratya who represents the seller and the dalal who negotiates on behalf of the buyer. The dalal will not deal direct with the seller. There seems little justification for the two middlemen. Thus potatoes in Poona, all the stuff must pass through the hands of five adtis.

(5) bargains are usually made secretly by manipulation under a cloth, and without the knowledge of the cultivator-seller. This is always brought up as a grievance by the people, but open marketing is usually resisted by the dalals.

How much weight must be attached to each of these criticisms must be different in each case and requires a series of special investigations. I have recently asked the Government of Bombay to allow me to organize two such inquiries, in the case of ground-nuts and gul (in the Deccan) in the coming year. We have now a Professor of Agricultural Economics (Rao Bahadur P. C. Patil) and the actual control of the work would be with him. It is intended that these should be the first of a number of such marketing studies.

In one case, namely that of mango marketing, we have recently had an investigation by a committee into the present organization and its report is attached, and gives a very vivid illustration of the difficulties in the way at present. It insists on the evils of the secret buying system above described, and on the careful grading of the produce, before being sent to market. The whole question of action under this report is now before Government.

*Welfare of Rural Population*

Under this head I propose to consider the problem of rural development as a whole, in what must be its ultimate object, namely the welfare,—the happiness and comfort of the rural population. At present in the Bombay Presidency the village population has rather been left in a backwater, but the time has come to look at the rural problems as a whole, and see to whether definite progress cannot be made not merely in the technical improvement of agriculture, or the organization of village finance, or in pushing village sanitation, or in increasing the number of villages with schools,—but in taking all these things as part of one end, the re-creation of the villages as self-conscious units, with the idea of progress and of improvement.

Such a conception, it will be said, requires a missionary —and so it does. It can hardly be created by a Government agency, though if such an agency can be developed there is no reason why Government should not aid it. And I would, hence, like to see an agency develop, more or less on the lines of a number of educational societies in the Bombay Presidency and elsewhere, which will provide the missionaries needed to carry out the idea. The general conception was suggested to me by the Hon. Sir Chunilal V. Mehta,—and seems a most productive one. It would work from the Taluka Development Association as a base, and would envisage the village problem as a whole. The link would be to study a typical village in a tract on thorough lines. My own village study of Jategaon Budruk in the Deccan would, perhaps, serve as a type. This would be done, as mine was, with the close and hearty co-operation of the people. Then the needs would be considered and the possibility and cost of satisfying them.

In most of our Bombay areas the needs are usually primarily economic. The land does not produce its maximum, even without greater current expenditure than is now made,—a too large proportion of the value is taken away by others than those who have had a share in producing it, and there is too much waste time. Once produce a hope of better economic position, and the desire for sanitation, for education, and for land improvement will follow. With the proper missionary, who has the resources and information of the agricultural, co-operative, sanitary and other Departments at his back,— the future is more than possible. To me this is no fancy sketch. I know villages where work could be taken in hand at once, *provided the men are available*, and here the scheme for a Society of Servants of Rural India suggested by Sir Chunilal V. Mehta seems to supply the missing link if only the men can be found who will accept the conditions and devote themselves to the work. I would strongly recommend any move in this direction, and be prepared to subsidise such missionaries,—without in any way limiting their independence.

Only by some such means can the all round progress be made which alone results in increase of happiness and welfare. It is useless to have technical improvement if all the extra produce is to be taken by unsuitable expenditure; it is useless to provide co-operative credit, if this means (as has often been the case in the dry tracts of the Deccan) that the extra credit is used to the full without care enough,— and is lost in the next year of scarcity.

My scheme would, therefore, be

1. A Society of 'Servants of Rural India' to inspire and supply men, and inspire the movement. It is hoped that the plan would attract large public funds, which with Government subsidies would in the first instance provide the men.

2. Local Development Associations, which would organise the area in which such men should be placed,—beside doing more general propaganda over the whole area to which they belong, and preparing the way for the intensive work in villages and groups.

Again I would insist on the consideration of the village problem as a whole,—and the attack on the present uneconomic, and despairing condition of the people of many of our villages, as perhaps the most serious problem in the country at present. Whether enthusiasm can be organised and whether men can be found who will devote themselves to such a purpose,—which can never be under the direct auspices of Government or even of semi-Government bodies, is a matter which can only be tested by trying. It is, however, I feel a matter to which Government resource might well be devoted, and to that extent the general creation of further extensive Government paid staff, whether for agricultural, co-operative, sanitary or other similar propaganda.

*Chapter* 20

# THE AGRICULTURE OF INDIA*

The cultivation and use of the land in so vast a country as India must necessarily show enormous diversity and it may at once be stated that the popular idea that a very large part of the sub-continent is like the land of the lotus eaters, where the necessities of life fall with little or no labour into the mouths of the indolent population is entirely false. Equally false, however, is the view which, largely on the basis of statistics, pictures India as a country largely desert, the remainder being badly cultivated and only needing the presence of the more knowing men of other lands to bring it to the agricultural level, say, of the countries of western Europe. As a matter of fact, India has areas, small though they be, where living is easy and where a livelihood, albeit a very poor one, can be obtained with the minimum of labour. On the other hand, there are desert areas, and large tracts where a meagre return only can be obtained for the expenditure of a very large amount of labour.

But two features are characteristic of the country from an agricultural point of view, almost from one end to the other. The first of these is the fact that all over (with the single exception of the planting industries of tea, coffee, rubber, etc.) agriculture is a peasant industry, conducted independently by small holders with their own hands, and

* Reprinted with permission from the **Annals of the American Academy of Political and Social Science**, Philadelphia, Vol. 145 (1929), pp. 72-81.

to the extent which their own labour, or little more, supplies for the exploitation of the land. In this it is, of course, similar to most of the agriculture of Eastern Europe and of China, but the predominance of such peasant holdings marks Indian rural life off from that of modern England or the United States of America, and gives it characteristics which must always be remembered in trying to picture India as an agricultural country.

*Rural Life*

The typical unit of cultivation in India is, therefore, small. Though 125,000,000 of the population in British India live (to use the terminology of the Census of 1921) on "the exploitation of animals and vegetation," or, in other words on agriculture, and though these form 73 per cent of the total population, yet they only actually crop 226 million acres (1924-25) or only 1.8 acres per head, or say seven acres per rural family. This at once, marks out the character of the work. It is cultivation on a very small scale, carried out with the simplest implements, with a minimum of machinery, capital, or hired labour. The results are those which are found everywhere under similar conditions, unless modified by the development of cooperation, that is to say, a self-reliant, self-contained rural population, intensely devoted to the land and unwilling to change it for industrial labour if this can possibly be avoided, but generally living near the limit of subsistence.

There are exceptions to these general statements. The great planting industries of tea, coffee, rubber and the like are the outstanding examples of the extension of large scale cultivation in India. These occupy nearly a million acres[1] under these crops in *British* India alone. These industries, which have been developed most largely in regions previously uncultivated, by foreign agency, are not purely agricultural in character. Each of them is associated with a somewhat complicated manufacturing procedure (more

[1] The figures for 1924-25 are: Tea, 715,836 acres; Coffee. 94,298 acres; Rubber, 80,807 acres.

marked in the case of tea and rubber than in the case of coffee), and each means, in addition, waiting several years for a return on capital. Each has, moreover, a market which it is difficult for the small producer to exploit. Apart from these planting industries, localised as they are almost entirely in the northeast and extreme south of the country, agriculture on an estate scale conducted with hired labour, has never developed in any part of British India, in spite of many efforts and the expenditure of much capital.

The other feature which is characteristic of rural life and of agriculture in India is the tenure of the land and the collection of the people in self-contained and largely self-governing villages. Hardly any of the land in the country is, in the full sense of the term, privately owned. The assumption, on the part of all governing authorities, at any rate, is that the land belongs to the State, but is placed in the hands of subsidiary owners who, though they have in most cases, power to sell it, yet they or any buyers hold it subject to the payment of a fixed assessment to the Government. In some cases, as in most parts of Bengal, this assessment is fixed for ever; in most of the other parts of India, it may be varied by the State at definite intervals, of twenty, thirty or forty years, according to considerations which are now standardised in each of the provinces of India.

While this alienable tenure of land, under the final ownership of the State, and subject to the payment of a fixed annual assessment, is practically universal in British India, yet the form it takes varies widely, but resolves itself into two essentially different types. In North India generally, including almost the whole of the great alluvial areas in the valleys of Brahmaputra, Ganges, and Indus, the land is held by landlords (termed *zemindars*) who then let it to the actual peasant cultivators. In the remainder of the country, embracing the Central Indian and Deccan plateaux, Burma, and South India generally, the ultimate owner of the land (that is to say, the Government) deals with the peasant cultivator himself (*raiyat*) who is responsible for the assessment, and, in the ideal, cultivates the land himself. In any case, the result as regards the agriculture is similar. Small holdings are the basis of cultivation, and are necessarily grouped into villages with a very strong bond of union between the people of the same village, either because the land belongs to the same *zemindar* or because such small holders have to provide for common services (carpenter, blacksmith, ropemaker, etc.) Thus there is much community in supplying the needs of the people, but each *raiyat* cultivates, in absolute or almost absolute indepen dence, the small amount of land which he holds, either as the tenant of a *zemindar*, or as a direct holder from the State. This *independence in cultivation* of each of the holders of land, however small the holding may be, is very deep rooted in the people, and though farming in co-operation *does* occur, it is relatively rare, and the pooling of holdings to make a decent-sized farm is equally little known.

We have, therefore, almost everywhere, a system of peasant agriculture, with all its disadvantages intensified by the smallness of the holdings, and by the fact that (even when the cultivators are tenants of one *zemindar*) they cultivate independently. This results in perennial shortness and extreme expensiveness of working capital, and of inability to use the capital so as to get the best result in production from the land. These disadvantages are least seen where the soil and climate, or irrigation arrangements, permit intensive cultivation, and the use of a large amount of hand labour, which, in certain agricultural conditions, obtains yields and results that no large-scale cultivation has yet obtained. The disadvantages are most in evidence in those dry and precarious regions where intense hand work yields no adequate return.

*Conditions of the Land*

Recognising that, with the exceptions noted, India is a country of peasant agriculture, conducted independently by very small holders with very limited working capital, we may now consider how the use of the land is affected by other conditions. First, as regards soil. In this matter one great distinction may at once be made. The great alluvial plains forming two thirds of the cultivable area of British India, and comprising the whole of the valleys of the Brahmaputra, Ganges, and Indus, and of the lower courses of all the rivers draining into the Bay of Bengal and the Arabian sea, represent as good agricultural soil as occurs anywhere in the world. Indefinitely deep, in many places without a stone, originally consisting of forest jungle and grass plains, they are capable, with adequate water supply and good cultivation, of producing crops (and in some cases do actually produce crops) as good as any found in the richest areas of the United States. The rest of the country, comprising almost the whole of peninsular India and extending northward in Central India nearly to the Ganges and Jumna rivers, consists of rocky high lands, sometimes hills and valleys, sometimes high irregular plateaux, always poor as a whole, with patches or strips of rich land which have the benefit of erosion from the higher lands. Some of the soils of these uplands have a high reputation, like the black cotton soil or *regur* found on the Malwa plateau in Central India, or the Deccan plateau further south; but the really good land is always patchy, and, taken generally, peasant agriculture in these areas is usually hard and yields small returns.

*Irrigation*

Much of the country included both in the alluvial plains, and in the rocky high lands and plateaux has a small and precarious rainfall. An uncertain rainfall is, of course, far more dangerous from an agricultural point of view, than a small rainfall; for the areas of very small rainfall, like Sind and parts of the Rajputana desert and of the Punjab, remain uncultivated and often without landholders, until irrigation is introduced, and when water becomes available the country blossoms as the rose. But in the precarious tracts, the position is far more uncertain. In a majority of years a reasonable crop can be obtained. In the remainder, the crop is small and incapable of supporting the peasant holder, and, in extreme cases, no crop at all is obtained. Under these conditions irrigation has seemed the only remedy, and though there has been some experiment in the direction of retaining water in the soil by improved methods of cultivation (the so-called "dry farming"), the extension of this method is a matter for the future.

But irrigation has been developed in the desert and in the precarious tracts of India on a colossal scale in recent years, particularly in the areas included in the great alluvial areas of North India. The construction of irrigation works, either by private individuals or by the State is, however, no new thing in India. All parts of the country where rain is not sufficient or sufficiently well distributed, or too precarious, and where water is available at a reasonable cost, abound in sources from which the land can be watered which are still used and are still being extended. Wells are employed in almost all parts, from the Ganges valley where they can be constructed for a few rupees, to many of the Deccan areas where a well capable of irrigating only five or six acres will cost a thousand rupees. Further, Madras and the South generally is the home of the irrigation tank or lake, made by blocking up the mouth of the small valleys so common in that region. The larger rivers had, even in pre-British times, already been utilised for inundation canals which flowed and fertilised the land during the flood season, but it has been reserved for recent years to undertake works on a scale not hitherto thought of, and so to extend enormously the area which has been made no longer solely dependent on the most precarious rainfall.

In some parts of India, this extension of irrigation has created an agricultural revolution. These are chiefly in the areas of the great rivers of the north, particularly in those watered by the Indus and its feeders, and to a less extent by the Ganges and Jumna. In what has been here termed the rocky peninsular areas, the opportunity is less, and only a small proportion of the cropped area can ever be irrigated, owing partly to lack of water and partly to the conformation of the country. But a general idea of the importance of irrigation to India as a whole can be obtained from the following statement showing the proportion of the area in British India sown with crops, which is irrigated.

NORTH INDIA AREAS, WITH BURMA

| | Per Cent |
|---|---|
| Bengal .. .. .. .. .. | 3.5 |
| Assam .. .. .. .. .. | 7.5 |
| Burma .. .. .. .. .. | 8.5 |
| Bihar and Orissa .. .. .. | 17 |
| United Provinces .. .. .. | 19.5 |
| Punjab .. .. .. .. .. | 40.5 |
| North West Frontier .. .. | 36 |
| Sind .. .. .. .. .. | 73 |

PENINSULA AREAS

| | Per Cent |
|---|---|
| Bombay .. .. .. .. .. | 4 |
| Central Provinces .. .. .. | 4 |
| Hyderabad State .. .. .. | 5 |
| Madras .. .. .. .. .. | 24 |

Excluding Madras, where the figure given is very largely increased by the inclusion of the irrigation in the lower valleys of the great Deccan rivers near the East Coast, it will be seen that the area irrigated compared to the area sown, is decidedly small—almost negligible in fact—in the Peninsula. In the North Indian provinces, however, when irrigation is needed at all, the proportion may reach a very high figure, as in the Punjab and Sind. In the Punjab, in fact, the recent development of irrigation (where it has now reached over 13,000,000 acres) has converted a poverty-stricken province into almost the richest region in the country.

*Primary Production*

The primary production of the agricultural land of India is food for its 300,000,000 of people, and it will not therefore be surprising that out of a total of 225,000,000 acres annually under crop in British India, no less than 200,000,000 are, at least partially, under various food grains, out of which by far the largest area is occupied by rice. Except for rice and wheat, little of all this production of grain is exported. Even of the two grains mentioned, under 8 per cent of the rice production leaves the country, and about 14 per cent of that of wheat, though this latter is one of the chief export crops and the amount sent from the country in 1924-25 reached a total of 1,100,000 tons.

These food crops, grown by India, thus, almost entirely for the feeding of its own people, on eight-ninths of the area actually cultivated, consists chiefly of eight grains. *Rice* heads the list, and occupies nearly 80.000,000 acres, being grown wherever there is a rainfall sufficient, (say over 40 inches per annum) or good irrigation, and where the soil is good enough. It is, *par excellence*, the crop of flat lands, and though in the regions suitable for them, varieties are known which flourish at almost any time of the year, the great crop is that grown during the rainy season, ripening in standing or flowing water, and grown, over most of the area, by transplanting the seedlings laboriously by hand into the fields where they finally grow. Rice is the typical crop of the small cultivators in the wetter parts of India in the rainy (*kharif*) season.

Next to rice stands wheat, grown entirely as a winter (*rabi*) crop and almost entirely in north and Central India, where 90 per cent of the area lies. Over one third of the land under wheat is irrigated. The average yield per acre compared with other countries, is small, being 17 bushels per acre under irrigation, and 11 bushels per acre without irrigation, in the greatest wheat growing province (Punjab).

After wheat come the group of millets, which make up together nearly 40,000,000 acres, almost exclusively used as a food by the people in India. These are chiefly sorghum (*jowar*) which is by far the most important, extending over 22,500,000 acres; pearl millet (*bajra*) with 12,000,000 acres, and *ragi* (*Eleusine corocana*) a small grain hardly grown elsewhere on a large scale, but occupying 4,000,000 acres in India. The two last are grown in the rainy season, on poor lands, while sorghum, the typical crop of heavier land in the drier tracts without irrigation, is cultivated both in the rains and in the winter season.

Maize is relatively a far less important crop than would have been expected, though it occupies nearly 5,500,000 acres. Most of this, however, occurs in the drier alluvial tracts of the Punjab, the North West Frontier and the United Provinces. Very little is grown in Peninsular India. Barley is likewise only important in the north where it is grown as a winter crop, but in spite of this restriction in area, it occupies nearly 7,000,000 acres.

The importance of Indian crops in world agriculture does not, however, depend on these food grains, but rather on those others which are classed by Indian cultivators as money crops. These are grown for sale, and in many cases for export. These include cotton, which occupies by far the largest area; jute and similar fibres, of which India has almost a world's monopoly; oil-seeds of various kinds, of great importance in the export trade; dyes like indigo whose cultivation is, in most cases, now dying out; sugar cane whose product is entirely absorbed in the country itself; tobacco, another large scale cultivation done now chiefly for Indian consumption: and the planting crops, tea, coffee, and rubber, previously referred to.

The cotton crop in India is only second in quantity to that produced in the United States of America, and furnished over 6,000,000 bales of cotton to the world's supply. It is a vitally important crop to the Indian cultivators through almost all the drier parts of the country with an annual rainfall of less than 40 inches, from almost the extreme north of the country to Cape Comorin in the South. There are over 17,000,000 acres under this crop in British India alone, out of which 3,000,000 acres are irrigated. This only very inadequately represents the importance of the crop, for at least 6,000,000 additional acres of cotton are grown in the various Indian States. The average yield per acre is low, being only 104 pounds, though there has been a steady increase in recent years. The characteristic Indian cotton is short in staple and rough, suited for coarse spinning and weaving, though there are now types grown in Bombay, the Punjab, and Madras which are of high quality. Over by far the greater part of the cotton area, especially in the black cotton soil regions of Central and Peninsular India, where the concentration of cotton-growing is the greatest, the crop is grown in rotation with sorghum (*jowar*), the latter giving food to the peasant population and their animals, and the former furnishing the money for their livelihood.

In its own area, the jute crop is equally important, but this is almost entirely in the northeast of the country, with a rainfall of over 50 inches per annum. The total land under the crop reaches nearly 3,000,000 acres in British India, of which 86 per cent is in Bengal. The crop rotates with rice, and is grown between April and October, being planted before the rainy season so as to get well started before the land is flooded by the heavy rains. The conditions required for successful jute cultivation are fairly narrow, and the retting of the fibre can only be done in a country of great rivers and abundant water. The lower Ganges and Brahmaputra valleys present these conditions and hence almost monopolise the production of *jute* in the world. Jute substitutes, like sann-hemp (*Crotalaria juncea*) or Deccan hemp (*Hibiscus cannabinus*) are fare more widely grown, to the extent of three quarters of a million acres, but the true hemp (*Cannabis indica*), though a common plant in many parts of the country, is never grown for fibre, but is cultivated to a limited extent for the intoxicating drug known as *hashish* or *ganja*.

If Indian agriculture is responsible for the world's supply of jute, and for a substantial portion of that of cotton, it furnishes also one of the most important sources of oil seeds and their products, including linseed, sesamum, castor, cocoanut, and, in recent years, groundnut, besides many others which are less well known. Together the area under these oilseed crops is over 15,000,000 of acres in British India, while a very large acreage is also found in Indian States.

The international importance of these cultivations is suggested by the fact that the export of oilseeds and oil-cakes was valued at nearly 24,000,000 pounds sterling in 1925-26. Naturally a variety of crops like these demand very different conditions of growth, but almost every part of India furnishes one or other of them. Linseed, for instance, which occupies over 2,500,000 acres, is concentrated in the Central Provinces, and the country to the north between them and Himalayas, and the estimated yield in British India only is between five and six hundred thousand tons. Sesamum, a still more typical Indian oilseed crop, occupies 3,500,000 acres of area. It is very widely grown, but is most concentrated in a belt through the central parts of the country from the northern parts of Madras, through the Central Provinces, to Rajputana—and in Burma.

Rape and mustard, on the other hand, form essentially the winter crop of Northern India. These seeds occupy an area of nearly 4,000,000 acres, and yield over 1,000,000 tons of seed per annum. The progress of the ground-nut (pea-nut) crop in recent years is most remarkable, and it now occupies nearly 3,000,000 acres and is calculated to yield about 1,500,000 tons per annum. The cultivation is rapidly developing in all the drier regions of the country, especially in the plateau soils of the peninsula and in Central India, as well as in central Burma.

Finally, castor may be mentioned, though the centre of its cultivation lies in the Hyderabad State. But its importance is likely to increase, and it occupies chiefly high, dry lands in the Indian peninsula, fit for little else. At present there are about 1,250,000 acres under the crop, of which three-fifths are in Hyderabad, and half the remainder in Madras.

*Other Characteristics*

It is only necessary, in a short sketch like the present, to refer to two or three other characteristic Indian agri cultural crops and products, for these are less numerous than they used to be. Dyes, especially indigo, formerly so important, have now sunk to insignificance, owing to their replacement by synthetic products. The cultivation of opium, formerly so important in the Ganges Valley and in Central India, is now very severely restricted, being only grown under the strictest Government supervision. Silk is a declining production, and, it seems, is fated to fall even below its present amount. In the case of lac, for the production of shellac, where India is the main producer for the world, the position is different, and despite the loss of the market for lac dye, the cultivation of the lac insect on local trees and cultivated plants is a matter of great importance to the peoples of several of the wilder areas of the country. The annual value of the Indian lac export is not less than 5,000,000 pounds sterling (1925-26).

Sugar cane cultivation in India is very widespread, but except in a few parts of North India has never become the basis of a sugar industry on modern lines, while the import of sugar is one of the biggest items in the external trade. The crop occupies 2,500,000 acres and, on the average, only yields under one and a half tons of crude sugar (*gur*) per acre. There is no crop with greater possibilities of development, and the yield per acre in northern India is certain to advance rapidly in the future as a result of the discovery and breeding of higher yielding canes.

Tobacco is another rapidly developing crop, and now occupies over 1,000,000 acres. Indian tobacco has usually been classed as inferior, strong and coarse; but its cultivation is nevertheless very widespread, reaching its highest intensity in northern Bengal, on the east coast, and in parts of Burma. It is essentially a crop demanding high class work on rich soil, and here, as in other countries, to be a successful grower of tobacco marks a man as a cultivator of the first rank.

*Animal Husbandry*

This partial enumeration of the chief crops of India does perhaps little to carry an idea of the agriculture of the country, and yet it *should* do so to a much greater extent than in most other countries of the world. For the Indian farmers are essentially producers of *crops*. Animal husbandry and the preparation of animals and their products for market take a far less important position in India than almost anywhere else. For a very large part of the Indian population consists of vegetarians, and, for the rest, the consumption of meat is very small per head. Hence the demand for animals for meat is very limited. Further, in India, the horse is not used for agricultural purposes, its place being taken almost entirely by the bullock. Therefore, the production of horses is very restricted: the production of cattle and even of sheep for meat is extremely small: and animal husbandry concentrates on supplying cattle for work purposes and for milk, and on sheep and goats for wool (or hair), skins, or other products—and only to a very minor extent for meat.

Nevertheless there are large areas which are almost purely pastoral, and a large part of the cattle and other animals needed in the life of the country are produced in a few thinly peopled areas where the growing of crops is almost impossible, but where the soil and rainfall ensure ample supplies of grass. In such tracts—like Rajputana, parts of the Punjab and Sind, Kathiawar and North Gujarat, or the high lands of Central India, the best breeds of working cattle are raised by semi-nomadic pastoral people, and sent to the more densely peopled areas. The milk supply is very largely obtained from the water-buffalo, which though nearly useless as a work animal, yet gives a fairly large amount of rich milk on a diet too coarse for ordinary cows.

The keeping of sheep and goats as a business (though many goats kept for milk will be found in almost any Indian village) is also very largely in the hands of semi-nomadic breeders and shepherds, though they are found in very many areas where cattle breeding under similar conditions is not possible. The sheep are poor, both in size and in production of wool, and, as a matter of fact, a *large* sheep is not desirable where meat is not a primary object. Many attempts to improve the wool have however been made by cross breeding, with a good deal of success in the Punjab and north India generally, and with little success anywhere else. The importance of the wool production can be perhaps, judged by the fact that, after supplying local needs, there is a net annual export of raw wool from India of 3,000,000 pounds sterling, besides an export of woolen goods of 1,000,000 pounds sterling (1924-25).

*Improvements*

The picture of Indian agriculture which has been given is, therefore, that of a vast mass of small holdings, worked as a rule on a family basis with a minimum of hired labour, and also with a minimum of capital. The farmer grows such crops as the land is suited for, chiefly for the feeding of himself and family and his working animals (the latter being invariably bullocks), but also growing a proportion of money crops which enables Government assessment to be paid and such necessities as are required to be purchased from outside.

Perhaps the item in this summary which affects the agriculture most is the fact that the capital possessed by the farmer is usually the minimum possible, apart from his land, and this absence of capital determines to an almost inconceivable extent the way in which agriculture is carried on. For the absence of capital—especially in the absence of a well-developed agricultural banking system, and still more when matters are complicated by an uncertain and precarious climate such as occurs in a very large part of India—means comparatively crude methods of work, primitive implements, hesitation to utilise fertilisers, whose purchase demands out-of-pocket investment, and generally, the impossibility of making experiments, or of making very radical changes in agricultural methods. As a result, the cultivators of the land in India are *supposed* to be extremely conservative, to refuse change when the advantage of change is clear, to refuse to take advantage of methods which have proved their value in other parts of the world.

After long experience of Indian farmers in many parts of India, I think that this idea of innate conservatism among the rural classes is not correct, and possibly they are really less averse to change than a very large proportion of the farmers of western countries. I have seen, again and again, within twenty years an old but less efficient implement replaced almost entirely, over large regions, by one more efficient, of an improved type of seed replace that in use for a hundred years, or the employment of artificial manure become general. And it would really seem to be true that readiness to adopt new methods is the characteristic of the Indian cultivators, provided they are *proved,* to their own satisfaction, to be of advantage, and provided they give a return which will warrant the borrowing of capital at high interest. To put it another way, economy of capital or out-of-pocket cost is more important than economy of running expenditure, where the labour is a man's own and has to be provided with food and maintenance in any case.

Hence, throughout India, implements will be found to be crude, rough, less economical in working than those which might easily replace them and which are available at a higher capital cost. The manures and fertilisers used are generally those which can be obtained locally, and with little or no cost except for labour. The most welcome improvements are those which, like an improved type of seed, will provide increased returns with little capital outlay, even if much more work is required in raising the crop. If the result is good and the returns are increased, then other improvements immediately become acceptable.

These considerations must be in the mind of anyone who studies the gradual but relatively rapid advance in the technique of Indian agriculture in recent years. After all that has been done, the outside observer will probably consider farming in India as being in a very primitive state of development, but I doubt whether it is so. Certain it is that the attempts at the wholesale adoption of western methods in Indian agriculture have usually failed, and I know no sadder sight than the museums one sometimes sees of large numbers of implements, appparently suited to the conditions, imported by some enthusiast into a particular area for use by the cultivators of the country—but which have never come into use.

And yet, as has already been stated, there has been a great advance in recent years in the agriculture of the country, as judged by the returns which the land can be made to give. Many of these advances have been made as the result of the work of the various agricultural departments in India, which, though founded many years before, were placed on a substantial basis in about 1905. These have very largely expanded since that time, and in most of the Indian provinces as well as in some of the Indian States very full advantage has been taken of the discoveries made and the experiments carried on.

These recent improvements have chiefly taken two or three lines. Of these the first is the production and extension of types of crops giving a bigger return than those actually in cultivation, and these have been taken up with enthusiasm by the smallholding cultivators of the country Improved cottons, sometimes higher yielding types, sometimes better quality types, are now in cultivation over millions of acres; higher yielding kinds of jute have already spread to 10 per cent of the total area under that crop. In the production of food crops like wheat and rice, particularly the former, very great success has been attained in getting higher yield per acre and better quality of grain and greater suitability of the plant to the conditions. Sugaracane varieties have been evolved suited for north India, which give vastly greater yields than were formerly possible in that part of the country. Tobacco strains have been isolated and are now in widespread use, which have enabled a much higher grade of product to be obtained.

Implements are rapidly being modified to suit the conditions. I remember the time when the iron ploughs in use in the Deccan could be counted on the fingers; now they are there by the hundred thousand. And where a cheap implement is found which will do better work under any particular conditions, then there is rapid adoption. But the problems of cultivation under the special conditions of Indian agriculture are only beginning to be really attacked by the agricultural departments.

The use of artificial fertilisers is spreading, particularly in irrigated lands. Indian soils are usually, though not universally, rich, except in nitrogen, and hence it is the use of sulphate of ammonia which has developed more than that of any other fertiliser. But others are coming, and, in certain cases, the proved value of green manuring has modified profoundly the methods and results of growing a crop.

These are some indications of the lines in which progress, in such an old agricultural country as India, is even now being made. I see little limit to what can be done, provided always the fact that the essential unit of agriculture in India is a small holder, with little or no capital beyond his land and bullocks, is kept in view. A rise of 50 per cent in the production of the land in India in the course of the next generation is not, in any sense, a Utopian ideal.

*Chapter* 21

# NOTE ON THE PLAN OF ECONOMIC DEVELOPMENT FOR INDIA*

When I first saw a summary of the plan devised by Sir Purshotamdas and others I felt the dawn of a new hope. Here for the first time in my experience were a group of men who saw that the rapid raising of the standard of life of the people of India was possible, that, in order to accomplish it, it was necessary to make a very large investment of capital, and that the scheme to accomplish it was worthy of being placed on the first priority list of matters to be dealt with by the national government of India when this is formed. These conceptions themselves form a great advance.

And I firmly believe that the vision represents something which can be made a reality. Too long have we been told that India is a poor and over populated country, that its resources are very limited, and that it is inevitable that its standard of life should remain very low. I have long held that India is potentially a much richer country than we have ever been allowed to think, and that if it were possible to make use of these resources there would be ample to raise the standard of life of the mass of the people to a very great extent. But there has always been something else that was considered by the authorities as more immediately important and the diversion of national

* In the Spring of 1944 Dr. Mann was asked by **The Hindu** of Madras to give his view of the recently formulated "Bombay Plan." In reply, he sent them this note.

funds into channels which would serve primarily this purpose has been on a meagre scale compared with the way in which money has been lavished in other directions.

Of course, in a scheme like this, nobody could be expected to agree with all the details of the plan. But seeing it is put forward as a basis for discussion only, I can welcome it with enthusiasm and leave criticism of details till later. But there are some propositions which, if accepted, would justify most of the proposals made in the scheme, and in which I have long believed. Some of these are as follows:—

1. India is large enough and productive enough to be almost self-contained if its resources were well utilised, especially if the economic unit included Burma and Malaya.

2. I doubt whether any country can give its people a proper standard of life if it is fully populated and yet 70 percent of its people are engaged in agriculture.

3. All the same, a very large increase in agricultural production is possible. The authors look for an early doubling of the produce of the land: I think this reasonable if irrigation is developed to its full extent and if the fragmented agriculture can be replaced by cooperative agriculture. I am a little alarmed at the proposal (page 31) that the latter (cooperative farming) should be established compulsorily, because I feel (from my experience with the cultivators of the Bombay Deccan) that it will be unnecessary.

4. The standard of life which is considered as a minimum in the proposals under discussion as a target seems reasonable though it is considerably higher than I found was expected or desired by the village people in the Bombay Deccan.

If the above propositions are correct, it means that in some way or other, the energies of a large number of the people must be diverted from agriculture to industry. I should prefer to see this done, as far as possible, by the development of cottage industry which would not mean the transfer of the population to great centres such as

Bombay or Calcutta. But somehow or other, a much larger part of the population must have industry as their main occupation if the country is to have the general standard of life raised to the extent mentioned in the scheme.

Of course, there are a good many questions which one would like to put to the authors of the scheme. The first, and perhaps the most important from my point of view is how to guarantee that the increase of production which would come about would not be absorbed by middlemen and financiers. I have always hesitated to urge a certain increase of production on villagers because I foresaw that the whole of the increased production would be taken by a money-lender or his equivalent. It is not, in fact, sufficient to increase production: you must also arrange that this increase should go to raise the standard of life of the worker. How is this to be done? I notice that Mr. Birla has partially dealt with this question in his speech of March 4th, 1944, and I am rejoiced to see that the matter has not been forgotten.

Let no one think, however, that the revolution which would be produced if the scheme were carried out, can be brought about without a good deal of self-denial by the people of the country. I was in Russia a number of times while the several five year plans were being carried through. And there is no doubt that, especially in the early days of the first five year plan, the people had to tighten their belts pretty much to carry it through. I do not see why so much self-denial should be necessary in India, for India starts in a far better financial position than Russia did and foreign capital is available to a far greater extent.

If the increased production which is anticipated—and, as regards agriculture, I believe, rightly anticipated—is obtained, I have no fear of the capital outlay which is suggested, namely the sum of Rs. 10,000 crores. This means an outlay of about Rs. 2500 per head of population, and this can easily be liquidated if the result, in increase of production, is anything like what we have a right to expect. There would seem no necessity for the raising and utilisation of this money to cause large inflation. In fact, inflation on a colossal scale has already taken place since the preparation of the scheme and entirely independent of it. With proper management there does not seem any need for the inflation which some people fear so seriously to take place.

But if a scheme like that under discussion is to be carried out, I think it is only fair to say that the money required must be raised by the community as a whole and not by private capitalists. That means that it cannot be carried out until there is a national government and until this is far more broadly based on the wishes of the people than any of the constitutional proposals that have been passed into law hitherto. I think too that in any authority that carries out the scheme the agricultural interest (not the landlords but the actual cultivating classes) must be very decisively represented—far more, indeed, than has ever been the case in any local or central governing body in the country.

*Chapter 22*

# VILLAGE BETTERMENT IN THE NEW INDIA AND PAKISTAN

The improvement of Indian villages and of the life which is lived in them is the improvement of India. This is as true now as it has ever been in the past, and the present moment is one when a consideration of what has been and is being done for the betterment of rural conditions is very much to the point. We can, in this country, now take a more objective attitude to what occurs in any part of the sub-continent than has been possible in the past, owing to the change in responsibility for events in that country since last August. But though we have less responsibility, there are many of us who are as interested in the progress of the village people as ever we were. Some of us have worked for many years among the rural population, and both admire and love them, and hence we follow all the efforts at the improvement of their lot with the greatest keenness and keep to a very large extent in touch with those who are now working to this end.

Until comparatively lately the information about the condition of villagers and of village life has been extremely meagre. Much information existed in settlement reports and similar literature, but a large part of this was never published, and most of what was published was rarely if ever read. In the middle of the last century Hunter wrote

* Reprinted with permission from the **Asiatic Review**, London, Vol. XLIV (1948), No. 158, pp. 154-170.

an account, chiefly historical, of the rural conditions in Bengal, and in the early years of the present century Jacks wrote on the economic life of a Bengal district in a book which inaugurated the modern study of village conditions. But the general public, both here and in India, was inclined to believe that the increasing Government revenue and income-tax returns, and similar indications represented increasing prosperity among the rural classes as well as among the commercial and administrative sections of the population.

*Below the Poverty Line*

The facts, as time and as many investigations have now shown, are far otherwise. In my own studies of Bombay villages thirty years ago, I found, rather to my surprise and certainly to the surprise of many others, that about 40 per cent. of the people of the Deccan villages where I worked were living below the poverty line even according to their own low standard of life and feeding. In the many studies done in other parts of the country since that time the situation (if we except some of the regions which have come under new schemes of irrigation) is very similar, whether they are in North India, Bengal, the coast districts of Bombay or the rice-growing areas of Madras. Even in some of the districts which have come under irrigation, the material prosperity has not been accompanied by the real betterment of the people, and has, in some cases, been bought by the disappearance of the previous peasant population and by the impoverishment of the surrounding dry areas.

The results of recent censuses tell the same story. The increase of industries in the larger centres of population would, one might have thought, have relieved the situation in the over-populated countryside. But this does not appear to have been the case. The percentage of rural as opposed to urban workers has hardly changed from 1881 to 1941, though the total population has increased by over 55 per cent. With the disappearance in large measure of the village artisan classes as well as the village industries, the people have on the whole not gone, as has been the case in Western countries, into the towns, but have settled on the land as peasant cultivators either on their own land or on land rented from others. The result has been that about half the members of the artisan and other non-agricultural castes have deserted their traditional occupations and become second-rate cultivators. Thus it has come about that rural India, and Pakistan as well, has, except in certain favoured regions, tended to be an agricultural slum, and the dependence of the people on agriculture has shown a steady increase from 61 per cent. in 1891 to 73 per cent. in 1931. I have not the figures from the 1941 census, but I am almost certain that they would show a similar trend.

*Conditions Still Unsatisfactory*

There have been many attempts to change all this and to increase the material prosperity and hence, it was supposed, the happiness of the rural population. The huge development of irrigation, particularly in North India, has as its principal end the prosperity of the country and, in the first instance, of the village population. Everywhere canal irrigation has gone, the material standard of the people has risen very greatly, but not always has the happiness and health increased to a corresponding extent. The introduction of co-operative societies for the supply of credit to the agricultural population was thought to herald a new day (at least by those who felt that the chief hindrance to comfort and material independence was the domination of the moneylender), when there was at least a chance for a larger part of the produce of labour to be retained in the hands of the peasants themselves. Now, forty years after the first co-operative Act was passed, the history of the movement is one of very partial success, and on the whole, it has not been able to attract the confidence of the rural population. Brilliant successes there have been, and that in almost every Province, but there has been little sign of the growth of a co-operative commonwealth, such as some of us looked forward to in the early days. I could go on citing scheme after scheme which were introduced as the cure for the misery and squalor of so many Indian villages, were worked by a limited number of people with enthusiasm, but have failed to move the apathy of the peasants, and then were dropped quietly in the background.

So that, at the time the British people have retired from responsibility for the rule of India and Pakistan, we have to confess that the conditions among the rural population in the greater part of the country are unsatisfactory, and I, for one, am inclined to be very depressed at the position, and sometimes I feel that a good many years of work among the villages of India and also of Pakistan have shown far less result than we hoped for and very little of which we can be very proud.

*Necessary Changes*

And yet nothing is more clear than the fact that a large part of the conditions which lead to this depression can be altered, even without great Government reform schemes. I do not think that all that is necessary can be done without such schemes however. There must be large changes in land tenures, in the law relating to debt, in the control or lack of control of moneylenders, in the subdivision and fragmentation of land, in the provision of money for improvements under better terms than have ever been available, and the like. But for the moment I want to consider the experience of many of us who have attempted to take a part in improving things even with Government and public control as they are, and see where the great difficulty lies even in those areas where it is clear that peasant health, wealth and comfort could be increased with comparative ease, resulting in a general raising of the living conditions of the people.

For this purpose I want to use not only my own experience, for that is rather remote, though I do not think that the outlook has changed very much since I lived among the villagers of the Bombay Deccan and of Sind, but also notes made by a large number of others who have devoted much time and energy to the service of rural advancement in almost every part of the country. I may mention some of these, though they are only people of whose work I have had pretty close knowledge. The first is Mr. Brayne, whose work in the Gurgaon district of the East Punjab is unique, and who, I know, retains his faith in the possibility of the villages becoming the homes of sweetness and light even without very radical changes in legislation. I was induced to give this lecture on account of papers which my friend Mr. S. G. Patil, of Indore, sent to me telling of his experiences in trying to follow Mr. Brayne's lead among the villages of his State. Then, all over the country there have been missionaries, co-operative workers, members of the agricultural departments, and hundreds of others who have in recent years made the improvement of villages their main enthusiasm. Towering over them all is that leader whom we have just lost, Mr. Gandhi, who drew public attention in a more powerful manner than anyone else had been able to do, and who lived among the villagers and looked upon their rehabilitation as perhaps the greatest of his ambitions.

In spite, however, of the work of all these and of many others, the results are meagre. Sometimes I am reminded of the proverbial result which one of our doggerel poets has enshrined in verse when St. Anthony preached to the fishes:

> The sermon once ended, the good man descended,
> The eels went on eeling, the crabs went on stealing,
> Much delighted were they, but they went their own way!

*Reasons For Slow Amelioration*

Why is it that so much effort has achieved such comparatively small results? I think that a frank study of the question is the only way to achieve what so many of us have at heart. Let us suggest several reasons now for the comparative failure.

First, I feel that it is partly because it has not always been realized that Indian villagers *are* peasants, and that peasants are the most immovable of all types of population. For to a peasant everywhere and it is essential in the peasant mentality—life on the land is a means of livelihood and not a business, and that accordingly the continuance of what has been done by his fathers is the most natural thing to do. Nay, it is almost a sin to change the method of doing anything. Now India is a country of peasants, with this type of thought very deeply ingrained. It is not limited to India, and I have met with exactly the same spirit in my travels and work in the Middle East—in Turkey, in Iran, and in Iraq. It would take generations to convert an Indian villager in most cases into a real town dweller who worked, say, in a mill. I have talked to many a peasant whose necessities have driven him to go to Bombay for a few months of work: in practically all cases they considered themselves as exiles and pined for their village in some remote corner of what might be a district in which scarcity perennially prevailed. This is the reason why there is really no drift to the cities as the figures I have given already show.

The second point on which I would like to insist is that except in certain areas and among certain classes and castes, the idea of making a profit, expressed in money out of their activities is secondary: the first is to make a living. Among the Maratha *kunbis* whom I know best, for instance, it is, of course, an excellent thing if there is a bumper crop, which can be sold at a good price, thus providing money which can be spent in jewellery for his wife and children, or in buying an extra pair of bullocks, but it will make very little difference in the style of living or in the food which he will use. To take another instance. When I was in Sind I had some extra land which I did not want to use, and made it known that I was prepared to let this out for a year till I was ready to use it for my experimental work. It was well supplied with water, and I had several applications almost at once. But it was at a time when we were all for increasing the acreage of cotton, and I made it a condition that one-third of the land should be put under cotton. To this all applicants demurred. "We will only put the remainder of the land under cotton when we have provided for our own food throughout the year," they said. They agreed that there was more profit to be made out of cotton, but that was a detail, and I had to let the land on their terms. The temptation of profit, in fact, goes a very little way among the peasants of the greater part of India, and if that is the only inducement to change of method of life, you will not get very far with them.

*"Pathetic Content"*

The third point, against which all workers among the peasants of India have come, is the extreme apathy which is found everywhere. I remember reading the very interesting little book of Mr. Brayne's *Socrates in an Indian Village*, and was amused at the very great amount of persuasion which Socrates had to use to get the people to take up even the simplest improvement, and the way in which this would tend to be neglected unless Socrates was continually pushing from behind. This apathy is perhaps justified in many cases, because in so many cases they have been victimized by unscrupulous outsiders, and they are apt to consider that any recommendation has a snag in it that they cannot see. "Better the devil we know than the devil we don't," they feel. The suspicion is very largely justified, and will never, I suppose, be met till they feel themselves on an intellectual level with those who approach them, or unless something has given them special confidence in the person with whom they are dealing. Socrates could get things done, because they had got confidence in him, and such confidence is a matter of slow growth.

The apathy is also caused by the feeling that any gain they may get, so far as money is concerned, will not come to *them*, but will be used either to pay off old debts to their moneylender or may be used as a justification for raising the rent of the land they occupy. As regards the former reason, there is good justification for the fear, for in many of the areas where famine may come at any time, the only security for the money lender, in keeping the peasants alive during a period of scarcity, is the idea of recompensing himself in a good season or when prosperity arrives, whatever be its cause. As regards the possible increase of rent as a result of village prosperity, this hardly applies to zemindari regions except so far as sub-tenants are concerned, but is ever before the eyes of cultivators in the ryotwari areas where the tendency in recent years has been for a larger and larger proportion of the people to be sub-tenants

Whatever the cause, all workers in Indian villages are agreed that the apathy is there, and none of the methods of overcoming it used in other countries have hitherto availed to break the vicious circle of listlessness, low production and squalor, leading to further listlessness, low production and squalor. No doubt there have been many individual cases where all this has been overcome, and I could quote many of them. Sometimes it has been by some form of cooperative effort which has appealed to some of the leading members of a village community, sometimes it has been the presence of some outsider in whom the village has confidence, sometimes by the return to a village of a member who has seen better things elsewhere, or sometimes the presence of an energetic and self-sacrificing village teacher or the like. In spite of the apparent apathy, and also the other hindrances, I am confident that the villages of India which I know are waiting for something which will sweep through the country and bring about the "will to improve," which is now and has been in the past so sadly lacking.

*Grounds for Hope*

Let us for a moment or two consider the advantages which Indian villages have, just as we have considered the difficulties in the way. The first thing is that nearly all the areas are, in a normal year, far less productive than they need be. I am not referring to famine areas in famine years, though there are one or two years at least of scarcity in every decade in many parts of India. But, apart from these, it is true that if the resources of manure and land were fully used, the production could be increased at least fifty per cent., without material change in the conditions otherwise. Now such an increase of production is the very first step towards a better life, and is vital to the future of the country apart from its effect on the individual villages. This reserve of possible production is a great asset if better living villages are to be obtained.

Next, there is now a large reserve of labour, unoccupied through a large part of the year, in the villages. This reserve of labour is increasing, and though most thinkers on India's future are greatly concerned at this excess of population, I am convinced that for the next few years it may be of great advantage in the rehabilitation of the countryside. If this reserve of labour can be harnessed in its spare time to the work of cleaning up the village, collecting the available manure, planting up the now waste ground, carrying out anti-erosion work such as bunding, digging wells where this is possible and the like matters, the chance of an early improvement of the crop yield, of stoppage of the waste due to erosion and similar causes, and the consequent bettering of the health of the people would be greatly increased. I have myself seen something of what can be done in one or two of the very doubtful areas of the Bombay Deccan, and there are no doubt many who have seen even more striking successes in other areas.

*Government Help*

There is also the very great advantage that a large number of Government departments are anxious to help. The agricultural departments, if they have done nothing else, have been able to supply seed of the main crops, which will give larger yields than those normally in use, have been able to give cheap methods of cutting losses due to diseases of crops such as smut among almost all the corn crops, and have been able to supply bulls which will in a very few years, if properly used, increase the capacity of the cattle to cultivate the land. Other departments can also help, and I am assured that under the new regimes will be even more ready to help than they have been in the past. I know that contact with Government departments has been dreaded by villagers generally in the past owing to a tendency to corruption and to demanding maintenance from the villagers when visiting any place. I sympathize with that dread, for I know that the charges have been true. But the more the villagers take matters into their own hands the less will be the danger in this direction, and the more the people used by the departments are really village-minded the less will be the trouble on this account.

And yet, with these advantages and others like them, it seems a very uphill task to rehabilitate the generality of Indian villages. As I look round the world and see what has been done in peasant countries to improve conditions, I am rather depressed. Apart from revolutionary conditions, there are few areas where a peasant economy has been raised from a low to a higher level and life to a more healthy and comfortable standard. Denmark is sometimes quoted, but that is not a peasant economy as we know it in India. In Russia there has been a change from what was at least as unsatisfactory a position as is India, but it has been done only by destroying the peasant economy to a very large extent. All over Eastern Europe efforts are now being made, though it is very difficult to find out exactly what is happening, but the impression is that here also the old peasant economy is being destroyed. In Western Asia, though I have myself seen the very great efforts that are being made in Turkey, there seems so far very little result. But I am not without hope in India all the same. Two things, however, must, I feel, be always remembered in this connection.

*A First Priority*

First, if real progress is to be obtained, the whole force of public and Government opinion must look upon the improvement of the countryside as being *a first priority* in the improvement of the nation. It must be a first charge on finance and on the energy of the authorities and of voluntary organizations. For the next few years, even if at the cost of postponing the beautification of the towns and expenditure on development in other directions, the rehabilitation of the rural areas must be done. This is the reason why so much has been achieved in Russia, though the methods by which it has been done there are abhorrent to most of us. Even if it be done in order merely to increase the land production, and not for the sake of the people themselves. I shall be glad to have it done whatever be the motive.

Second, I feel that the whole question of better living for the villagers must be considered as one whole. In the past we have had so much of piecemeal work. There would be a visit from a vaccinator one day; just after, a health visitor would arrive; then he would be followed by an agricultural fieldman, and he again by a co-operative agent, and each of them would see only the success of his own work. The only way is to treat the villages as a whole with one agent, who will help (if necessary with the help and guidance of others) the better living of the people. One of the most striking successes in this direction has been the rehabilitation of a group of villages as a whole in a very depressed corner of Kathiawar, initiated by Nanavati in 1912 and still prospering. The idea of Mr. Brayne's village guides is a right one, and, provided that the proper people can be obtained and properly used, they may form the key to the betterment which we are discussing.

*A Vision of the Future*

I am afraid that this lecture has been on the whole depressing. But I am not by any means hopeless. I can see in my mind's eye a Deccan village as I should like it to be. It is not one which is favoured by an irrigation scheme, though there are a number of wells which give water to a limited area for, say, ten months in the year. The village is clean, and all the rubbish and nightsoil is put into the village manure pit, which is used to fertilize the crops, especially the irrigated crops. All cattle manure is also there, for fuel is being grown in part of the village area where formerly only thorns and bushes grew. Every area not flat is carefully bunded so as to preserve as much water as possible. The two most prominent buildings on the village site are the co-operative centre, where in addition agricultural stores are kept, and where crops are collected for co-operative marketing, and the school where both boys and girls are trained, at least up to the age of twelve, not only to be literate but also in village patriotism. When I can see this, and a thousand things that are not visible today, I shall be satisfied that India is on the march, not only to economic security, but also to having a happy and contented peasantry such as Mahatma Gandhi, whom we mourn today, would have been proud to see.

*Chapter 23*

# THE GENERAL REPORT OF THE ALL INDIA RURAL CREDIT SURVEY*

I well remember the enthusiasm with which the report of Sir F. Nicholson was received in 1904 when, as a result of his visits to the Far East and specially to Japan, he formulated schemes for agricultural co-operative credit societies in India. All over the country there was a feeling, in governmental and other educated circles, that at long last a method had been found under which the squeezing of cultivators by moneylenders could be made to cease and to that extent at least, the position of even the poorer classes of cultivators could be improved. In almost every province of India, after the passage of the first Cooperative Societies Act, there was a vigorous effort to found and multiply rural cooperative credit societies, and the effort has gone on ever since. Various methods were adopted in different parts of India, the extent of state intervention of supervision was not the same, the stress on different aspects of cooperation varied a good deal but a vast number of societies was established almost everywhere.

But, almost from the beginning an evil fate seemed to follow the movement. Its success varied very much in different provinces. Perhaps the most successful societies were in Madras, Bombay, and the Punjab. Elsewhere there were serious breakdowns. I remember the great increase in numbers in Bengal and the rapid dying down of the whole

* Reprinted with Permission from the **Indian Economic Review**, Delhi. Vol. III (1957) No. 4. pp. 76-80.

movement there. I have also seen loss of confidence in Bihar, in the Central Provinces, in Hyderabad, and in many other parts of India. And now, after fifty years, the Rural Credit Survey speaks of the failure of the movement to do anything more than an almost negligible amount towards the solving of the problem of rural credit. "Cooperation has failed, but Cooperation must succeed", says the Survey Report.

This is not to say that there are not many very successful cooperative societies, There are. I have known many, though my own experience has been almost exclusively in Bombay State. My first contact with village societies was with one or more in the irrigated tracts of the Bombay Deccan and there the results were apparently very good in some cases, though by no means in all. When the scope of the cooperative movement was widened, great success was achieved in cotton sale societies with which I was closely in touch.

But, generally, the results have been so disappointing that now, the whole question of rural credit has been again examined, and this time, more thoroughly than on any previous occasion. Taking a series of supposed typical villages in almost every part of India, the Rural Credit Survey has collected data which will serve as raw material for many a long day. Nobody will ever read the three mighty volumes which have been published, but the second of these which contains the report and conclusions is undoubtedly the most thorough-going examination of the economics of rural life in India to-day that has ever been made. It seems to have been made by fair-minded workers and its contribution to the question is of the greatest importance. If in the rest of this review I venture to criticise, it is not because I have less admiration for a wonderful piece of work well done.

On two points in the Report I must insist at once. First, the question of rural credit cannot be separated from the development of an economically sound agriculture. If the produce of the land cultivated does not suffice to maintain the workers and dependents who have produced it, then no scheme of rural credit can possibly be successful in the sense of leading to an independent and self-respecting population. In my own inquiries in villages in the Bombay Deccan forty years ago, I found that 40 per cent of the population were living—and necessarily living—below even their own meagre standard of life, and a recent re-survey of one of the villages studied seems to-day to justify my then conclusions. The village in fact, only succeeded in carrying on, even apart from the exactions of money lenders and the like by reducing their standard of living to below their recognised needs. Such a situation is wholly uneconomic, and no system of supplying credit can meet the situation. For the other sixty percent of the population the question of credit presents little difficulty from an economic point of view. They are the ones who are creditworthy and who join a cooperative credit society. For the rest whether you have large societies or small ones, or however they may be organized, the situation will remain unsatisfactory.

A second point which I feel is insufficiently stressed in the Survey Report is the very great variation in the seasons in many parts of India. In England we consider the agricultural seasons a failure which do not give 80 per cent of normal corps. In parts of the Bombay Deccan, I calculated some years ago that, in 1918, there was only 9 per cent of a normal crop, and in 1920 very little more. No normal banking system, as usually understood, can face such a situation and the cooperative societies have never done so. Previous to years quoted I risked some money in two cooperative societies in the Ahmednagar district and, of course, lost my money. This difficulty is hardly mentioned in the Report, but it is vital. For irrigated areas or others, which are more or less independent of such variations, the banking system represented by the cooperative societies may do, but no amount of state partnership nor other banking procedure can meet this situation. In such areas, it is one of the great strengths of the local resident moneylender that he has devised a system which does meet it. It can be and is oppressive but it works, and it enables the small cultivator to continue to exist.

The Survey attempts to deal with the first of these vital points by insisting on the improvement of agricultural methods and yields at the same time as credit facilities are being increased, by the greater and greater use of other forms of cooperation so as to eliminate the many middlemen who fatten on the rural population, by the consolidation of holdings and other similar procedures which have been recommended in the past, and by the consolidation of village interests in the development of multipurpose societies. So far so good. But I would suggest that such agricultural improvement must precede a satisfactory system of credit if the latter is to apply to a large part of the medium and small cultivators of the country. I say with all solemnity that no system of credit can be satisfactory while the produce which a full time cultivator is able to get from the land is not enough to maintain him and his family at the standard of life to which he considers he is entitled. So I venture to say again that a vastly increased amount of land produce must precede a sound system of rural credit whether this is given by cooperative societies or by any other means. In other words, progress in increasing the produce of the land must always be a little ahead of credit improvement.

The actual positive proposals made by the Survey, from a village point of view, are really two in number. The authors of the Report come down solidly in favour of cooperation as the only sound system of rural credit for the future. But they want larger societies based not necessarily on single villages but on groups which would allow of more competent management, and they want the State, through a rather hierarchical arrangement of banks, to take a much larger and more intimate part in the organisation of rural credit by placing much larger funds at the disposal of the appropriate authorities. The arguments adduced in the Report in favour of these proposals are very weighty and, so far as they go, valid. But to some of us who are conversant closely with conditions in the villages both of them seem to have dangerous implications.

The larger society is much better from a purely banking point of view but it would certainly be less cooperative, would almost surely be dominated by the larger or more active villages in the group, and would tend to be more like a local bank than a cooperative society. The elasticity of the village society, which itself is very much less than that of the local village moneylender, would be very largely lost. The whole question has been very competently dealt with by Sir Manilal Nanavati in a recent number of the Journal of Agricultural Economics, and, on the whole, I agree with him in his contention that it would be with very great regret that the idea of a village unit for a rural banking system should be given up for a larger primary unit for the supply of credit.

I am also a little timid about the large part which is suggested for the State in the scheme developed in the Report. I have spent some time in Russia on several occasions where the so-called cooperatives have become little more than departments of State which do as they are told by the powers that be. If this were to occur in India—and I am very doubtful if it could be avoided if the scheme in the Report is carried out—it would mean more or less the abandonment of the idea of a peasant India and the substitution of one more or less like the collective farm system of Russia. This may seem very far distant at present, but I think that this might be the final result.

Perhaps my most serious criticism of the Report is that it is essentially a banker's treatment of the subject. I wonder what would be the reaction of a small or medium, but honest and progressive, cultivator to the proposals. Such a man has been accustomed to rely on the village moneylender or shopkeeper for his finance. He has a small surplus in a good year when the moneylender takes the greater part of that surplus, but in a bad year he is helped through. The system is oppressive. Everyone acknowledges that. The advent of more abundant funds at a lower rate of interest would be welcomed But some of the elasticity would be gone, and he would tend to be under the domination of people of perhaps a rival village who might or might not understand his problems. I would at any rate like to have the feelings of such a cultivator before me when taking action on the recommendations in the report.

In spite, however, of any criticism that may be made of the proposals in the Report under review, I welcome very deeply the publication of the volumes of the Rural Credit Survey. They furnish data which will be invaluable for a good many years to come for action and, though the ultimate solution of the question of Indian rural credit may be on somewhat different lines from those recommended in the Report, yet the facts that have been collected and summarised must furnish the materials for any serious consideration of the subject.

*Chapter 24*

# THE SIGNIFICANCE OF LAND TRANSFERS*

Almost all the early information about the social and economic conditions in rural India during the British period is contained in settlement reports and other similar documents whose first and often only purpose has been to see how far the area in question is capable of being taxed, and it is only in recent years that such studies have been undertaken by those who have no ulterior purpose to serve and whose sole object has been to ascertain how the people live and how their conditions of life have been affected by circumstances beyond their control. One of the first of such studies was by Mr. Jacks in the Faridpur district of Bengal and was published in 1916. This district is a typical area in Lower Bengal, right in the heart of the Ganges delta, low-lying and swampy for a large part of the year, highly populated, growing rice and jute, with the usual social arrangements in Bengal of large and largely non-resident zamindars under whom there is a whole population of land-occupying peasantry. It is in this same district that Dr. Mukerji has made a study of land transfer during a long series of years in the present century.

One of the key questions in such a district, which may reveal changes in the condition of the people, is the fre-

* Dr. Mann's discussion of **The Problems of Land Transfer** by Karuna Moy Mukherji was written in January, 1958 in response to a request from Dr. Mukherji for an introduction to his book. As it happened, the volume was printed without Dr. Mann's foreword. The final four paragraphs have been added from a subsequent (June, 1958) letter of Dr. Mann to Dr. Mukherji, who had asked for further comment on his concrete proposals for dealing with the land transfer situation revealed by his study.

quency and character of land transfer, that is to say how far the peasant landowners (or land occupiers) are retaining or losing their land by sale or even by mortgage. If they are losing their land it will at once be asked who is taking over the properties which the peasantry are losing and what is becoming of the former owners. This question is treated in the present book with a wealth of detail, based on a full study of the records of land transfer in the past and on a special residence on the spot among the people which, as I know full well, is the only way of really getting at the heart of the matter. Dr. Mukerji's method was to take a number of villages in the district, work among the people whose land was in question, using in addition all the materials which he could obtain from local government sources. Then he went further in respect of one or two of the villages so as to get details of every piece of land which has changed hands between 1922 and 1944. This is a type of work which has very rarely been done in India in respect of any social factor and certainly in no place with regard to land transfer.

Now the question of the way in which ownership (or working occupation) of land is changing is one of the most important means of determining the future shape of economic conditions in India. Up to the present, the country everywhere, and in no area more than in Bengal, has been in the occupation of a land-owning or at least of a land-occupying peasantry. This has been independent of the existence of landlords (zamindars), and the efforts of most governments for the last hundred and fifty years has been to strengthen the hold of the cultivating peasant on the land and to reduce the chance of his being dispossessed by any superior holder. For this purpose laws have been passed in almost all Indian provinces which have, on the one hand, given the cultivating occupier the right to raise money on the value of his lands, and, on the other, to restrict the power of creditors to take over the fields on the security of which they have lent money. Many types of legislation have been passed in different provinces, and in no area more so than in

Bengal, culminating in the Bengal Agricultural Debtors' Acts of the nineteen forties.

It is clear that with the unbiased study of local records and with notes taken on the spot the results are bound to be important. And so they are. Dr. Mukerji finds that during the period of his investigation lands were rapidly passing out of the hands of the smaller occupiers into those of the larger agriculturists, and, to a still greater extent into the hands of non-agriculturists. There seems to be no doubt about the fact in the Faridpur district, and I see no reason for doubting that the same thing is happening in other parts of Bengal and possibly of the whole of India. This means the gradual creation or at least the very large increase in the landless proletariat and so a comparatively rapid change in the character of the rural population of India. One of the chief features brought out by Dr. Mukerji's study is this increase in the landless population and the tendency for the rich (whether agriculturists or not) to become richer and the poor poorer.

As to the cause of this change, Dr. Mukerji's investigation is very revealing. Any period of scarcity or famine such as that in Bengal in 1943 leads at once to a very large increase in land transfer. The same is the case when there is a large increase in price of necessities of life such as there has occurred since the war. In the latter case, the present study brings out clearly the close correlation which exists between rise in prices and amount of land transfer. When one comes to think over the matter perhaps the result in both cases is not surprising when the vast mass of the landholders have only enough land to fill partially the needs of themselves and their families for food and other necessities. If an agriculturist is substantial enough to provide, from his own area, enough for his family and only sells what he does not himself require, a rise in prices may improve his financial position. If, on the other hand, his holding is only able to supply a part of his needs, every rise in price will tend to push him nearer to bankruptcy and hence to realising his only valuable property, namely his land. Only forty years ago I drew attention to the effect of a rise in prices on rural prosperity in regard to the economic condition of the majority of the people, and the evidence which Dr. Mukerji has collected in regard to land transfer confirms the conclusions which I then drew.

The changes revealed by Dr. Mukerji's study will generally be considered as undesirable, at any rate by those whose ideal conception is that of rural India as essentially the home of a landholding peasantry, and who believe that all efforts ought to be made to secure this end. The question of how far restriction of free transfer of land should go is, as Dr. Mukherji remarks, one of the most controversial matters both in the past and at present. His study gives a final blow to the idea that free transfer will result in a happy peasantry. He has, in fact, made out a good case for limited restrictions on land transfer. But a concomitant of this is a method of obtaining cheap and sufficient credit in spite of the reduction in the sole thing the peasant has which is valuable, namely his land. How is this cheap and sufficient credit to be made available, particularly to the peasant who is not economically viable because of the limited amount of land which he has to cultivate. This is the problem which has defeated all administrators. It was hoped that cooperative societies would meet the situation, but they have not done so except to a very small extent, and I see little prospect of their being able to do so.

The real difficulty is that no scheme of financing cultivators who are not self-supporting can be successful, and in a period of rising prices their position will become worse and worse. Until you can get rid of the peasants who are not viable, that is to say, who do not work enough land to supply their needs, you cannot have a sound agricultural economy. So that while there should be a ceiling for landholding, there should also be, for each tract and each type of land, a minimum holding to which finance should be available. This is a rather cruel conclusion, but what can one do? In the Deccan where my own inquiries were conducted, I found that 40 percent of the cultivators were below this minimum. Dr. Mukherji's work suggests that in Lower Bengal the proportion is not far from the same.

This is the crux of the Indian rural problem, namely the very large proportion of landholders who have not sufficient land to support themselves and their families, and who, at every special stress, have to part with their land,—a process which Dr. Mukherji has shown is going on at an increasing speed during the last few years when prices have been rising rapidly. The question is how to deal with this problem. In Russia, which I know well, they have dealt with it by converting individual holdings into large scale farms and, at the same time, both absorbing the redundant rural population by a very rapid industrialisation and also settling large new agricultural areas in Siberia and elsewhere. I do not think this method is applicable in India. India does not have the large new areas to settle that existed in Russia, so that the sole method for absorbing the redundant rural population would be their transfer to manufacturing industries, and I do not believe India can do that.

The problem of how to deal with these non-viable cultivators will eventually have to be solved if a happy and satisfactory India is to come about. Land transfer is a symptom of the underlying disease, and Dr. Mukherji's work is valuable in showing how it is related to the disease.

*Chapter* 25

# THE ECONOMIC RESULTS AND POSSIBILITIES OF IRRIGATION*

The primary problem in the economic development of India is the increase in production from the land. At present there seems a race between the increase in population and production, and one often feels that the former is going ahead at a greater rate than the latter. In fact, in the latest study of the subject with which I am acquainted, that of Thorner and Blyn[1], it is suggested that the annual production of food crops is tending to decrease, and though this is compensated for by greater yields from commercial crops, yet the position is that what is obtained from the land *per head of population* is much less than fifty years earlier. These figures only took the records to 1945-46 and there has been a substantial Increase in production since then, yet the fact that in the present season there has been a large shortage of home produced foodgrains must give rise to much anxiety for the future. It is idle to talk about raising the standard of life of the people when the actual production of the country per head of population is stagnant or declining.

Increase in production from the land can, as has been emphasised by many authorities, be reached by many methods. It can be achieved by extension of the area of

* Reprinted with permission from the **Indian Journal of Agricultural Economics**, Bombay, Vol. 13 (1958), No. 2, pp. 1-6.

[1] Thorner, D. and Blyn, G.: "Long Term Trends in Output in India" (contained in **Economic Growth: Brazil, India, Japan**, Durham, N. Carolina, 1955).

cultivation, by improved seeds, by the more extensive use of fertilisers, by improved methods of cultivation, and by irrigation. In the present article I propose to consider the question of benefits which accrue in the matter of increase of crop yields by the use of irrigation, and also of other favourable or unfavourable results which may follow the introduction of any system of using water on the land.

### *Mean Effect of Irrigation on Crop Yield*

At first sight it would seem obvious that in a country with such a variable rainfall as occurs in the greater part of India, yields must increase to a very considerable extent if the supply of water can be made secure. It is difficult to determine to what extent such an increase actually occurs, but perhaps the general result can be best judged by figures given in the reports on the average yield per acre of the principal crops. The latest edition of these figures relates to 1946-47 and a summary of the results there shown for some of the staple crops in some of the principal provinces is given in the next page.

I am not sure how far these figures can be relied upon, but as they stand they suggest that the effect of irrigation is far greater in Madras and Bombay than in the provinces of northern India. Taking all the above quoted crops together they would make the increase in crop due to irrigation, on the average, as follows:—

| | |
|---|---|
| Madras | 164% |
| Bombay | 159% |
| Uttar Pradesh | 44% |
| Punjab | 53% |

Thus, we may say that, if these figures are at all reliable, in South India irrigation will at least double the output of our staple crops while in the Indo-Gangetic plain it will only increase crops by 50 per cent. But the figures given are only averages and they confound together results which may be great in some parts of a province with what may be a minimum advantage in another part of the same area.

YIELD PER ACRE

| Crops | | Madras | | Bombay | | Uttar Pradesh | | Punjab | |
|---|---|---|---|---|---|---|---|---|---|
| | | lbs. | Increase % | lbs. | Increase % | lbs. | Increase % | lbs. | Increase % |
| Rice | Unirrigated (No irrigation) ... ... | 873) | | —) | | 800) | | 806) | |
| | Irrigated ... ... | 1202) | 38% | 1250) | ? | 1050) | 27% | 1170) | 47% |
| Wheat | Unirrigated (No irrigation) ... ... | — | | 502) | | 750) | | 614) | |
| | Irrigated ... ... | — | | 1238) | 147% | 1100) | 47% | 1037) | 69% |
| Jowar | (No irrigation) ... ... | 480) | | 602) | | — | | 307) | |
| | Irrigated ... ... | 1247) | 160% | 1389) | 131% | — | | 553) | 80% |
| Gram | No irrigation ... ... | — | | 408) | | 750) | | 520) | |
| | Irrigated ... ... | — | | 1153) | 183% | 900) | 20% | 806) | 52% |
| Cotton | No irrigation ... ... | 78) | | 113) | | 120) | | (D) 130 | |
| | Irrigated ... ... | 286) | 295% | 311) | 175% | 220) | 83% | (D) 196 | 51% |
| | No irrigation | | | | | | | (A) 115 | |
| | Irrigated ... ... | — | | | | | | (A) 135 | 17% |

D is **Desi** cotton. A is American cotton.

*Effect of Irrigation in Marginal Tracts*

The figures, in fact, very much over-simplify the situation, for it seems certain that under normal seasonal conditions there are many areas, both in South and North India, where irrigation will not increase the yield at all. In fact, in such tracts the most that can be said for irrigation is that it is an insurance against drought and would be rejected in a normal season by the cultivators at least for staple crops. This is certainly the case in the eastern Bombay Deccan and I believe it is also the case in the lower reaches of some of the canals in Uttar Pradesh. In fact we may divide, as the early irrigators in India did, irrigation schemes into those which are used to get a crop where such a crop could not be otherwise obtained and those where the means of irrigation are merely an insurance against drought. The tendency of those who finance work on irrigation canals, whether these be from rivers, from reservoirs, or from tanks, is to concentrate on the former type of irrigation where large yields of special crops can be certainly obtained but which do not do very much to raise the general level of land production. The tendency in fact, has been even to convert what were originally designed as protection against famine into sugarcane canals where water is very largely used for cash crops which could not be grown without canal or other irrigation water. Such crops are profitable to the providers of water as well as to the cultivators of such crops, but they furnish only a small contribution to the question of the increase of the produce of the country.

I want to make a special point of the fact that in many cases irrigation is likely to make only a small contribution to this increase because it has been ignored by most of the writers about the benefits of irrigation. In one most valuable report, however, by Gadgil[2] the conditions prevailing in a Deccan district are discussed and it is claimed that, as a group, the owners of dry farms were distinctly better off than those with irrigation. The

2. Gadgil, D. R. and Gadgil, V. R.: Survey of Farm Business in Wai Taluka, Poona, 1940.

authors say, in fact, as follows:—"The non-irrigated farms have not only a better representation than the average in the net profit classes but have also a much smaller proportion of their numbers in the largest loss classes than the average." The writers of this report deal with the profit to the cultivators and not with the effect of irrigation on the production of the land, but my own experience in the eastern Bombay Deccan has been that there, in the growing of *rabi* crops, the people have often, if not usually, stated that the application of water for the staple crops generally grown (*jowar*, etc.) is of no advantage and may actually diminish the yield.

Such conditions may be and probably are exceptional and would prevail most largely in the black cotton soil areas, particularly where this is deep and not early drained. But, in general, I think that in what may be called marginal areas, the benefit of irrigation for the usual crops which are required for food or for sale has been exaggerated. It is very much to be desired that in such areas irrigation should be regarded chiefly as an insurance against drought and so the charge for water should be very low or only made when the water is actually used. This would mean that in such places water would not be used in many years for the staple crops and, to that extent, the capital invested in making water available would lie idle or be only used to a minimum extent. In a famine year, on the other hand, such possibilities of irrigation would be invaluable. When I was Director of Agriculture in Bombay during the famine of 1918, the existence of sources of irrigation, even when the reservoirs only filled to a very partial extent, was beyond price. In other seasons water was hardly wanted except for garden crops or, to a limited extent, for those like sugarcane which cannot be grown without irrigation water.

### *Effect of Soil on Irrigability*

The extent to which irrigation water *can* be obtained and used varies very much in different parts of India, owing to the very different conditions of the water supply and, perhaps, even more to the character of the soil. It is generally recognised, for example, that deep black cotton soil cannot be profitably irrigated owing to inevitable waterlogging even when every effort is made at drainage, In consequence the black cotton soil areas are those where the proportion of the cropped land which is under irrigation is the smallest in the country. The actual proportion of the cropped land which was irrigated in 1953-54 in several of the Indian States is shown below.[3] The figures given include the area under rice cultivation, though this stands on a rather different footing from that for other crops. I have, however, left the figures as they stand.

Proportion of Cropped Area—Irrigated

| | Per Cent |
|---|---|
| Bihar | 16.7 |
| Bombay | 5.3 |
| Madhya Pradesh | 5.9 |
| Madras | 26.7 |
| Punjab | 35.8 |
| Uttar Pradesh | 24.9 |
| West Bengal | 18.6 |

It will at once be seen that in the States which have predominantly black cotton soil, the areas now irrigated are far smaller in proportion than elsewhere where other soils prevail. Such States are Bombay and Madhya Pradesh as they were before the re-arrangement of State boundaries. In Bombay, for example, the proportion of the cropped land which is under irrigation is only 4.9% in Gujarat and 4.3% in the Karnatak, while in the Deccan in spite of the great extension of canal irrigation in recent years it is only 7.4%.

As already stated, this deficiency in Bombay and Madhya Pradesh is partly due to the difficulty of getting water on to the land and partly to the unsuitability of the land for irrigation. As regards the former it is perhaps

3. Abstract of Agricultural Statistics, 1953-54, Delhi, 1955.

not realised that in the Bombay Deccan, if all known sources of possible water supply were utilised, about 86% of the cropped area would still have to be cultivated under dry crops. This is not to suggest that there is not very great scope for the extension of irrigation whether from wells, tanks, or canals, but to indicate that such extension must take account of (1) the requirements of water for the staple crops in normal as well as in famine seasons, (2) the character of the soil, its depth and its drainability, and (3) the accessibility of water.

*General Irrigation Policy*

From the point of view of increasing the total production from the land, the tendency to make canal irrigation primarily irrigation for sugarcane and garden crops has perhaps been a mistake, at least in part. Areas which have been successfully converted to sugarcane and similar crops have become prosperous, and have been a source of great profit to the cultivators of those crops and also to the irrigation departments concerned. My own connections have been with the Deccan canal areas and notably those under the Nira and Mutha canals. The economic effects of irrigation in such areas have been vividly described by Gadgil[4] and no one who has visited them can help feeling that the supply of water there has been to bring prosperity to districts which were formerly very poor, and so far as the watered areas are concerned, to increase vastly the production per acre. But this local prosperity has been achieved with both agricultural and social consequences to which comparatively little attention has been hitherto given.

I do not wish to stress here, however, the danger of the development of waterlogged and salt land as a result of canal irrigation in such areas. A good deal of attention has been focussed on this point in the recent past, and it is almost a criminal offence, with existing knowledge, to

4. Gadgil, D. R.: Economic Effects of Irrigation, Publication No. 17, Gokhale Institute of Politics and Economics, Poona, 1948.

put in canal irrigation without arranging for adequate drainage *at the same time*. This was neglected in the early days of canal irrigation, with the result of making a good deal of the area, which gave the best results when water was first given, becoming permanently barren. I say *permanently* barren because in the Nira Valley we found that the subsoil of such water-logged patches was stinking with sulphurated hydrogen owing to the expulsion of oxygen from the lower layers of the soil by excess of water.

But another agricultural result has followed in these Deccan canal areas in the draining of the manurial resources of the surrounding dry country into the watered region. The growing of sugarcane demands a very high degree of manuring, and every source for manures must be tapped, apart from the oilcake and artificial manures which are bought by the sugarcane growers. For quite a large region round the Nira canal area (and I doubt not that the same has occurred round the Godavari and Pravara canals) cattle manure and other similar materials have been drawn into the watered zone with the result that the dry crops there have been to that extent starved of the manures which they might have had. There are no available data to show to what extent this has taken place, but if the high productivity of the irrigated land (chiefly with sugarcane) is secured by a reduction of the yield of staple crops grown without water, it must be placed to the debit of the canal irrigation.

The social result of the introduction of canal irrigation in the Deccan areas is a very striking one, at any rate under the older canals. The lands which were formerly owned and worked under dry cultivation by landowning Maratha cultivators has to a very large extent passed into the hands of invaders who, used to irrigation practice, have taken over the suitable plots, while the former owners have largely disappeared as landowning cultivators. What has become of the former owners who worked the land, I do not know. Have they become landless labourers or have they moved to other parts of the country? On this point we have no evidence and the census returns do not help us much in the matter. But if the phenomenon is more widespread it becomes important. I can quite believe that a cultivator who has been accustomed to grow *jowar* or similar crops on unwatered land, may not be able to accommodate himself to the more arduous work of growing sugarcane and its associated cultivation. Even in the dry regions, there is usually a certain growing of garden crops irrigated from wells, but this is a very different matter from the large scale cultivation of sugarcane and the change over seems to have been too much for the average Maratha cultivator. I am not sure how far the same change has taken place in the more recently irrigated tracts, but it is a social development which cannot and should not be ignored.

*Summary*

To summarise the points which I have tried to make in this paper, it would seem that

(1) The increase of crop yield which may be expected as a result of the introduction and extension of irrigation varies a good deal in different parts of India and is probably a good deal less in the alluvial tracts of northern India than in the Peninsula areas.

(2) Such increase as is obtained will vary very much with both climatic and, soil conditions. When climatic conditions are favourable for crop growing, and soil conditions are unfavourable for irrigation, the introduction of canal irrigation may even lower yields and water must be looked upon only as an insurance against drought, with the consequence that, as regards staple crops, the capital invested can hardly give a regular annual return but may be invaluable when a famine occurs.

(3) The tendency to encourage the use of irrigation water for the intensive cultivation of sugarcane and garden crops to the detriment of its use for staple crops, while leading to very great increase in production and in profit in local areas does very little to increase the total production of the land, taking the country as a whole. It is a convenient method of using water and a profitable one for both the sugarcane growers and for the providers of water, but its

effect on the national production per head of population is only small.

(4) Without laying stress on a subject which has already been much debated, namely the effect of heavy irrigation on the development of waterlogged and salt land, I would urge, as I have often done before, that no new irrigation scheme should be even entertained which does not provide adequate land drainage at the same time as a fresh supply of water is given.

(5) There is a case for the close study of the effect of the introduction of a new source of irrigation on the agricultural results in the surrounding dry areas. It is almost certain that the demand for manures in the irrigated tracts has, in the past, tended to lower the standard yield of crops in such dry areas.

(6) There has been little study of the social results of the introduction of canal irrigation on the inhabitants of the area to which water is supplied. In certain irrigated tracts these have been very considerable with the displacement of the traditional landowning cultivators by invaders who are more used to the growing of intensively irrigated crops.

*Chapter* 26

# THE CENTRAL MECHANIZED FARM AT SURATGARH — REPORT 1956-58*

Everyone is agreed that the economic future of India depends largely on such an increase in the agricultural production of the country as will raise the possible standard of life of the mass of the population. It is clear that this means that the total yield of food crops from the land must rise more rapidly than the number of people who have to live upon them. This rise in production can be obtained either by an increase in the crops obtained per acre (whether this is achieved by improved agricultural methods, by irrigation, or by the extensive use of fertilizers) or by an extension of the area which is actually under cultivation.

The same problem has faced a large number of other countries in recent years, especially in Asia. Most of them have depended largely on extension of the cultivated area for the necessary increase in production. This has been, for instance, in Russia where in spite of the most vigorous efforts, the actual yields of food crops per acre have been disappointing. In China it has been reported that there has been a very large increase in production per acre but the figures given are so extraordinary that I venture to doubt whether the yields claimed have been obtained on any large scale. In Turkey where I have figures before me for the last twenty years or more, the

* Reprinted with permission from the **Economic Weekly**, Bombay, Vol. XII (1960), pp. 1827-1828.

results show that the large increase in total yield have been almost entirely the result of extension of area rather than of gain in yield per acre. This fact is so striking that I venture to quote official figures for several of the most important crops. The following are the kilograms per hectare in 1936-40 as against those obtained in 1956-58.

| | Yield per hectare in kilos. | |
|---|---|---|
| | 1936-40 | 1956-58 |
| Wheat | 1059 | 1059 |
| Barley | 1170 | 1277 |
| Maize | 1383 | 1184 |
| Rice | 2105 | 1959 |
| All cereals | 1098 | 1117 |
| All pulse crops | 859 | 903 |

In fact, it would not be unfair to say that in the countries of Asia, with the exception of China where I think the figures now given are unreliable, any increase in production has been almost entirely obtained by increasing the area under crop. I do not doubt that this is probably a temporary phase and that a big future increase will be obtained by irrigation, by improvement in the type of crop grown, and by the more extensive use of artificial fertilizers.

The possibility, however, of getting the necessary increase in production by an extension of the cultivated area in India is far less than in Russia or China. The boundaries of India are more or less fixed by ranges of mountains or deserts while Russia had the whole of Siberia and most of central Asia to use. Though such use has not been easy, yet a very large part of the available area has been taken in hand. China can extend her cultivation, again not without difficulty, into the vast areas to the north west and the production can be largely increased by such extension. But in India the vacant areas are limited. There is a certain amount of suitable land, in the Himalayan Terai and large areas in Rajputana and

Central India, which could be used if adequate water was available. Apart from these regions, I doubt whether there are many large spaces available for agricultural extension, for the largely jungle tracts of the Eastern Ghats have such poor soils and such irregular formations that it is very doubtful whether it will ever be possible to use them to a much greater extent than is the case at present. Except for the Himalayan Terai it would seem likely that extension of area depends on increase in the supply of irrigation water, for Rajputana is at best an arid tract.

This makes the work at Suratgarh Central Mechanized Farm of very special interest because, with the recent increase in the land in Rajputana commanded by irrigation schemes, it has indicated a method of getting a return quickly from an area where otherwise development might take a long time to occur. The establishment of this farm has been made possible by the gift of agricultural machinery by the Soviet authorities for a 30,000 acre farm,—the largest farm in Asia. In a desert area of Rajasthan, where water has recently become available and where there was little or no population, the establishment of this farm may be expected to show what can be achieved in places which have been almost completely desert. The report of what has been so far done at Suratgarh up to 1958 which now lies before me gives promise of a very valuable contribution to our knowledge of the best way to make use of the land under similar conditions. Here there was previously no cultivation, partly owing to lack of water and partly to the way in which an arid area such as this tends to develop sandhills and a general topography which makes it almost impossible for an ordinary cultivator to bring the land into use. The presence of water for use either by flow or by lift irrigation has made agriculture possible and the large scale implements have allowed the rapid preparation of much of the area which would otherwise have been almost useless. The report under review tells of the area cultivated and the crops grown in 1956-8. There were actually 3,000 acres under crop in the *rabi* season of 1956-7, nearly 1,900 acres in the *kharif* season of 1957 and 2,500 acres in the *rabi* season of 1957-8. The yields obtained are not outstanding, but it must be remembered that they were obtained in a new area and by pioneers in the use of the equipment provided.

The actual yields per acre of the more important food crops were as follows compared with the official averages for the Punjab.

| | | Suratgarh | Punjab (average, irrigated). |
|---|---|---|---|
| 1956-7 | rabi | | |
| | Wheat | 12.4 to 22.8 mds. | 12.6 mds. |
| 1957 | kharif | | |
| | Jowar | 5.4 mds. | 6.7 mds. |
| 1957-8 | rabi | | |
| | Wheat | 5.4 to 8.8 mds. | 12.6 Mds. |
| | Barley | 9.2 mds. | ? |

There is no serious indication that the yields will be greater than are obtained in the already cultivated parts of the Punjab except with paddy, and the area of this crop grown was so small that I do not think anything can be concluded from the figures given. But the results do show that in this desert land, which has not been cultivated before, crops can at once be obtained which at least compare with those which are being obtained in the same part of India under irrigation.

It will be seen that the development of the farm is not taking place very fast, and so far the chief crop grown has been mustard in the *rabi* season and *Jowar* in the kharif. But I think that if the area under crop can be rapidly increased and if, as should be the case, the yield per acre can be raised substantially, the farm will have at least provided a valuable guide to what can be done in a previously desert tract.

But the question now arises as to how far Suratgarh may be a guide for the future in India. And I think that the inevitable conclusion is that the farm does show the way to a *rapid* increase in the area under food crops in *unoccupied* regions where water has recently become available. The key words in this conclusion are *rapid* and *unoccupied.* I have no doubt that when water becomes available there will be, in any case a slow extension of cultivation into such areas by energetic peasants like the jats of the Punjab. But if an area is already largely under cultivation no one would think of removing the present peasant population in order to create large state farms such as that at Suratgarh. In such a case the only way to use the mechanized equipment available, as at Suratgarh, would be to create large co-operative farms in which, as in the Russian collectives, the peasants would be essentially labourers on a large estate in whose management they would have little part. A few of the larger owner-cultivators would be influential enough, owing to their large holdings or other factors, to demand an active part in the running of the farm, but the largest part of the owners would tend to become labourers who would draw their share of the profits but whose pride in the ownership of land would tend to vanish. Even if such a scheme could be pictured on a large scale and that this would be acceptable to the present peasants it would mean such a revolution in the relationship of the people in the land that it would change the whole economy of the country.

The only justification for the adoption of such a policy would be an assurance that it would lead to a large increase in the production of the land. I own that the yields at Suratgarh given above do not suggest that large increases in yield would be obtained above those which are now being got by the Punjab peasants, but they are pioneer attempts and it is likely that much better results will be reached when Suratgarh is well established. But, in my mind, these results are doubtful and the Russian experience is not very encouraging in this matter.

In fact, while I think that the establishment of the Suratgarh mechanized farm is of great value in showing a *quick* method of developing a *new* area where water has been recently supplied, its value as a specimen of what can be achieved over the greater part of India is limited. I do not think it can show the way to the great increase in agricultural production that is now so essentially required. I am very glad that the Suratgarh experiment is being made, and it is both possible and probable that in the end a higher yield will be obtained there than is customary in the adjoining regions of the Punjab and that it may lead to a generally higher standard of cultivation in that region, largely by the economical use of water and by the demonstration of higher yielding varieties of crop and of the value of artificial fertilizers under the conditions of irrigation cultivation. But that is for the future.

**PART V**

**THE USSR AND THE MIDDLE EAST**

*Chapter* 27

# THE COLLECTIVE FARM SYSTEM IN RUSSIA*

The story of co-operation in Russia has been different, both in its history and in the final form which it has now reached, from anything that has happened in any other country in the world. It began about the end of the last century on lines similar to those in other countries and although it was distrusted by the Government of Imperial Russia, yet it developed very rapidly. Before the great world war the co-operative societies numbered at least 27,000, chiefly for the supply of agricultural credit. The Communist Revolution in 1917 destroyed almost the whole of the previous structure, especially so far as credit societies were concerned, and what was left became for the time being practically a State Department.

A reaction took place in 1921 with the re-creation of the Co-operative Movement. So far as agricultural co-operative societies were concerned at this period, the members were individual peasants, and they were very largely used for the purpose of combined purchase or combined sale. In the palmiest days of the 'New Economic Policy', as it was called, the number of members of such societies became over eleven millions or about half the peasant population of the country.

A new epoch, however, opened in 1928, and, since that time, while ordinary co-operation, as we have known it

* Reprinted with permission from the **Indian Cooperative Review** (Madras) Vol. IV (1938), No. 2, pp. 149-155.

elsewhere has tended to decline, the organisation of communal agriculture has gone further than in any part of the world and has been conducted on co-operative lines. The extent to which other forms of co-operative work have declined can be judged by the abolition of the urban consumers' movement in 1935—a form of co-operation which, when I first went to Russia in 1930, seemed universal and seemed likely to absorb all shopkeeping in the country.

But rural co-operation, centred round the collective farm, has developed marvellously,—though with many changes of policy, many set-backs, and at one stage a narrowly-avoided collapse,—till it has become almost universal, either in the form of the collective farms themselves or in the rural consumers' movement which acted as a buyer of agricultural produce. In the present article I will limit myself to an account of the organisation and working of the collective farms themselves, both as recorded in published reports, and also as I have seen it myself in the country.

It is well known that the ownership of land in Russia has been in some sense communal throughout the whole of its history. It was only in the second half of the 19th century that the idea of peasant ownership came to the front. This being the case the transformation to a collective system, though full of difficulty, was not nearly so revolutionary as it would be, for example, in India, where individual possession of land has a very long history and is very deep-seated in the minds of the rural population. All the same, one of the first results of the Communist Revolution in 1917 was the seizure of almost the whole of the agricultural land by the peasants. The large estates of the landlords were split up and, in spite of Communist theory, it looked as if the whole land would become a mass of individual peasant holdings, though of course the community, as represented by the State, demanded a much greater share in the produce than the State had ever taken before.

This continued, as I have already said, until 1928, when a genuine attempt was made to reconvert the mass of individual holdings in a village or similar area, into a communal holding. At this time, however, there were a good many different views as to what the nature of the communal holding should be, how far it should go in taking account of all the possessions of each peasant, how far the area should be treated as one, and also how far the management should be unified. There were some, in fact, who wished that everything possessed by the peasants should be communalised, that is to say, should be treated as belonging not to individuals but to the community as a whole. Some went so far as to consider that no peasant should hold individually even his house, his poultry, or a milking cow. There was a time when it appeared as if this view was likely to dominate the situation, but it was always resisted by the peasants over the greater part of the country, and the result of attempts to enforce it led to the slaughter of a very large proportion of the livestock in the country in 1932 and the early part of 1933. I fancy, indeed, that the whole Soviet system came more nearly to a breakdown at this time than at any other period of its history,—largely owing to the resistance of the peasants to the complete collectivisation of property which was demanded.

As has so often happened during the re-organisation of Russia since the Revolution, the authorities realised the dangerous position of things in the early part of 1933, and a very famous decision was made at this time, largely under the inspiration of Stalin, whereby individual ownership was recognised over a very large part of a peasant's possessions. He was encouraged to have a milking cow of his own; to possess chickens and small animals of similar kind; to have a cottage and surrounding garden which belonged to himself; while at the same time the general agriculture of each collective unit continued to be run on a communal basis. The result is that at the present time it is estimated that four per cent of the rural land in Russia (1935) form the private gardens and other lands belonging individually, to the members of collective farms.

From 1933 the organisation has become fairly stable, and I think it may be said that the collective farm system of Russia is now well understood by the peasants, and works (at any rate in a fairly successful season) with no more friction or trouble than is the case with the agricultural organisation of a more old-fashioned country. This result has, however, only been achieved by a very close co-operation between the State and the collective farms. Unless the State had been prepared to invest an enormous amount of capital in making it easy for the collective farms to work, and unless it had, moreover, been prepared to make the success of the collective system one of its primary purposes, I am sure that the present organisation could not have reached success. In fact, I feel (as a result of my Russian observations) that, if any country wishes to develop a collective agricultural system, its Government must make this a primary purpose of its existence and must be prepared to invest capital and energy in the struggle to make it a success, which is far beyond anything which seems to have been realised by any Government in the world outside Russia.

Let me now try to describe the organisation of a typical collective farm in Russia at the present moment. The area of such a farm may include one or more villages and its area may run from say, 200 acres up to 12,000 acres, though over a greater part of the main corn-growing tracts of South Russia a collective farm will usually contain somewhere from 2,000 to 6,000 acres. Usually the collective farm is made so as to include one or more villages and there has been very little cutting up of the old villages in order to make more convenient collective farms. In 1934 the average area per household in collective farms amounted to nearly 16 acres of which it was stated that nearly 13 acres were under grain or similar crops. Naturally this figure would vary enormously over different classes of cultivation and over different parts of the country, but the figures I have given give an idea of what is aimed at in these farms. Owing to the demand for labour in the rapidly developing industrial system in the towns and also for extending agriculture into new tracts in Siberia and elsewhere, the area per household has tended to increase since the collective farm system was first organised about 1928. It must be realised at once that in a collectivised village a peasant must either join the collective farm or cease to hold any land at all beyond the garden attached to his own house. Membership, however, is not a right of any inhabitant of a village and members can be expelled by the general meeting of the collective farm and new members likewise admitted even from other villages.

Such a collective farm as I have described is now a legal constituted body and will have received grants of land from the State, which are to be devoted to its use for ever. Its property includes the arable land and the common grazing ground of the village, central farm buildings, work animals, implements and the common herds of cattle, pigs, sheep or poultry.

The collective farm is governed by a small committee elected from their own members for two years. The land is worked without reference to its previous ownership in the most economic way and the actual work of the farm is carried on by 'brigades' each under a leader appointed by the Committee. These 'brigades' contain both men and women, while the care of livestock is usually a matter for the men.

But it must be clearly understood that the cropping of such a collective farm is not entirely at the option of the Committee. Each collective farm receives each year from the Government a plan specifying how much grain and similar crops is to be grown, while for cash crops they are expected to grow a definite minimum of each of the products for which the place is suitable, for sale to the State factories. In fact, a contract has to be entered into at the beginning of each year with the State factories specifying the minimum amount for which any individual collective farm will be responsible.

An agricultural expert appointed by the State is usually, though not always, attached to a collective farm and his advice will normally be taken, though the Committee are not compelled to do so. Further, in a greater part of Russia the State has established what are called Machine Tractor Stations, and, from these, implements and tractors can be hired which are beyond the capacity of each collective farm to purchase for themselves, and seed and manures can also be obtained from them. The use of all the implements for the production of a crop, including the use of tractors, will usually be charged at about one-sixth to one-fourth the normal yield of the crop grown and this will be paid for in kind.

The crops are, therefore, grown by the members of the collective farm with the assistance of seed, manure and implements as well as advice obtained from the nearest Machine Tractor Station. When each crop is ready for harvest the whole is reaped and remains at the disposal of the Committee of the collective farm. From the products so obtained the charges incurred in its production, including the land revenue payable to the State, have first to be set aside. In fact, the order in which the produce is disposed of will normally be as follows:—

(1) Delivery of a fixed quantity of grain or other crop per acre amounting to about one-sixth of the average yield to the State. This might be considered as a form of land revenue, but it is rather in the nature of a forced sale at a price which is only a fraction of that obtainable in the market. In the great grain-growing area of the Ukraine the amount of grain delivered under this arrangement amounts to about one to one and a half maunds per acre. It is a fixed amount based on the average yield.

(2) The payment to the Machine Tractor Station for supplies of the manure, seed and advice, together with the hire of implements during the growth of the crops.

After these two payments have been made the rest of the produce belongs to the collective farm. Such a proportion of it as is necessary to pay the cash requirements is sold in the market collectively, and out of the money so obtained the following items are paid for (1) the subsistence wages of the members of the collective farm which have been paid during the growth of the crop, or other wages due to the workers. I will deal with these wages later; (2) the agricultural tax payable to the State and amounting to about one-fortieth of the cash income; (3) all costs of production other than those due to the Machine Tractor Station; (4) the administrative costs of the collective farm which must not exceed 2 per cent of the cash income; (5) the purchase of equipment designed to increase the production of the collective farm, and (6) any other expenditure for the improvement of the collective farm which is decided upon by the general meeting.

The whole organisation is simple except so far as concerns the wages payable to the workers on the collective farm during the production of the crops. To subsistence wages paid during the growth of crops, is added the amount which is earned by the worker according to the quality of his work, so that a small amount will be payable to each worker during the growing season and, it may be, a larger amount at the end of the year when the produce has been sold. The calculation of the amount due finally to any individual worker is one of the most interesting features of the collective farm system. All farm labour is divided into seven groups according to the quality of the work done. In the highest of these groups, which includes the best workers on the farm, each day's work is counted as two 'labour days'. In the lowest grade, on the other hand, a day's work is only counted as half a 'labour day'. This unit of labour represents the amount which might be expected from an ordinary good labour without special skill. If the amount of work done or the quality of work improves, a worker will be advanced to a higher group. If, on the other hand, it declines a worker may be put down to a lower group. The whole decision in this matter being made by the Committee of the collective farm.

After the payments which have been mentioned have been made the whole of the remaining produce is divided among the members of the collective farm, who can store it or sell it or use it at their own free will. Most of it is, I fancy, sold to a rural co-operative society and this is the method which is generally favoured by the State authorities. Though there is no restriction on private sale of produce, yet it must be understood that this cannot lead to the growth of a private merchant's business because any such business is not permitted. If the sale, in fact, is not made to a co-operative body, it can only be made, at least in theory, to an actual consumer.

The general system which I have described has not been reached all at once, and it is quite possible that the organisation has not reached its final methods. Up to the present there has been such a demand for labour in connection with the new industrial development in Russia that there has been no pressure of population in the village but, on the other hand, the tendency has been for the number of members of any particular farm to decrease. This may not always be so and if the amount of land per member decreases it may bring back poverty again where the collective farm system has temporarily removed it. The safeguard against this is represented of course by the enormous underdeveloped areas in Russia, to which any excess population can go and will be able to go for many years to come, where every assistance will be given to new colonists who try to make an agricultural country from an undeveloped waste.

It will be seen that the system, if properly worked, should lead to the most efficient working of the land within the area of a collective farm, and as a result the actual production from the land is now tending to rise quite rapidly. To obtain this result, of course, means an enormous State organisation to assist the collective farm to make the best use of its land. Experience in Russia, as elsewhere, has shown very clearly that there is no magic in the conversion of a large number of small scattered individual holdings into a collective farm. In fact, experience has shown that the production of such a collective farm may be less than that obtained by the members working each on his own holding. It is only where there is good leadership and satisfactory knowledge that the desired end can be achieved. And it would astonish anyone who has not been to Russia to see the amount of energy and training which has been expended in order to obtain the necessary amount of village leadership, and to make available the requisite skill to the collective farms. In the absence of these two necessary things the collective farm system almost collapsed. With their gradual improvement and the consequent increase of production, the whole seems to have passed beyond the critical stage and is now likely to become a permanent feature of rural economy in Russia.

I have no time in the present article to discuss the extent to which any similar scheme of collective farming is suitable for India. There are many difficulties in addition to those which have been met in Russia. Some of these are the intense pride in land ownership and all that this involves, the absence of undeveloped areas to anything like the same extent as in Russia into which the excess population could be drafted, the absence of industrialisation of the country to anything like the same extent as has been attempted in Russia, and finally, the existence of a large 'middle man' class whether existing as landlords, private merchants, money-lenders, etc., all of whom had to be 'liquidated' in Russia before the collective farm system could be a success. Unless these difficulties can be met and got rid of I do not think that the success of any collective farm system in India is likely to occur. Whether they can be got out of the way, except in a revolutionary period, is a matter on which opinions may differ. But if these hindrances can be met, I think there is something to say for the collectivisation of agriculture as a means of improving the standard of life of the rural population in India, and the raising of the rural communities out of the abyss of bankruptcy in which they now lie.

*Chapter 28*

# AGRICULTURAL IMPROVEMENT IN TURKEY*

When I was in Turkey in 1940, I was very deeply impressed with the agricultural possibilities of the country, at any rate in the parts I was able to visit. Whether I was staying in the western valleys of Anatolia, or on the high plateau of Ankara, or whether I was travelling in the Pontic area near the Black Sea coast to the north or in the Adana plain in the south, the impression left with me was that while much of the cultivation was excellent and the products first class, yet there was much land that could be used which was not used, the crops that were usually grown were capable of much improvement, and a good many things could be grown which are now almost unknown in the country. A long experience of India and the Middle East confirms the impression that if advantage is taken of investigations made and experience gained in India, in Egypt, and in the Soudan, as well as in the subtropical regions of America and in the southern part of the U.S.S.R., there should be at least the possibility of increasing the value of the agricultural products of the country by fifty per cent if not by more.

In any country, as Ataturk repeatedly affirmed, a well conducted agriculture is the foundation of national economy. In saying this I do not in the least wish to suggest

* From the first of a series of radio talks made during the latter part of the second world war, exact date uncertain.

that in working for the advance of agriculture, one should attempt to hinder the development of any industry which utilises to the full the resources of the country, whether derived from mines or from any other source. I should like to see all the rich mineral resources which are known to exist being utilised to the greatest possible extent. I should also like to feel that the products of the land, such as cotton, or jute, or sugar cane and sugar beet were being worked up in the country itself into commercial materials whether for home use or for export. But the fact remains that for a good many years to come, agriculture is likely to play a predominant part in Turkey. This was fully realised by Ataturk who, only a few months before he died, expressed this belief very strongly. He had seen the revival of his country, the rise of pride among his people and of confidence in her future. And yet, in his last days, he expressed the conviction that unless the agriculture of a country is sound and the land is yielding its increase in great and sufficient quantity, a satisfactory economy for a nation cannot be built up.

This idea had pervaded his activities for a good many years. I have myself visited and seen the large agricultural estate which he established a few kilometres from Ankara and which he designed as a model to show what could be done and which he wished to see duplicated in many centres in Turkey. On this estate can be seen to-day the very large increase in the production of the land which follows a properly organised system of farming and also the way in which the products of the land can be made the basis for rural industries. I have also seen the estate near Trabzon on the Black Sea coast and the efforts which he made to show what could be done for agricultural development in that favoured region. I believe that he also laid out his own money in the cotton producing areas near Adana, as a pioneer in the introduction of more valuable varieties of cotton than those previously in use.

If one may judge by figures the inspiration which he gave has already had great results in extending the cultivated area in Turkey and in improving the crops grown since the new Turkey rose just over twenty years ago. Only fifteen years ago a careful estimate showed that out of 22 million hectares of cultivable land, only 4 to 5 million hectares were actually in cultivation, and the cultivated area including fallow land was only about 7 to 10 per cent of the total area of the country. By 1938 this proportion had been nearly doubled, which is a testimony to the enthusiasm with which agricultural development was being carried out and to the energy of the peasants as soon as the peace and safety of the land was secured.

The position of Turkey at the junction of the east and the west gives it peculiar advantages. It has been the source and origin of the cultivation of many of the world's most important products. It was in Anatolia, most probably, that wheat cultivation originated. Here also other grains like rye first reached the status of cultivated crops. Many pulse crops like peas and lentils also came first into general use here, and some of our most important crops for feeding cattle, like lucerne, were developed in parts of Anatolia. There is thus a strong historical basis for the belief that Turkey can again acquire a far richer and more powerful agricultural economy than it has to-day. To do this, however, it will be necessary to utilise the rich variety of materials in the form of plants and animals which can be found in the country, and also to use the great amount of knowledge and experience obtained in other countries similarly situated in recent years.

In the British Empire and Commonwealth under conditions which have some similarity to those of Turkey, there are many crops grown which might well be of great use here. In India, especially, great discoveries have been made which would probably be of much importance. Varieties of wheat, of other grains, of cotton and other fibres, of oilseeds and of other crops have been evolved there which may be of considerable value in Turkey. Similar results have been obtained in Egypt. All the information collected in these and other eastern countries is at the disposal of the leaders of Turkish agriculture, and no-one would be happier than the workers in India or Egypt if the results of their work led to successful developments in Turkey.

In the past, the whole of Western Asia has suffered very badly from the erosion of the land. This applies to India as well as to Palestine, Syria, Persia and the adjoining countries and to none more than to Turkey. But while in some of the countries mentioned, almost the whole of the soil of the uplands has been washed down into the lower lands and into the sea, in Turkey a much larger amount of the forests have been preserved and this has saved the plateau lands of Anatolia from the worst effects of erosion. Yet the first impression I got when travelling on the plateau is the bareness of the country except that at the bottom of valleys and the almost complete absence of trees on the upper slopes. Hence there is a large amount of land which at one time could have been cultivated but which now seems destined to be waste and only to furnish meagre pasturage for sheep and goats. Much has been done in recent years to ascertain the best ways of preventing at least the extension of the losses caused by erosion in the United States, in the drier parts of South Africa and Australia and India. I would particularly dwell on the work done in India where the improvement of eroded lands and the prevention of further erosion in a peasant economy has been carried out on a large scale.

Already in some directions of agricultural production, Turkey leads the world and dominates the markets of Europe and elsewhere. When we in England speak of figs, for example, we mean the dried figs of Izmir. In the case of dried raisins and currants we have in mind the products of Greece and Western Anatolia. Turkish tobacco, though grown and cured in South Russia and the Levant, is dominantly obtained from northern and western Anatolia. The world's supply of hazel nuts (*findik*) is very largely obtained from the Pontic region where they are a principal peasant crop on land which would probably be uncultivated if it were not for this crop. Anatolia is the home of the mohair goat and the remainder of the world depends on Turkey for its supply. All these products and some others have been produced to perfection here and the rest of the world has copied what has been done for many generations by the Turkish peasants.

We may take figs as an example. The dried figs of Smyrna are the best the world produces. When it was desired to introduce the cultivation of figs into the United States, the growers there failed for a long time to get a product which was in any sense equal to that obtained in Turkey. They sent experts to examine the conditions on the spot who found that the secret was partly in the variety which was grown but also in the fact that unless there were a limited number of male trees in an orchard—the so-called capri-figs—from which insects could come to fertilise the flowers of this variety, no figs at all could be obtained. These male trees were taken to California and the methods for getting the flowers of the Smyrna fig tree fertilised by the insect, employed by the Turkish peasants, were introduced there, and now California can and does give figs that are equal to those obtained from Western Anatolia.

I might also mention the almost unique importance of Turkey in the production of tobacco. The variety used is different from that used in America or in the Far East, and the method of curing is again different. Both these factors give a special quality which seems peculiar to the tobacco grown and made round the Black Sea and Egean coasts. From here, the industry has spread to adjoining countries like Soviet Russia, Bulgaria and others, but it is the plant and the methods of curing which were developed in Turkey that have given Turkish tobacco its secure place in the world market.

The vine is one of the national plants and its dried products, raisins and currants, form one of the principal articles of export. It is cultivated everywhere where such cultivation is possible. It is equally a field and a garden crop, and all the cities of Anatolia where it can be grown are ornamented with vines. It makes the charm of small towns everywhere and the street are often galleries overgrown with grapes. The whole area under vines amounts to about 350,000 hectares and the best varieties (which may originally have come from Persia) are as good as any in the world and are peculiarly suitable for drying for the purposes of international commerce.

But while in some ways, Turkish agriculture has a good deal to teach the rest of the world, yet, in its larger aspects, there is a great deal that it can learn from what has been done elsewhere. In the following talks in this series I propose to speak of some of the special agricultural difficulties which have come to my notice in Turkey and to indicate what experimental results obtained in the United States, in Soviet Russia, in Egypt, and in the British Commonwealth (especially India) can do to help in the solution of the difficulties. I hope also to describe some of the new cultivations which may be found suitable for Turkey, such as tea and jute, and to describe in more detail the methods of checking soil erosion and other destructive incidents. My notes will be purely suggestive and will be made with a full realisation of the great progress that has already been made in recent years.

*Chapter* 29

# THE NEW AGRICULTURAL LAND LAW IN TURKEY*

When the Royal Commission on Agriculture was sitting in India in 1926, it was a source of very deep disappointment to me that it was forbidden to consider the question of land tenure as a factor in agricultural production. The result has been that while a great deal has been done since then towards the discovery of the best ways of utilising the land so as to produce the most profitable results, very little has, in comparison, been achieved in getting the present cultivators, whoever they may be, to put these improvements into practice, and this failure is, I believe, very largely wrapped up in the question of the relationship of the cultivators to the land which they work.

In nearly all the countries of the Middle East, where the anxiety to increase agricultural production is as great as it is in India, this fact is now realised, and two lines of attack of the problem are being constantly discussed. One school of thought favours the adoption of a system similar to that now worked in Russia, where individual ownership and individual cultivation have been very largely replaced by collective cultivation, enabling large scale production to be carried out, and resulting, with proper guidance from above, (it is hoped) in a very large increase in the production of the land. I have recently read a very

* Reprinted with permission from the **Bombay Cooperative Quarterly,** Bombay, Vol. XXX (1946), No. 2, pp. 77-81.

careful and thorough examination of the present situation in India by Baljit Singh, who takes the view that nothing short of this will bring production to anything like the point which is necessary in view of the rapidly increasing population. He says also, quite wrongly I believe, that small scale agriculture is necessarily uneconomic and always gives a smaller yield than work on a larger scale may be expected to produce. This last proposition is, of course, by no means always true, for perhaps the most intensively cultivated parts of Europe like Flanders and the Lombardy plain are cultivated by peasants in small holdings.

The other school of thought, which has been adopted in nearly all the countries which border on the Soviet Union, is that the best way to increase agricultural production is to give the peasants an individual ownership of the land, to see that each owner has enough land to give him a good living, to see that this unit of land ownership is maintained, and to see that he is adequately financed and provided with the best expert advice. The line of action which is being taken is well illustrated by the new agricultural land law in Turkey passed on June 11th 1945 of which I will try to give a short account.

Like most of the other countries of the Middle East, the land system of Turkey has been essentially one in which the cultivation has been done by peasants, but the land has been generally owned by landlords whose only connection with their property has been to draw their rents and live far away from which they get their livelihood. In addition, the rents have usually been rack rents, that is to say, rents as high as the land will bear after providing a low subsistence level for the people who grow the crops and cultivate the ground. In India, the worst effects of such a system have been mitigated by the regulation, in the *zemindari* provinces, of the rents which the zemindar could charge for the use of the land, but in the *raiyatwari* provinces, where the land has been gradually passing out of the hands of the cultivators into those of non-cultivators, there has been no such mitigation. The position in Turkey has been similar to that in the *raiyatwari* tracts of India in this sense, but the one difference between the conditions here and in Turkey has been the fact that in the latter country there is a very large reserve of cultivable land while in India the amount of such land is relatively small.

The whole area of Turkey is about 19 crores of acres. If we deduct the mountains, waste land, and the forests, there remain about 14 crores of acres available for agricultural use. Out of a total population of 1.8 crores, from 80 to 85 p.c. are engaged in agriculture, about the same proportion to the total as occurs in India. Thus we get between 1.4 and 1.5 crores of people dependent directly on the production of the land, which allows between 9 and 10 acres per head. Nevertheless, between 0.4 and 0.5 crores of the agricultural population are either landless or are considered to have insufficient land to provide a reasonable subsistence for themselves and their families. 5.7 crores of acres are now under arable cultivation, but more than double this area is described as under grass and pasture. Some of this is really waste land and may be worthless, though this cannot be stated for certain. But it contains very many areas which would be greedily grasped at if it was available, while the present land system has prevented its being used. The general conclusion is that while there is enough land to provide about 40 acres per family, and thus, for a peasant country, there is no real land shortage, the use of that available is very poor and much remains unused. The primary purpose of the new land law is to secure the cultivation of as much of the country's surface as is possible by providing landless persons and those who have too small holdings, with land suitable for cultivation. Incidentally, it also provides, for people who have land but have not the capital to work it, funds for both capital expenditure and for working expenditure also.

The land which is to be made available under this law is either state domain which is not used for public purposes, land which belongs to local authorities but is not required by them, land of which the ownership is not known, and also land in private hands which is excess of what is required by the owners. Land which has not been sown, ploughed or planted for three years without good reason can also be taken for the purpose. For the present purpose, lands are classified as to size into three groups, small, medium and large. Small properties are under 125 acres in size: medium range from 125 to 1235 acres: large are in excess of 1235 acres. As there is not, in Turkey, a record of rights in land officially kept, the owners will be held responsible for establishing the boundaries of their property to the satisfaction of the authorities.

When land is to be obtained from private owners, the latter are to be paid a price which is determined by the value of the land as declared in connection with the land tax in 1944. For large estates, the amount so declared is that which is to be paid to the owner. For medium estates, the amount to be paid is double this figure, and for small properties, it is three times the value so determined. The land will be paid for by the State in 'Soil Bonds' which bear interest at 4 p.c. and will be paid off in twenty equal annual instalments.

The persons to whom the available land is to be distributed are in a definite order of priority. This order is as follows:—

(1) Those who do not own land but work as tenants or partners in agricultural undertakings.

(2) Those who have not enough land to make a livelihood.

(3) Agricultural labourers of long standing.

(4) Graduates of agricultural and veterinary schools.

(5) Those not farming, but who are believed to be willing to farm.

All persons who receive land must be Turks, must be farmers, and must cultivate the land *themselves.*

When land is distributed, accounts will be prepared showing what the new owner has to pay, and this amount will have to be paid in twenty annual instalments without interest, the first payment only to be made in the sixth year after the land has been taken. An interesting provision is that there will be a deduction of 5 p.c. of the remaining instalments for each of the owner's children as they reach primary school age.

The new owners of land will not be left without finance. For capital expenditure, such as for buildings and equipment, and for land improvement, money will be advanced by the agricultural bank, on the basis of repayment within twenty years, interest not to be higher than 5 p.c. Credit for money needed for current working of the land can also be obtained from the agricultural bank, with credit for one year. It will thus be seen that the financial burden which the State is prepared to take upon itself is very great, and this is a measure of what is considered to be the urgency of getting new land under cultivation and so increasing the produce of the country.

Other sections of the law deal with the terms under which the land distributed will be held and with arrangements for dealing with it on the death of the owner. I am not very clear as to what will ultimately be decided on with regard to the last point, but it is certain that there is no intention to allow the land again to be, on the one hand, fragmented, and, on the other to be again absorbed into large landlord estates. Very large powers are granted to the Ministry of Agriculture in the matter of making regulations in regard to most of these matters.

It will be quite clear after this very meagre summary, that the new law may mean a revolution in Turkish agriculture. By itself it does not necessarily mean very much, but when it is considered as one of a number of factors in agricultural development in the country, and when other factors such as technical improvement and efficient expert advice are being pushed ahead with great energy, it is possible that in a very few years, the rather depressing picture of cultivation in Turkey may be radically changed. I know enough of the Turkish peasants to feel that they are likely to respond to the possession of land and the relief from the bitterness of private finance, and, if they do, I cannot see the end of what is confessedly the first step in a long series of measures which may make them among the best cultivators in Western Asia.

*Chapter* 30

# WHEAT IN THE MIDDLE EAST*

During the last four or five years the Middle East has attracted attention, not only for its political importance, but also for its potential value as a source of food supply for the external world. By the Middle East is meant the countries which lie round the eastern end of the Mediterranean and extend farther east to the confines of India. On the north the boundary of our area is the southern coast of the Black Sea, and on the south the mountains of Abyssinia. Within these limits it is probable that agriculture first developed, and wild grasses were first cultivated to give crops of cereal grains. It was here that, probably about 15000 to 10000 B.C. (I),** the earliest wheat crop was reaped, and during almost the whole of the classical period parts of the Middle East were looked upon as the granaries of some at least of the political units of the west. Now, however, the situation is vastly different. Hardly any of the countries are much more than self-sufficient in bread grains (though several are exporters of barley), and they make very little contribution to the world's supply of its staple human foods.

*Varieties.*—All authorities agree that wheat originated in the area we are considering. In the ancient world Diodorus Siculus tells us that wheat was discovered by the goddess Isis in the country of Nysa, a high mountain in Phoenicia.

* Reprinted with permission from the **Empire Journal of Experimental Agriculture,** Vol. XIV (1946), No. 53, pp. 31-42.

** Notes may be found at the end of the chapter.—Editor.

Aaronsohn (2) claims that he has found a form of wild wheat on the hill-sides of northern Palestine. The earliest type of wheat grown seems certainly to have been emmer (*Triticum dicoccum*) or some progenitor of this variety. It is still grown by backward peoples in many parts of Europe and most parts of Western Asia as well as in Abyssinia. In South India, under the name of *khapli* wheat, it is quite extensively cultivated, largely because it is fairly resistant to diseases, notably to rust; but its cultivation is tending to contract and it will probably soon have little more than historical interest.

In later times, including the classical period, emmer was gradually replaced by two or three other types which still form the main wheat production in the Middle East. One of these, *Triticum vulgare,* common or soft wheat, which according to Vavilov (3) probably originated in the valleys of northern Iran (Khorassan), Afghanistan, or northern India, forms the basis of by far the greater part of the wheats cultivated at present in the world. They do not appear to be really adapted to a Mediterranean climate and have by no means so completely dominated the wheat position here as they have done in most of the temperate regions. They are, however, the natural wheats of the north of Iraq and the eastern regions of our area where the climate largely depends on the monsoon. Here, also, *Triticum compactum,* or club wheat, is largely grown.

The typical wheat of the Mediterranean region, and of all areas where the climate is similar, is *Triticum durum* or hard wheat. It appears to be the main wheat of cultivation in North Africa including Abyssinia, in Asia Minor, and in the countries adjoining these centres. It is not very susceptible to the more common forms of rust, which is, as we shall see later, a very important matter in the Middle East. Hard wheat is essentially a crop of semi-arid climates and will stand a good deal of salt in the soil. In Iran and Iraq it is at least equally important with soft wheat, and it forms by far the most common variety in Egypt. In Syria and

Palestine it is almost the only variety grown. It is taller than soft wheat and is always bearded. It is said (4) that the root-system is more extensive than that of *T. vulgare*.

In fact, in the Middle East, the present evidence seems to indicate that, whilst in Iraq and to the east of it, where, however indirectly, the influence of the monsoon prevails, the cultivation of *T. vulgare* does and should tend to be the most important, there is little doubt that in the remainder of our area the future lies with some form of *T. durum*. Unfortunately, except in America, until very recently little work was done in breeding improved forms of the latter variety, and the chance of large developments in wheat cultivation would seem to depend largely on such improvement in the Mediterranean area.

For, after all, the character of the wheat that can best be grown is determined by the climate, and in the Mediterranean area this is so definite and characteristic that it is always likely to cause an equally definite selection among the numerous kinds of wheat that are available. As applied to the whole of our area, the remarks of the Howards regarding the varieties of wheat cultivated in India are very much to the point. They say (5):

> "The wheats at present in cultivation in this vast Empire, in which a civilised agriculture has been practised from time immemorial, represent the survival of types most fitted for the conditions of the various tracts. Nature has eliminated the unfit, and the experience of past centuries, handed down by tradition, has taught the cultivator what soils and what tracts are suitable for this crop. Varieties of wheat introduced by sea from western countries have, in recent times at any rate, had no influence on the crop and have not been adopted by the cultivators."

This is equally true of the wheat-growing regions of the Middle East. Attempts to introduce American varieties and hybrids from Arizona into Egypt failed completely, and after three years' work Dr. Freeman discarded all American wheats in favour of Egyptian (6), though some Australian types have more recently been fairly successful. It would seem that any development in wheat-growing in the future will have to depend chiefly, if not entirely, on the utilisation of forms already growing in each section of the area, and improvement will mainly result by selection from them or by hybridization among them.

*Influence of climate.*—According to ancient writers in the Mediterranean region, wheat requires a rich humid earth, preferably deep valley alluvium, with plenty of ground-water, or fertile volcanic soils retentive of moisture (7). It is also fastidious as to rain-supply at germination, and if the autumn showers hold off for a few days after sowing, there may be failure (8) as a result of the character of the Mediterranean climate; in this the great bulk of the rain falls in the winter months from November to March, whilst the summer period, May to October, is practically rainless and generally very hot. As one proceeds east and south, the total rainfall tends to decrease till in Egypt, Arabia, and the Syro-Arabian desert, only irrigation cultivation is usually possible. There seems some reason to suppose, with Ellsworth Huntington (9), that within this general character of climate the Graeco-Roman civilization grew up in a period of increased total rainfall which lasted from about 500 B.C. to A.D. 200 and extended not only over the Mediterranean but also over most of south-western Asia. After the latter date it is suggested that there was a decline in the amount of rainfall, and especially in the autumn rainfall on which the grower depends for the sowing and germination of his crop, or that the latter, at least, became very much more uncertain. Though this theory of Huntington has not been accepted by most authorities, who consider that there is little evidence of any such change of climate (10), it seems certain that after about A.D. 200 the Middle East lost its character as the granary of the western world and was replaced at least partly by the supply of grain from North Africa. At any rate, from about this time, the Middle East tended to become less important as a wheat-growing region, till now the whole area is barely self-supporting in wheat production.

*Present production.*—The position at present is that the Middle East, including Turkey, Syria, Palestine, Iraq, Egypt, Iran, and Transjordan, contains units which are constant importers of wheat, and others with a small export. But the general picture is well shown by the following figures which indicate for the period 1934-7 the production and net export or import relating to each of the states. The figures, representing thousands of short tons, are approximate, but they tell the story (11).

| | Total production | Exports | Imports | Net utilization |
|---|---|---|---|---|
| Egypt | 1,305 | .. | 8 | 1.313 |
| Palestine | 97 | .. | 75 | 172 |
| Transjordan | 89 | 25 | .. | 64 |
| Syria and Lebanon | 549 | .. | 2 | 551 |
| Iraq | 527 | 58 | .. | 469 |
| Turkey (1937) | 3.600 | 65 | .. | 3,535 |
| Iran (1934-8) | 2,000 | ? | ? | ? |
| Total import or export | | 148 | 85 | .. |

The import and export figures for Iran are not available. Most of the smaller sections of the Middle East, however, absorb the whole of the net export shown above, and it is not too much to say that the area we are discussing has little or no wheat for export to other countries.

And yet wheat is almost universally grown. The writer has seen it in what would appear to be the most unlikely places, such as the wetter parts of the Turkish Black Sea coast, and it certainly forms one of the staple cultivations in all the countries. Cereals everywhere are the main crops grown, occupying up to 90 per cent. of the cropped area. The percentage of the cereal acreage occupied by wheat varies from 29 per cent. in Egypt to 64½ per cent. in Transjordan. In Turkey it is 50 per cent. and in Cyprus 55 per cent. It is not, therefore, that wheat is not now grown, but that the total area under arable cultivation forms only a fraction of the available land, and that the general yield per acre is small almost everywhere. Of course, where irrigation is used, the yield *can* be high, as in Egypt, but even without irrigation, where the rainfall is sufficient, very large yields are actually obtained, particularly near the Sea of Màrmora and in north Smyrna in Turkey and a few other places, where in fact they may be as good as anywhere in the world. In spite of the limitations caused by the Mediterranean climate, there is no doubt that wheat will grow and grow well over most of our area. With irrigation some of the greatest difficulties in wheat-growing are removed and in such cases yields may be obtained that are phenomenally high.

*Areas of production.*—Except in Egypt, and to a less extent in Iraq, most of the wheat is grown without irrigation. Among cereals, under such conditions, its only serious competitor is barley, which usually replaces it where the rainfall becomes too small or too uncertain for wheat; where the rainfall is satisfactory, wheat is almost always preferred as a cereal crop. In Anatolia (Turkey), though wheat is grown almost everywhere, there are three special large centres of production of wheat grain. The first lies in north-west Turkey, the second lies on a stretch across the plateau of central Anatolia, and the third between the above, stretching from Ismir northward to the Black Sea. In 1938 Turkey was credited with a wheat production of 4,300,000 m.t. as compared with 2,500,000 m.t. of barley. In Iran the valleys and basins between the ranges of the northern mountains produce heavy crops of both wheat and barley, with a centre perhaps most characteristically at Meshed. But the whole fairly well-watered fringe of the Iranian plateau is well adapted to wheat-growing, which of course disappears equally on the damp coast of the Caspian and in the dried-out deserts of the south. In Iraq the northern plains grow the wheat, which disappears in the purely irrigated tracts farther south. In Transjordan and Syria the crop struggles against drought, against a vicious land system, and against the continual fear of attack from the nomads of the south. But here, in the Hauran, there lies one of the ancient world's sources of wheat. This is a high plateau, east of the Jordan rift, about 2,500 ft. high, where west winds from the Mediterranean give a fairly constant rainfall of about 400 mm. in winter and early spring, supplemented by considerable dews. This is an old volcanic district with lava soil, rich and retentive, overlying limestone, which at one time supported a large farming and live-stock-breeding population. It is this region which causes Transjordan to be an exporter, albeit a small one, of wheat to the adjoining areas of Palestine. For Abyssinia and Arabia we have no data as to the yield usually obtained, but we do know that there is no export from either of these countries.

The other reason why the production of wheat in the Middle East is so far below what might have been expected, even when full allowance has been made for the fact that it is grown mainly on non-irrigated land, is because there are very large areas of possible cultivation which are not cropped or are cropped only at intervals of years. Even in Turkey, only 14 per cent. of the land is cultivated every year, 14 per cent. is never cultivated, and nearly 60 per cent. is pasture. Similar figures probably occur in all the countries, and even if we exclude the regions which are too dry or where the rainfall is too uncertain, only a very small proportion of the remaining area is actually under crop in any one year. The system of a bare fallow every 2 or 3 years is very common. In view of all one knows it is legitimate to feel that the possibilities of largely increasing the production of wheat in the Middle East are considerable. Among the countries which could contribute to such an increase, one of our greatest authorities (12) specifies Abyssinia, Iran, Arabia, and Afghanistan, and to these we could add Turkey at least.

*Difficulties.*—Leaving aside such semi-political questions as the existence of a vicious land system almost everywhere (by which the peasant cultivator is taxed and rented so highly that he is left with only a bare subsistence from the land, and under which any improvements made by him lead only to higher taxation and a higher rental), and the baneful influence of moneylenders and merchants, there are special agricultural difficulties that make wheat-growing a far more uncertain proposition even in the irrigated areas of Egypt, Iraq &c., and, to a much greater extent, in the regions which depend on the natural rainfall to grow and ripen the crops, than in the wheat-growing areas of western Europe. The extreme riskiness of this crop is well shown by the production in Palestine in two successive years. In 1937 the recorded yield was 127,250 t.: in the following year it was only 44,000 t., so that in the latter year the wheat crop was only 35 per cent. of that reaped in 1937, apparently entirely caused by meteorological conditions or diseases induced by them. The report to the League of Nations on the crops of 1938 states that although there were good rains up to the New Year, 'which caused heavy and rapid growth on the coastal plains, the spring rains of early April, followed by a rise of temperature, brought a universal attack of rust (*Puccinia graminis*) which affected the wheat from Dan to Beersheba'. I quote this case as it is one about which we have full data, but similar experiences are found wherever the Mediterranean climate prevails.

Over almost the whole region we are considering wheat is planted in autumn, except at high elevations, and in some regions, like Abyssinia, with a summer rainfall. Although the durum wheats are by nature a spring crop, even they must be sown in the autumn owing to the climate. The first difficulty is, therefore, the uncertainty of the autumn rains. The sowing is determined, of course, by these, and they prevail from the latter part of September to November. The rain then increases usually in December, January, and February and begins to fall off in March and finishes in April. The early cessation of the rain is an even more serious likelihood than its late beginning, and is almost always considered as perhaps the most serious risk in the cultivation of wheat. Associated with the latter is the early arrival of the hot dry weather, which produces what the French call *echaudage* or scorch, causing the grain to shrivel either at the time of its formation or as it ripens. This often leads, in the regions of the greater summer heat, to a loss of from 25 to 40 per cent. of the crop. It is usually produced as the result of intense sun accompanied by dry winds following a period of damp weather favouring vegetative growth. It is especially prevalent on soils that are badly worked, shallow, impervious, or wet (13). The damage from this cause is specially severe where the early spring temperature is high, some authorities stating that it becomes serious only when this exceeds 32°C. Frosts, especially late frosts in the spring, often cause serious demage, and may lead to a violent attack of rust, whilst excessive rain during the winter may lead to the same result. In fact the frequent occurrence of rust is perhaps the most important factor in determining whether wheat can or cannot be grown and the type which is most suitable for any particular set of conditions.

*Wheat Cultivation in the Individual Lands of the Near East*

*Turkey.*—Turkey is by far the largest producer of wheat in the Middle East. On the average of the years 1934-8 there were 8.5 million acres with an estimated production of 125 mn. bu. of grain. There is no country in the Middle East where more energetic measures have in recent years been taken to raise the standard of cultivation of the crop to a much higher level than prevails at present. All parts of the country grow wheat except a few areas on the Black Sea coast and some of the high level districts like Kars in the eastern highlands, but the concentration is greatest in the west of Anatolia and in the central plateau. Wheat accounts for about a quarter of the value of the agricultural produce of the country. The area under wheat increased by 25 per cent. between 1929-33 and 1934-8.

The dominant wheat is *T. durum* or durum wheat, but there are also grown large areas of *T. compactum* or club wheat, which occupies about 13 per cent. of the area under the crop. In the eastern regions large areas of *T. vulgare* are grown, which is nearly all a genuine spring wheat and a few years ago occupied one-third of the area covered by other types. One of the greatest hindrances to the development of wheat cultivation in Anatolia is the very great mixture of types that are grown together. The fields of wheat are nearly always mixed with rye, often with barley and oats, whilst the mixture of durum wheats, club wheats, and soft wheats is universal. Improved strains of wheat have been discovered by the Government research stations, and their use is gradually being extended. Great efforts have been made to replace spring wheats by winter wheats where this is possible on account of the greater yields thus obtainable. This work is not easy in a peasant country. There is no shortage of land, though the holdings are small, and the usual method of cultivation in the past has been to leave the land fallow every second year.

Among wheat diseases rust, both *Puccinia graminis* and *P. rubigo-vera*, is common, especially after cold rain in winter if this is long continued, or if the crop is grown in wet lands, whilst a damp spring is fatal owing to the increased incidence of this disease. Smut (*Ustilago tritici*) is almost universal, especially on spring wheats.

It is difficult to see how very much more can be done than is being done to extend the cultivation of wheat in Turkey and to increase its productivity, without very far-reaching changes in the organisation of agriculture. But everything that the writer has seen seems to show that there is no *agricultural* reason why Turkey should not again be one of the granaries of the world. The fact that in certain areas a yield of 25 to 30 bu. per acre is common (in the Van district it is stated that over 60 bu. per acre have been reaped) shows what can be done and how the yields obtainable are comparable with those secured in the best parts of western Europe.

Ataturk was especially identified with attempts to popularise the best methods of cultivation, and the peasants have been freed of tithe and some other taxes. But the burden of taxation is still very heavy. On the other hand, the Agricultural Bank guarantees a fixed price for wheat by its own purchases, and there are arrangements for credit. The wheat so bought is stored in silos and sold later on the Bank's account.

*Syria and the Lebanon.*—In this area there are mountains close to the coast, and an undulating plateau to the east running down to the Tigris-Euphrates basin. The amount of winter rain decreases rapidly from Lebanon to Syria, but the Bekaa valley between the Lebanons grows large areas of winter cereals and pulses. Syrian agriculture is primarily cereal production, and a characteristic feature is the long fields cultivated in strips by communal villages. The usual rotation is wheat, fallow, and lentils or vetches. The north-east corner of Syria is fertile with good winter rains, sufficient for wheat, and was one of the granaries of Rome: now only relatively small areas of it are farmed. As one proceeds from the irrigated areas round Damascus into the deserts at the east and south, the sequence is dairying, olive and fruit plantations intercropped with cereals, vines, cereals alone, and desert.

The cultivation consists almost exclusively of durum wheat, and the most frequent type is a reddish one termed *haurani*. It is sown in October and November and reaped in May or June. In the north the chief difficulties encountered are: 1—a low winter temperature, often as low as —10° C.; 2—drought during March and April; 3—frosts during March; and 4—hailstorms in spring. Of these by far the most important is the lack of rain in March and April, with drying winds and high temperature. The other difficulties are comparatively rare, but there have been cases, even in recent years, when almost the whole of the wheat crops were lost owing to frosts both in winter and in spring. The uncertainty of the climate is shown by an estimate (14) that in central Syria out of 10 years there will be 3 bad on account of drought and *echaudage*, 4 years about average, and 3 years with a good and satisfactory yield.

Most of the wheat is grown without irrigation, but where water is available it is often used, as in the neighbourhood of the Euphrates and in the Lebanon valleys.

*Palestine*.—Passing from north to south in Palestine, the rainfall decreases in the coastal plain from 20 to 10 in., in the Jordan valley from 20 to 2 in., and in the central range from 35 to 25 in., but the annual variation is considerable and ranges from half to double the average. The most general system of agriculture is a 2-year rotation. The winter crop is wheat or barley, and is rotated with a winter leguminous crop or with a summer crop in the succeeding year such as *durra* (sorghum) or sesame. A 3-year rotation is also in use. There is little manure applied and what might be used is often sold to citrus plantations.

In this typical area of the Mediterranean climate we have two well-marked periods of precipitation. The 'former rain' in October and November is not usually large, but it increases in December, January, and February, begins to fall off in March, and finishes in April. There is a moist W. or SW. wind in winter, dry N. or NW. wind in summer. The dreaded sirocco from south and east comes in April and May and sometimes in September and October. But both at the beginning and end of the wet season the rain becomes doubtful, and it is not so much a deficiency in quantity as this fact which causes the very great differences in its effectiveness. Thus, for example, in 1938 although the aggregate rainfall was somewhat above the average, the irregularity in its incidence together with a serious and widespread outbreak of rust resulted in yields of both wheat and barley that were most disappointing both in quantity and quality.

Most of the wheat grown is durum, as this is more resistant to rust and also to drought. It is, as usual, hopelessly mixed and the agricultural department has made great efforts to obtain and spread purer stocks. In 1937-8, 187,700 kg. of such seed were distributed. The sowings begin in November and continue up to the end of January. The harvest dates from about the middle of May, being later in the north than in the south. Complaints are frequent of failure due to lack of rain in the autumn. If there are less than 70 mm. of rain in the two sowing months, there is likely to be failure. This occurs in 40 per cent. of years at Jaffa. Equally difficult is the lack of rain at the critical period of March and April, and if this is accompanied by the sirocco winds, there is bound to be large-scale complete loss, and what is not lost by shrivelling of the plants may be lost by scorching of the ears (*echaudage*).

*Transjordan.*—This country, including the portion in Syria, was once one of the granaries of Rome and supported a highly developed and advanced population of about a million people—three times its present primitive and backward population. The region's agricultural and cultural development reached its greatest height under the Nabateans before the second century A.D.; subsequently it was dominated by the Roman and Byzantine empires, and finally blotted out by the Arab invasions beginning in the seventh century (15).

The climate is essentially subtropical and steadily gets more and more arid as one proceeds to the east and south. In the Jordan valley wheat and barley are grown under irrigation, and a rotation of wheat in the winter and white *durra* (sorghum) in the summer is common. The plateau region to the east of this was where the great supplies of wheat were raised at one time, the climate being typically Mediterranean, having a rainy season lasting from October to May and a long dry summer season from May to September. But it is very irregular, and Lowdermilk (15) notes that 'reliance of farming and grazing on regularity of rainfall has often brought disaster to man and beast alike'. It is evident that this irregularity must be taken into account in any attempt to regain some of the wheat-growing capacity that the region formerly possessed.

*Iraq.*—The great possibilities for wheat culture of the upper valley of the Euphrates in Turkey have already been mentioned. This area extends far to the east right up to the edge of the Iranian plateau, and wheat is a large crop as far south as the 36° N. parallel. To the south of this wheat gradually disappears. In all this area, with a strongly continental climate, including the possibility of very great cold in winter but a certainty of great heat in the spring, and with a likelihood of severe drought when the corn is ripening, wheat is cultivated under great difficulties and only varieties which have been developed in the area can stand the conditions. Even as it is, the spring drought and the accompanying heat often lead to shrivelling and *echaudage*. The possibilities of irrigation in these regions are rather limited, and in any case it seems likely that durum wheats will be of the greatest importance, or some types of common or club wheats which have been developed for similar climatic conditions elsewhere.

*Iran.*—Except in the higher mountain regions, where spring wheats are grown without irrigation, wheat in Iran is almost entirely an irrigated crop. Hence, though wheat and barley are the predominant crops in the country, the normal system of cultivation is to leave a considerable proportion of the land in fallow largely because water is limited. Only from 10 to 15 per cent. of the land is cultivated, but a further 20 to 30 per cent. could be cultivated if water were available. The extent of wheat cultivation is therefore very largely dependent on the abundance of the autumn and winter rains in the mountains bordering the Iranian plateau giving a sufficient amount of water for adequate irrigation. At the same time, if water is abundant and applied too freely, there ensues a bad attack of rust.

Winter wheat is usually sown in the latter part of October: if delayed till late in November, there is danger on the high plateau of Iran that the winter frosts will damage the wheat. Irrigations are further given from February onward and harvest takes place from June onward.

In Iran we are approaching what appears, according to Vavilov (16), to be the centre of origin of *T. vulgare* and *T. compactum*, and hence these are more commonly grown than is usual in the Mediterranean climate of the other countries of the Middle East. But they are restricted by the extreme susceptibility to rust of most of their strains, for in such areas as Fars and Kerman, where rust is very common and serious, they have to give place to less sensitive types.

*Afghanistan.*—The cultivation of wheat in Afghanistan has been studied in great detail by Vavilov (17), who finds here the centre of distribution of *T. vulgare* or common wheat, and of *T. compactum* or club wheat. Durum wheat does not occur even in admixture (Vavilov). Dry wheat, i.e. non-irrigated, is composed entirely of common wheat; irrigated wheat consists almost entirely of club wheats. Both soft and club wheats as grown in Afghanistan are very subject to brown, yellow, and black rust, and smut is very common indeed. Spring wheat is grown up to an elevation of over 10,000 ft. and winter wheat up to over 9,000 ft. It is impossible to give any idea of the yields usually obtained, but they are almost certainly small, and the wheats, as grown, are almost hopeless mixtures of varieties.

*Egypt.*—The cultivation of wheat in Egypt is among the most interesting in the whole of the Middle East. It is all irrigated either by basin, water being applied during the river flood only, or by canal, 2, 3, or 4 irrigations being given during the growth of the wheat. The chief wheat grown is *beladi*, a durum wheat, occurring in both red and white forms. The so-called Hindi wheat, which is grown only on canals, is a soft wheat brought from India and is less mixed than those of local origin, but it has proved very susceptible to rust. Of late years some varieties have been introduced from Australia, and have given excellent yields under canal irrigation. Most wheats are sown from October to the first half of December, and are reaped in May or the first half of June.

There is no doubt that the great enemy of wheat-growing in Egypt is rust. Foaden and Fletcher stated that rust was rarely serious, but that is hardly so now. Balls also states that rust does little damage to cereals, as the local wheats are naturally immune. Nearly all more recent writers lay stress on its importance, and it is said that even local Egyptian wheats are severely attacked by yellow rust. Azzi (14) states that rust and other maladies (chiefly smut) cause 80 per cent. of the serious damage to the wheat crop in Egypt.

*Abyssinia and Eritrea.*—In Abyssinia, which Vavilov considers as the most probable centre of distribution of durum wheats, and where 75 per cent. of the wheats cultivated consist of this variety, we have a region of summer rains where, after slight falls in March and April, the main rainfall is received in June to September. The country is generally well watered and is of high fertility, and the standard of peasant cultivation is high. There is little doubt that Abyssinia could be a large producer of wheat for world use if transport facilities existed for its movement in and out of the country. All the wheat is grown as a summer crop, and is usually sown in May or June and reaped from October to December.

In Eritrea there is a summer rainfall of 20 in. on the north and south plateaux and about 12 in. on the western plains. A recent observer informs the writer that although there is a genuine agriculture on the south plateaux, the land is not fertile except in spots, and most of the people are warriors and not farmers. On the northern plateaux, on the other hand, the people are nomad pastoralists. The western plains are the most immediately capable of development, and this is being done largely by Sudanese and West Africans.

*Conclusions*

This survey of the cultivation of wheat in the various countries of the Middle East indicates that so far as climate and soil are concerned wheat-growing could be greatly extended, and that there is little to prevent its becoming again an important factor in the world's supply of this cereal. But to do this several things are needed.

First, unmixed varieties of wheat must be grown. Everywhere it is complained that the peasants are using seed which is a hopeless mixture of types, even when it does not contain rye, barley, and oats. This necessity has been recognized by the agricultural departments in almost all the Middle East countries, and their energies have been concentrated on producing and spreading selected seed among the wheat-growers. Such work is slow and wearisome, but considerable success has been achieved in Turkey, Egypt, Palestine, and several other countries. At the Karej station in Iran they have as many as 1,500 varieties on trial, with facilities for multiplying those which prove themselves valuable.

Not only is it necessary to provide unmixed varieties, but they must be at least fairly resistant to rust, that curse of wheat-growing in the whole of the Middle East. Resistance to rust depends also on the prevalence of good agricultural conditions and particularly on the absence of waterlogging of the land owing to lack of drainage or to excessive winter rains. Rust may also follow late frosts in the early spring, a contingency which can hardly be provided against except by the development of varieties which ripen early.

Early ripening is also important as a protection against *echaudage* due to hot dry winds at the time of grain formation or ripening. The writer is not aware that very much attention has been given to this matter, but in view of the great amount of damage said to be produced in the greater part of the area every year, it would seem to be one of great importance.

Manuring is important if more than 11 or 12 bu. per acre are to be obtained, even with the best of climatic conditions. It is one of the dangers of the extension of intensive cultivations, like that of citrus in Palestine, that it drains the surrounding country of its manure, and such crops as wheat suffer in consequence. But unless a very frequent fallow is given (and this means the waste of much of the land from the point of view of production) high yields are impossible without the use of a good deal of manure, especially nitrogenous manure, whether derived from organic or other sources.

In view of the paramount importance of water-supply both in the autumn and in the early spring for wheat cultivation in a Mediterranean climate, and of the irregularity of rainfall in most of our area, the importance of water conservation can hardly be overstressed. It is perhaps the uncertainty of the rain in the spring that is the biggest cause of failure of the wheat crop, and so the use and development of early maturing strains of wheat affords the most likely method of meeting the difficulty. How far the method of vernalization worked out by Lysenko and other Russian investigators has been applied to wheat in Turkey and other Middle East countries is unknown to the writer, but if even a fortnight could be gained in the spring it would be of enormous importance in securing a far more certain crop than is now possible. Apart from these methods, all arrangements for checking soil erosion and retaining water in the soil should be of considerable value, and what has been already done on the Jewish colonies in Palestine gives a lead to workers in other parts of the Middle East. The way in which the methods of so-called 'dry farming' have been applied in India are well worth study by those who wish to increase the chance of high yields in the more doubtful areas of the countries under study (18).

The more the situation is studied, both in the Middle East and in the other peasant countries of the world, the more one feels that if agriculture, and specially wheat-growing, is to be pursued on a more reliable basis than it is now or has been in the past, one of two things must happen: either peasant cultivation must be abandoned, or a thorough-going system of co-operative work by the peasants must be substituted for the purely individual action that now exists. This is not the place to develop this thesis, but a long experience in several peasant countries of Asia has convinced the writer that there is hardly any third alternative if a sound agriculture, with a basis of wheat-growing, is to fit into the world-system of the future.

## REFERENCES


1. SCOTT ELLIOTT, Prehistoric Man and His Story, London, 1915.
2. A. AARONSOHN, Agricultural and Botanical Explorations in Palestine. Bur. Plant Ind., U.S.A., 1910. Bull. No. 180.
3. N. VAVILOV, Studies on the Origin of Cultivated Plants. Bull. Appl. Bot. & Pl. Breed., 1926, **16,** 155.
4. DONDLINGER, Book of Wheat, London, 1908; C. L. ALSBERG, Durum Wheats and their Utilisation. Wheat Studies, 1936, **15,** 337.
5. A. and G. HOWARD, Wheat in India, Calcutta, 1909.
6. R. H. FORBES, Sultanic Agric. Soc., 1923. Bull. No. 10.
7. VERGIL, Georgics ii, 217.
8. A. PHILLIPPSON, Das Mittelmeergebiet, Leipzig, 1907.
9. ELLSWORTH HUNTINGTON, Climatic Changes New Haven, 1922.
10. V. CONRAD, The Climate of the Mediterranean Region. Bull. Amer. Met. Soc., 1943, **24,** 127.
11. A. I. TANNOUS, Foreign Agriculture, Washington, 1943, **7,** 243.
12. T. W. GRINDLEY, Proc. World's Grain Exhib. & Conf., Regina, Canada, 1934, p. 119.
13. F. and P. BERTHAULT, Le Blé.
14. AZZI, Le Climat du Blé dans le Monde, Rome, 1930, p. 819.
15. W. C. LOWDERMILK, Palestine, Land of Promise, London, 1944, p. 142.
16. N. VAVILOV, Studies on the Origin of Cultivated Plants, Bull. Appl. Bot. & Pl. Breed., 1926, **16.**
17. ——and D. D. BUKINICH, Agricultural Afghanistan, Leningrad, 1929.
18. See N. V. KANITKAR, Dry Farming in India, Simla, 1944.

*Chapter* 31

# MILLETS IN THE MIDDLE EAST*

Among the cereal crops, it is rather astonishing that so little attention has been devoted to the study of the growth and improvement of the various kinds of millet that are cultivated in many countries of the old world. And yet perhaps it is not really astonishing. The millet grains take no considerable part in international commerce; they are, in all countries, the food of the poorer classes, and are little cultivated in the more advanced agricultural regions. On the other hand it must never be forgotten that they form the principal food of perhaps a quarter of the world's population. They are the dominant cereals in the whole of Africa except the north. They are grown in India on about 60 million acres with an estimated production of 12 million tons of grain. They are the mainstay of the people of northern China and Manchuria. They are specially important among the nomadic people of central Asia, and are a large factor in the cereal production of SE. Russia.

Moreover, the millets have certain advantages over any of the more usual forms of cereal. Some of them, notably the various Panicum millets, grow very rapidly and ripen more quickly than any other cereal crop. They are, hence, suitable where the rainy season is very short. Some of them resist drought in a remarkable manner and can be grown

* Reprinted with permission from the **Empire Journal of Experimental Agriculture,** Vol. XIV (1946), No. 56, pp. 208-216.

where no other summer crop can be profitably cultivated. They are less exhausting than maize and some of them will flourish on land which is too thin or too poor to grow any other cereal. They require less attention than maize, though they will pay for attention. They are attacked by few insect pests. Most of them flourish under high-temperature conditions—a very important matter in many parts of the Middle East. All these factors determine the position of the millet crops and their maintenance as important sources of human and animal food in the hotter and drier regions of the globe.

In their present distribution, however, the Middle East has only a minor place, though they are of very ancient use there. Unlike wheat and barley none of the millets seems to have had its origin in this region. Sorghum appears to have come either from India or Africa (1). The various Panicum millets are of central or eastern Asian origin. Bulrush millet comes almost certainly from Africa (2). *Eleusine coracana*, the *ragi* of India, which is cultivated on 6 million acres in that country, has very probably been developed there. But although all the millets come from outside the Middle East, they have been known and used there from the very earliest times. Sorghum has been found in Sennacherib's palace at Nineveh, dating from about 750 B.C. (3), and though the suggestion that it has been found in Egyptian tombs dating from 2200 B.C. is probably incorrect, yet this millet was certainly known in Egypt from a very remote period. In the prophecies of Ezekiel a grain known as *dukhn* is mentioned which may be either *Setaria italica* or *Pennisetum* sp. It is difficult to get complete data about the present status of their cultivation with the partial exception of sorghum (*durra*), as in no case is the area under each of them separately recorded. In Egypt the area under millets, including sorghum, is between 200,000 and 300,000 acres, almost entirely in Upper Egypt. In Turkey, they occupy quite a small area compared with maize, the mean for 1940-2 being 186,000 acres with an estimated yield of 64,000 m.t.

of grain, whilst for the same years over 4 million acres were under maize with a yield of 600,000 t. of grain. In most of the other countries of the Middle East, the data needed to get an exact idea of the present status of millet cultivation as a whole, relatively to other cereal crops, do not exist, though in most of them sorghum cultivation, at least, is important.

Two or three of the millets that are grown on a very large scale in other parts of the world are hardly found in our region or only on its edge. These are (*a*) Pearl or Bulrush millet (*Pennisetum typhoideum*), (*b*) *Eleusine coracana,* the *ragi* or *nagli* of India, and (*c*) *Eragrostis abyssinica,* the *Am tef* of Ethiopia. The first of these (*a*) is very largely cultivated in Nubia and the Sudan and is also grown in Arabia and Abyssinia. Beyond this it is hardly grown at present though at one time it seems to have been produced extensively in Mesopotamia. It has great resistance to drought and grows in regions with less than 15 in. of annual rainfall. It is a cereal for warm areas with low rainfall. The second (*b*) (*Eleusine coracana*) is still less common. It is occasionally met with in Abyssinia, but only the white kind is used for food, the other varieties being used for making beer. The *Am tef* of Abyssinia (*c*) is locally important, and forms a very usual article of diet in the country. It grows on thin and dry soils and ripens in 50 days. Both brown- and white-seeded types are cultivated; but this grain has not spread beyond its borders in the Middle East. On the other hand, the crop has been taken to South Africa where there are now 500,000 acres under cultivation, almost entirely as a fodder crop (4).

In view of their limited distribution, I shall, however, say no more of these three millets in the present paper.

### *Sorghum or Great Millet*

By far the most important of the millets is *Sorghum vulgare* or the Great Millet. The two grand regions for sorghum culture are Africa and India, with a subsidiary but important area in north China and Manchuria. But in most of the countries of the Middle East, sorghum is grown and valued as an irreplaceable element in the cultural system. It disappears in the extreme north of the area, and is also hardly found at high elevations. This may be due to the absence of a sufficiently high temperature during the growing season, for the Americans have found that in the U.S.A. it is successfully grown only where there are at least 160 days free from frost, where the rainfall is 18-40 in., and the average July temperature is above 76° F. (3). It is hence unimportant on the Persian plateau and in Afghanistan, whilst in Turkey it is rarely found in the northern part of Asia Minor or in the high lands of Armenia.

The main advantage that sorghum has over maize is that it requires less rainfall. Where there is abundant water, the tendency is for sorghum to be gradually replaced by maize, as has happened in Egypt. But in a country liable to small or poorly distributed rainfall, the sorghum plant has the faculty, not possessed by maize, of remaining dormant during short periods of drought, whereas maize would be killed or greatly damaged. This is a very important quality in countries where such poor distribution of rain is likely to occur during the growing period. Further, sorghum is less exhausting than maize; for example, it has been found in Palestine that although legumes must be grown after maize, a second cereal crop can be taken after sorghum (7). The superiority of sorghum to maize in a region of limited rainfall is shown by the fact that the water-requirement, per pound of dry matter produced, is less than for any of the common grains; for example, to give one pound of dry plant wheat requires 513 lb. of water, barley 434 lb., maize 368 lb., but sorghum only 322 lb. (6). Further, long ago Hilgard (8) reported that whilst maize is rather sensitive to alkali in land, sorghum will succeed on white alkali soils. This is important in regions where the concentration of salt in the soil is bound to be considerable even if it does not appear as an incrustation.

The sorghum grain is not capable of making good bread with yeast when used alone, though it can be mixed to the extent of 20 per cent. with wheat without spoiling the bread. On the other hand, it grinds easily, is economical in use, and can be eaten as a porridge or in unfermented bread, as is common in tropical countries. It is an excellent food for animals and in some of the countries of the Middle East (Palestine, Transjordan, Syria, Lebanon) it is said that 50 per cent. of the crop is so consumed (9); as animal food, tests indicate that it may be about 89 per cent. as efficient as maize (10).

One of the great advantages of sorghum as a crop is that the straw, even after the grain is ripe, remains a good fodder—far better than that produced by maize. Sorghum straw is, in fact, the principal fodder of cattle in a very large part of India and in every country where sorghum is grown for grain. The leaves remain green much later than do those of maize. None of the other millets is as useful for this purpose as sorghum and, though there is danger in feeding fodder from immature or stunted plants (owing to the possibility of poisoning), there is nothing against the mature straw. Of course, sorghum can be grown for fodder only, when it is cut while the grain is in the milk-stage. Special varieties are kept for this purpose in India and also in Egypt, and it has been stated with truth that no other crops can compare with the sorghums in yielding a heavy weight of green fodder of good quality (11).

Except in NE. Africa and in Arabia, practically only two types of sorghum are grown extensively in the Middle East. These belong to the species *Sorghum durra* and *S. cernuum* (1). In N. Africa and the Sudan many other forms occur and are cultivated, but these two (which are very closely related) seem to be almost the only ones which have penetrated, as cultivated plants, into the greater part of SW. Asia. In both these types the panicle is shorter and more compact than in most other varieties. The grains are not large and are flattened, whilst the tip of the stem is often curved so that the head hangs down in a way that is called 'goose-neck'. One of the reasons why these types alone have continued in our region is that they ripen fairly quickly, i.e. in 80-110 days, if grown in the hotter part of the year. This is very important, for most of the other common sorghums of NE. Africa and SW. Arabia are much later in ripening (12). Except in Egypt, where irrigation facilities are universal, the qualities mainly desired are: 1—short plants, not more than 2 m. high, 2—early ripening plants, not needing more than 3 or 4 months to mature, 3—very dense heads with grains full at the top and the bottom, and 4—a high yield of grain.

We may now consider the status of sorghum cultivation in each of the more important countries of the Middle East.

*Egypt.*—The cultivation of sorghum for grain is of very old standing; now there is a tendency for its decline and for it to be replaced by maize; it is almost restricted to Upper Egypt. There are two common types, both of them forms of *Sorghum durra,* the first of which is known as *sefi* and is grown in summer, being sown in March or April and reaped in August, and the second, known as *nabari,* sown in August and reaped from October to December. The yield in both cases may go up to over 2,000 lb. of grain and about 8 camel-loads of fodder per acre. From figures given by Dudgeon, the mean yield of the sorghum crop for the whole of the country would appear to be over 2,000 lb. per acre. During the recent war, however, the yield per acre went down by over 18 per cent., chiefly due to the shortage of fertilizers.

The crop is not badly attacked by pests. Borer (*Sesamia*) may be serious but is not increasing. *Aphis sorghi* has been described as the most serious pest of sorghum, but in Egypt it is only of local importance (13). Among diseases, grain smut is by far the worst and is almost universal though it can easily be checked by seed-treatment. Birds are perhaps the most troublesome enemy of the crop, which has to be carefully watched when it is near ripening.

The grain is very little used as food in Lower Egypt, but is the main cereal of the people in Upper Egypt. It is generally used mixed with wheat, barley, or beans. Half sorghum and half beans makes the food of the poorest people (13).

Sorghum forms 10 per cent. of the cereal acreage of Egypt (9), where the estimated total production is about 300,000 tons. A small amount is exported, and about 20 per cent. of the production is used for feeding animals.

*Palestine, Syria and Lebanon,* and *Transjordan.*—In these countries sorghum is a very important crop, but in none of them is it so vital as in Palestine where it forms 23 per cent. of the cereal acreage (in Transjordan 5½ per cent. and in Syria and Lebanon 8½ per cent.). The total production is estimated to be about 51,800 tons in Palestine, 7,700 t. in Transjordan, and 86,000 t. in Syria and Lebanon. Of the amount available for use in these countries, about one-half is used for animal feeding (9).

In all these areas the goose-necked form of sorghum is almost universal. In Palestine, in the majority of the areas where summer crops can be grown without irrigation, sorghum is probably the only remunerative crop. In years of sparse spring rainfall, *durra* (i.e. sorghum) can be grown where maize would fail, and south of a line from Jaffa to Nablus it is probably the only paying cereal crop (7).

In *Turkey,* sorghum is cultivated chiefly in western Anatolia, especially in the valley of the Buyuk Menderes, in the vilayets of Aydin and Mugla. In this region it is one of the essential cultivations. In other parts of the country, it is only sown locally, but in the Taurus the fields of sorghum become frequent. The most common type of sorghum is that already described as being usual in SW. Asia, but kafir corn (*Sorghum caffrorum*), both with a red and white pellicle, is common. Sorghum grain is ground and made into cakes, and is also largely used as a food for stock.

In *Iraq,* sorghum is cultivated extensively as an irrigated crop in summer in the riverain areas between Basra and Mosul, especially along the Euphrates. Cultivation increased considerably in the years before the recent war, as the grain is 'well liked by the people as food for human consumption' (16). It forms about 9 per cent. of the cereal acreage and the production before the war was estimated at about 140,000 t., about 25,000 t. of which were exported. About 30 per cent. of what is produced is fed to animals, and the remainder is used as human food.

In *Persia* and *Afghanistan* the crop is of very minor importance. There is a small amount of sorghum grown in most parts of Persia, but especially on the plateau, though it is not in any way a staple of the country. In Afghanistan it is still less important.

*Arabia* is one of the very ancient homes of sorghum cultivation, and the crop is grown everywhere, both for grain and as a fodder crop, in the latter case under irrigation. In Aden a recent authority states that the main dry-land crop is sorghum, and that it gives from 400 to 600 lb. of grain per acre; for value as food and fodder it is unlikely that any crop can compete with it (17).

### *Panicum Millets*

Though sorghum may perhaps be considered as the most important of the millets, yet there is another class which is at least as widespread in the Middle East, but cultivated in a very different part of the area and under very different conditions. This is the class of the Panicum millets, which were probably developed in eastern or central Asia, but which have spread into all tropical and semi-tropical countries where the rainy season is short, where the soil is thin and poor, and especially where a crop is wanted that with very little effort can be relied on to give a fairly large amount of fodder as well as grain. The northern limit of the cultivation of these essentially semi-tropical crops appears to be about a June isotherm of 63° F., or possibly a July isotherm of 68° F.

There are four important species of this type of millet:

1. *Panicum miliaceum*, the common *proso* millet of Russia, has now spread far, not only in the Middle East but also in India and in the U.S.A. where it was probably introduced by Russian immigrants. The head of this millet consists of a large open-branching panicle somewhat like that of oats. It has coarse, hollow stems usually about 30 in. high. It is sensitive to frost and so should not be sown till late in the spring when the ground is warm and if possible moist. It grows slowly at first, but with increasing heat comes on very rapidly and ripens in from 50 to 60 days from sowing under favourable conditions. It seems to have the lowest water-requirement of any grain crop (310 lb. of water per pound of dry matter produced) (6). Its roots are very shallow. It grows well on any kind of soil except coarse sand, and is well adapted to plateau conditions and high elevations. In dry areas only 22 lb. of seed per acre are needed for sowing, with a little more in damper conditions. It is not usually considered suitable for irrigation, though in central Asia it is irrigated (18). The straw does not make the best of fodder and even the hay made from the flowering plants is poor. The ripe crop easily scatters the seed and hence it is usually cut when the upper part of the ear is still green.

The yield of grain under the conditions where it is usually grown—on poor thin soil without manure, after another grain crop—is from 400 to 750 lb. per acre, though as much as 2,000 lb. have been obtained under favourable conditions. When used for human food, it has an agreeable taste with a slight nutty flavour. The meal is cooked into a mash or porridge. Mixed with four times its weight of wheat, it makes a good substitute for wheat alone. For animal feeding it appears to be rather better than oats and not quite so good as barley.

The best types of *proso* millet for the Middle East are probably those from SE. Russia, where alone, till very recently, any effort has been made to improve the type of crop grown. The Russian varieties are said to be much earlier than those from China or India, and this is a great advantage with a crop whose value lies largely in its early ripening.

2. *Panicum miliare* or the Little Millet. This millet is often confused with 1, and hardly deserves separate mention, for it is apparently grown in the Middle East on a very small scale, though it is extensively cultivated in India, chiefly in the south. It is noteworthy, however, for its reputation of growing on the poorest, otherwise worthless lands. It can withstand both drought and waterlogging. It grows in India up to 7,000 ft. elevation, and, there, needs 3½ months to mature (20). The yield is not widely different from that of *proso* millet.

3. *Setaria italica,* also known as *Panicum italicum* or *Chaetochloa italica*, the well-known Italian millet, is grown on almost as large a scale as the *proso* millet, either as a crop or an admixture. In Japan it is the most important millet in cultivation. Its chief advantage over *proso* millet is that it not only gives a good yield of grain (though the grain is not so good for any form of bread as is *proso*), but also the straw forms an excellent fodder. If grown exclusively as a fodder crop and cut when in flower, it makes excellent hay. It is also not so exacting in the matter of climate as *proso.* Chiefly for these reasons it has enormously extended in the U.S.A. (21), almost entirely as a heavy yielding, rapidly growing forage crop. The hay is stated to be better than timothy hay.

Like *proso* millet, it should not be sown till the land is warm, and a little later than maize. It usually ripens somewhat later than the *proso* millets, but can be cut for grain in from 70 to 90 days after seeding. A good crop will yield about 800 lb. of seed and 2,500 lb. of straw per acre, and may stand 5-6 ft. high; heads of grain 7 in. long have been recorded, though the average is much less than this. It will grow on very poor land, but will not stand waterlogging or a long drought. It lacks the ability possessed by sorghum of recovering after a period of drought.

4. *Echinochloa frumentacea*, also known as *Panicum frumentaceum* and *Panicum crusgalli*, called by some authors Poor Man's Millet, can hardly be considered as a crop in most of the Middle East, though it occurs almost everywhere where the other Panicum millets are found, often as a weed. I mention it because it is the quickest growing of all the millets and has been known to give a crop of ripe grain in 45 days after sowing, under favourable conditions of moisture and temperature. When grown it is sown in the middle of June when the soil is moist. Rain is needed till the middle of August, when it is cut. This presupposes a monsoon climate with rain in summer, and, apart from irrigation, it seems very doubtful whether there is much place for this millet in the greater part of the Middle East.

Such are the chief varieties of Panicum millet to be found in our region. What is their status at present and their future possibilities? In Egypt, the so-called Poor Man's Millet, i.e. *Echinochloa frumentacea* (Arabic *dineba*), is used as a reclamation crop on lands too salt for rice. It is chiefly found in Behera and Fayum, and is rarely grown except for the above purpose. It is usually considered to need plenty of water, but to have the capacity of standing drought. It yields about 10 bushels (say 510 lb.) of seed per acre, but is generally treated as a fodder crop and cut for this purpose (22).

In Turkey, almost the only one of this class of millets is *proso*, though it is often very much mixed with Italian millet. This is met with in many districts but most largely in SE. Anatolia, though it is also cultivated in the south-west. In central Anatolia (Konia), *proso* is likewise grown, and, generally, it is used where the soil is very shallow (15). In the areas between Turkey and Russia, both Italian and *proso* millets are grown. In Georgia, for example, *proso* is by far the most common though it is commonly very much mixed with Italian. Even now, the chief food of the people in W. Georgia is a thick porridge cooked from one of these two millets (23). In the high lands of Armenia, Italian millet supplants *proso* in dry years and in exhausted soils and becomes an independent crop.

In Iraq, on the middle Euphrates zone south of Baghdad, *proso* millet is cultivated along with cotton, sesame, and gram as a hot-weather crop between March and September (24). It is specially prominent along the Shatt-el-Hai river. Other forms of millet are generally looked on merely as forage grasses or as weeds, though *Echinochloa frumentacea* is occasionally cultivated for its seed.

In Persia, most of the Panicum millets are found, particularly in the north-east of the country, though they do not appear to make a very large contribution to the food supply. In Afghanistan, *Panicum miliaceum* is grown all over the country by the sedentary population as well as by nomads, and quite a large number of its varieties are found there. It is usually mixed with Italian millet and, especially in the high mountain districts, it is difficult to tell which forms the principal crop. *Echinochloa frumentacea* is found only as a weed in rice fields (25).

### *General Conclusions*

This summary of the present position of the millets in the agriculture of the Middle East would not be completed without some consideration of the future. There seems little doubt that there is scope for a much more extended use of sorghum as a crop in areas where maize can hardly be grown owing to the shortness and uncertainty of the rain. I doubt whether sorghum has much future under irrigation-cultivation except possibly as a pure fodder crop, for the only advantages that sorghum possesses are its capacity to flourish where water and manure are short and where the attention that maize requires cannot be given. On the other hand, the market for maize is a better one, for its grain is usually preferred both for human food and for animals.

But where irrigation is not available and rainfall is short, sorghum seems the food crop *par excellence* for the summer in the hotter countries, just as wheat and barley are the natural autumn and early spring crops. But in order that the extension of which sorghum is capable should take place, there will have to be a great deal of selection or production of the most suitable types with a capacity to ripen early, with a somewhat dwarf habit, with dense heads and consequently a high yield of grain. There is an opening for the development of such types in all the countries of the Middle East, excluding those which are entirely in plateau country. Efforts in this direction have been made in Egypt, but they are still more required in other parts of the region where dry cultivation prevails necessarily over the greater part of each country.

The place of the Panicum millets in the agricultural systems of the future seems still more certain if only the best types are evolved and brought into general cultivation. There is real need of food and fodder crops that grow under high temperature conditions, ripen quickly, and grow on poorer land than the alternatives available. The Panicum millets have too often been, in the past, considered as nomads' crops which would tend to disappear as nomadic life became less usual. But the experience in SE. Russia shows how they can be a dominant feature in agricultural life under conditions where it would be almost impossible to grow profitably any other cereal crop. The experience in the U.S.A. also goes to prove that they have their place in advanced agriculture, though, there, they are chiefly forage crops. But to achieve their place on the plateau lands of SW. Asia, these millets will have to be much more closely studied than they have been in the past, and the very great admixture of one type with another (as of Italian with *proso* millets) will have to be reduced or entirely eliminated. But I can picture millions of acres in the Middle East, now looked upon as useless even to a peasant agriculture, being brought at least into subsistence cultivation by the proper use and extension of the growing of the Panicum millets.

## REFERENCES


1. J. D. SNOWDEN, The Cultivated Races of Sorghum, London. 1936.
2. C. R. BALL, Pearl Millet. U.S.A. Farmers' Bull. No. 168.
3. U.S.D.A. Yearbook, Washington, 1931.
4. G. W. GRABHAM and R. P. BLACK, Report on Mission to Lake Tana in 1920-1. Cairo, 1925.
5. A. PIEDALLU, Le Sorgho. Paris, 1923.
6. J. W. SMITH, Agricultural Meteorology, 1928.
7. Palestine Monthly Agricultural Bulletin, March, 1940.
8. U.S.D.A. Yearbook, Washington, 1895.
9. A. I. TANNOUS. Foreign Agriculture (Washington), 1943, 7. 243.
10. R. C. TOM, Colorado Farm Bull., 1943, **4**, 14.
11. J. MOLLISON, Textbook of Indian Agriculture. Vol. III. Bombay, 1901.
12. A. I. IVANOV, Bull. Appl. Bot. and Plant-Breed., 1930, **24**, 273.
13. G. C. DUDGEON, Egyptian Agricultural Products. Egypt, Min. Agric. Bull. No. 1a. Cairo, 1915.
14. A. I. TANNOUS, Foreign Agriculture, Washington, 1945, **9**, 38.
15. P. ZUKHOVSKY, La Turquie Agricole. Moscow, 1933.
16. E. GUEST, Notes on Plants in Iraq. Dept. Agric. Baghdad, 1933, Bull. No. 27.
17. B. J. HARTLEY, Rept. Middle East Devel. Conf. 1944. Cairo, 1945.
18. N. I. VAVILOV, Bull. Appl. Bot. and Plant-Breed., 1929, **20**, 80.
19. J. H. MARTIN, U.S.A. Farmers' Bull. No. 1162. Washington, 1924.
20. G. N. AYYANGER and N. A. WARRIAR, Madras Ágric. J., 1941, **29**, 461.
21. H. N. VINALL, U.S.A. Farmers' Bull. No. 793. Washington. 1917.
22. G. P. FOADEN and F. FLETCHER, Text-bk. Egypt, Agric. Cairo, 1908.
23. N. A. MAISSURIAN, Bull. Appl. Bot. and Plant-Breed. (Leningrad) 1929, **22**, 149.
24. W. D. GARBETT, Rept. Middle East Devel. Conf., 1944. Cairo, 1945.
25. N. I. VAVILOV and D. BUKINICH, Agricultural Afghanistan. Leningrad, 1929.

# APPENDIX TO PART V

## FOOD GRAINS IN THE MIDDLE EAST AND INDIA

(Outline of Proposed Book)

**Chapter**

I The Middle East and India as Sources of Food Grains in the past.

a) Centres of origin of the chief food grains.

b) Early large scale growth of food grains in Asia Minor, Syria, Mesopotamia, and Egypt, and consequent growth of trade.

c) Food supply of the ancient world and its centres of trade in the Middle East.

d) Decline of production and trade in the early Christian centuries and its causes.

e) The part played by India in the ancient trade in food grain.

f) Collapse of trade and also of production for export in the whole of the Middle East and India after the Mohammedan conquests and especially after Turkish power.

g) Modern developments. (25 pages)

II Wheat in the Middle East and in India.

a) Extent of production in ancient, medieval, and modern times.

b) Varieties.

c) Influence of climate on the types of wheat grown.

d) Areas of production.

e) Difficulties of wheat growing, including pests and diseases.

f) Wheat cultivation in individual countries.

g) Future of wheat cultivation and trade in the Middle East and India. (25 pages)

III Barley in the Middle East and India.
**(same headings as under wheat)** (20 pages)

IV Rye in the Middle East and India.
**(same headings as under wheat)** (10 pages)

V Oats in the Middle East and India.
**(same headings as under wheat)** (10 pages)

VI Maize in the Middle East and India.
**(same headings as under wheat)** (15 pages)

**Chapter**

VII Rice in the Middle East and India.

a) Probably the most plastic of all food crops.

b) Extent of production in ancient, medieval, and modern times.

c) Varieties, and their adaptability to very varied conditions.

d) Chief areas of present and possible production.

e) Difficulties of rice growing, including pests and diseases.

f) Rice cultivation in individual countries.

g) Future of rice cultivation, and possibility of its great extension. (25 pages)

VIII Sorghum and the Millets in the Middle East and India.

**(same headings as under wheat)** (25 pages)

IX Pulse Grains in the Middle East and India.
**(same headings as under wheat)** (25 pages)

X Possibilities of the Middle East and India in the world supply of Food Grains in the future. (25 pages)

Total: 200 pages.

There would be about 15 to 20 small outline maps in the text, and probably about 12 half-tone plates.

**PART VI**

**AGRICULTURAL SCIENCE AND METHOD**

*Chapter* 32

# RAB: A UNIQUE SYSTEM OF CULTIVATING RICE IN WESTERN INDIA

(*As summarized by Sir Edward Buck*)*

For some hundreds of miles along the western coast of India runs a range of mountains precipitous towards the Arabian Sea, but broken and undulating on the Eastern side of the ridge. Against this mountain wall, called the Western "Ghats" (or Passes), beats the torrential downpour of the summer monsoon, giving to the inland tracts on or beyond the summit rains which are heavier or lighter according to the varying amount of protection afforded by the different elevations. It is in certain extensive districts of the mountain area, covering some hundreds of sq. miles, where the rainfall is considerable that the unique method of rice cultivation known as the rab system prevails. The meaning of the term "rab" as used in this article is manure burned on the seed beds from which rice plants are transferred to the field. At one time the manure thus burned consisted mainly of branches and twigs of the trees and scrub bushes of the surrounding jungle. It has been suggested by WALLACE and others that the rab system might be considered as a civilised

* This paper is reprinted with permission from the **Monthly Bulletin of Agricultural Intelligence**, Rome, International Institute of Agriculture, 1915. It affords a convenient summary of a much longer paper by H. H. Mann, N. V. Joshi, and N. V. Kanitkar, for which a detailed reference is given below, P. 398.

adaptation of the custom practised by many aboriginal tribes in various parts of the world of making successive clearings in hill forest lands by burning the trees and vegetation in limited areas, passing on to new patches when the earlier clearing has been more or less cultivated. This custom, technically known as "Jhumming" in the eastern hills of India, as "Chena" in Ceylon and "Ladang" in Malaya, is also practised by the Maories of New Zealand and is common in some parts of Africa. Whatever its origin, the rab system is now firmly adhered to by the rice growers of those regions in the Western Ghats where local conditions are favourable.

It was not unnatural that the apparent wastefulness of a system which involved the cutting and lopping of trees in wooded areas should call forth a protest from the newly organised Forest Department in India, and in 1856 the Government of Bombay resolved to forbid the use of their woods and forests for such purposes, arguing that, in as much as the rab practice did not exist in the rice lands of other parts of India, it was an unnecessary waste of the natural resources of the country. Then came an answering protest from the cultivators who pleaded the beneficial effects of the system to their rice fields. The controversy led to a serious investigation by the Agricultural Department of the Bombay Province, the Director of which, Mr. Ozanne, instituted what were considered decisive experiments. He found that rab was better than unburnt manure; that burnt cowdung gave better results than burnt vegetation but that the latter was far superior to unburnt vegetable matter and even gave better results than unburnt cowdung. There was therefore something to be said for the burning. At any rate the system was pronounced to be justified and "rabbing" was upheld as the best known method of manuring under the particular conditions of the country where it was practised.

No attempt, however, was then made to explain the process so as to account, on scientific grounds, for its superiority. This has been the subject of more recent enquiry. Moreover in view of the fact that rab material in the form of branches and loppings of trees and other vegetable matter is becoming increasingly difficult to obtain in sufficient quantity, while in the form of cowdung it is costly and also limited in supply, the attention of the scientific investigations has also been given to the question whether cheap and efficient substitutes for rab can be found.

Before entering upon the scientific issues which have been raised, by subsequent investigations, it seems desirable to draw attention to some general features of the whole system.

Three points deserve notice:

1) that the rice is *transplanted* from a seed-bed to the field;
2) that the *seed-bed*, not the field, is *manured;*
3) that *burning* the manure increases its effectiveness.

Taking these points in order, it may be said at once that the practice of transplanting seedlings from the bed to the field now prevails over most parts of India for the growth of the better classes of rice, especially where conditions of climate and rainfall are not favourable for broadcast sowings on the main field.

Even in India, however, the farmers in some districts are said "to shrink from the transplanting system because of the cost and the heaviness of the labour" and no doubt experiment is everywhere required to ascertain how far the results of transplanting justify the cost.

The next point noted for consideration is the *manuring* of *the seed-bed*. In districts in India where the use of burnt manure does not prevail, it is not unusual to put on the seed-bed ordinary unburnt manure.

Another advantage of transplanting seems to be the possibility of economising in the use of manures since applying the manures to the seed-bed instead of to the field gives a greater efficiency for a given quantity of manure and incurs less expense in the application. This point would appear to be worthy of more attention in those countries in which rice is sown broadcast after manuring the field.

The third point for consideration is the *burning* of the manure. The practice of clay burning exists in many countries[1] as a means of ameliorating heavy clay soils, but apparently rather with a view of bringing the soil into a more workable condition than with regard to its manurial effect. The superiority of burnt manure is attributed by WALLACE to his belief that dung unduly exposed to heavy rain becomes sour and injurious to vegetation. Assuming that this may be the case, still recent investigations by Professor KNIGHT and Dr. MANN, of the Agricultural College, Poona, have tended to prove that there is an independent positive advantage in the heating of the soil on which rab has been burnt. This indeed has become the most interesting question taken up in the later inquiries.

During 1904 to 1910 KNIGHT[2] conducted experiments with a view to finding cheap and effective substitutes for rab. In the course of his experiments he showed:

1) that the greater part of the beneficial effect of the rab treatment might be attributed to the heating of the soil;

2) that the effect of the ashes of the rab materials, though considerable, was of minor importance compared with that of the heating.

3) that results as good as those obtainable with the rab treatment could be obtained by manuring with a considerable number of nitrogenous unburnt manures, the most effective at lowest cost being oil cake and particularly the cake of safflower (*Carthamus tinctorius*).

In 1909-1912 MANN together with JOSHI and KANITKAR[3] investigated the several factors concerned in the rab

[1] Prof. A. Bruttini, **Dizionario di Agricoltura**. Casa Editrice Vallardi, Milan.

[2] "Substitutes for Rab"—Bulletin No. 63; **Department of Agriculture**, Bombay, 1914.

[3] The "Rab" System of Rice Cultivation in Western India Mann, H. H., Joshi, N. V., and Kanitkar, N. V.—**Memoirs of the Department of Agriculture in India**, Chemical Series, Vol. II. No. 3. February 1912.

system, *viz,* the effect of heat on the chemical, physical and biological character of the soil; the effect of the ash constituents on the plants; the effect of heat alone compared with that of the ash and of each with that of the complete rab. Their results indicated that of the total increase of 182 per cent in dry weight of the seedlings at the time of transplantation, 44 per cent was due to the ashes and 56 per cent to heating. It was noted that the duration of the heating process does not exceed 1-1/2 to 2 hours and that the soil temperature at half an inch below the surface does not exceed 110° C., and at 1 inch 75° C. to 80° C. The effect of heating the surface soil to various temperatures was then determined in pot cultures. It was found that the increase in the growth of the crop was proportional to the increase of temperature up to 125° C., and that the effect was greatest when the heating immediately preceded the growing of the crop. This effect passed off within three months after heating when the soil was kept dry before sowing, but disappeared in six weeks if the soil was wet, thus supporting the native belief that rain before sowing impairs the efficiency of the rab treatment.

A detailed analysis of the heating effect in its various aspects was then made.

1) *Chemical Effects*: Analysis of the water-soluble constituents of the soil before and after heating showed a steady increase in the amounts of soluble minerals and organic matter with increase of temperature.

2) *Physical Effect*: The effect of heat on the soil texture was determined by sedimentation and percolation methods and the increase in permeability was found to be proportional to the increase of temperature. A similar increase in permeability brought about by the addition of gypsum was only able to effect an increase in yield of about 1/4 the amount produced by heating the soil. Safflower cake, which had proved to be most effective as a rab substitute, was found to have a greater flocculating power on a soil suspension than any other cake.

3) *Biological Effects*: With regard to the biological effects of heat on the soil, it was found that the activity of the aerobic organisms, as measured by the rate of absorption of oxygen, was very largely reduced, on heating to 125° C. for half an hour, but never wholly destroyed, and that after 7 weeks it greatly exceeded that in the unheated soils. Since the growth of the seedlings is greatest immediately after heating when the aerobic activity is lowest, it is concluded that the fertility of the soil is not dependent on the presence of large numbers of soil organisms during the growth of the plants.

Of the three effects due to heating, the chemical and physical effects were considered to be more important than the biological. This conclusion, classed as only provisional by the writers, may have to be modified in the light of recent progress in soil biology and the partial sterilisation of soils; and it is possible therefore that the heating effect in the rab system is capable of more complete explanation on biological grounds. Thus RUSSELL and PETHERBRIDGE[1] have shown that temperatures between 55° and 100° C. cause considerable changes in the bacterial flora and protozoa of the soil, resulting in a great and rapid *increase in ammonification* and promoting a greater development of the fibrous roots of the plants. In this connection it is interesting to note that KELLEY[2] in Hawaii and MENOZZI[3] in Italy have shown that nitrogen in the form of *ammonia* is the most effective manure for rice and that greater yields are obtained by applying it before planting than at intervals during the growth of the crop. On this view the heating effect of the rab system would seem to produce ideal conditions in the seed-bed for the growth of the seedlings, as it causes the

[1] **The Journal of Agricultural Science**, Vol. V, part I, Cambridge 1912; and **B** Jan. 1913, No. 14.

[2] "The Assimilation of Nitrogen by Rice".—**Hawaii Agricultural Experiment Station Bulletin**, No. 24. Washington, 1911.—Also **B.** Aug.-Sept.-Oct. 1911, No. 2513.

[3] "Les Engrais dans la Culture du Riz," by Prof. ANGELO MENOZZI; paper read at the International Rice Congress, Vercelli, 1912.

temporary increase in ammonification of the soil at the most favourable period in the development of the rice plant.

On the question of the wastefulness of the present system of using forest material as rab, the following remarks by the writers of the Pusa Memoir[1] are interesting: "The principal conclusion we have so far been able to reach is that no method not involving a greater actual out-of-pocket expenditure seems likely to yield the results which are obtained with the combined application of heat and of wood ashes to the soil which is given in the rab process.

"This being the case it would seem that attention should be largely concentrated on obtaining the application of these by a less wasteful method than the ordinary burning of rab. Our experiments clearly indicate that if the heating effect is obtained, combined with the application of wood ashes, it matters little how it is brought about. No one who has watched the actual practice in vogue can doubt its wastefulness of both heat and fuel, and we feel confident that the quantity of fuel (i.e. cowdung, branches and other materials) could be reduced to, at most, one quarter of the amount at present used if, instead of its being spread over the surface of the soil, the latter could be burnt in heaps. The area to be burnt is small and the soil to be heated consists only of the surface layer to one inch deep. Thus to plant one acre of rice would require the heaping up and burning of a comparatively small amount of soil.

"It would not seem difficult to devise a reasonable method requiring far less fuel and trouble in collecting fuel than is submitted to now. That soil, especially a heavy sticky, clayey soil like that in these regions, can be so burnt in heaps is proved by the former common practice of so burning it in Europe, and by the practice in the fen districts of England at the present day. And in this direction the solving of the difficulties, both of the cultivators and of those who are anxious to prevent the damage to the forests and trees in the rice tracts of the Deccan and Konkan, seems most probably to lie."

[1] Mann, Joshi and Kanitkar—(**Ed.**)

*Chapter* 33

# THE EARLY HISTORY OF THE TEA INDUSTRY IN NORTH-EAST INDIA*

## I

During the period of my engagement as Scientific Officer to the Indian Tea Association (1900-07) I had unrivalled opportunities to collect materials concerning the establishment of the tea industry in north-east India, both by having access to old reports which were placed in my hands, more particularly by the courtesy of the Superintendent of the Assam Company, by conversation with people now no longer with us who remembered the early days of the industry, and by examining the files of daily newspapers and weekly and monthly periodicals which exist in Calcutta. This being the case, I collected together a large number of copies of some documents, and notes from others bearing on the subject, for I felt that as the tea industry is practically the only successful Indian industry in the establishment of which Government took any large part, a study of it would probably be very useful in these days when so much is being stated about industrial development. For ten years these materials and notes have remained with me unused. Their interest has, however, by no means diminished in the interval,—and I trust that the record of the pioneer labours, often against the strongest opposition and most disheartening circumstances, will be of some advantage and

* Reprinted from the **Bengal Economic Journal**, Vol. II (1918), pp. 44-59.

encouragement to other pioneers in connection with the agricultural and industrial development of India.

From its original introduction into use in Europe the supply of tea had been a Chinese monopoly, and the trade in it to England had been a monopoly of the East India Company. In the early part of the nineteenth century, on the renewal of its charter, the East India Company lost its trading monopoly, and as the trade in tea was one of the most valuable parts of its activities, it became anxious to obtain a rival source of supply entirely under its own control. Moreover, especially in the thirties of the last century, Japan broke off all trading connection with the West, and suspicions were rife[1] that China would do likewise, and so at once cut off the source of supply of tea from England.

As a result of these political changes and suspicions, great anxiety arose for the production of tea in India, if such production were by any means possible. It was already known that the tea plant would thrive under very widely varying conditions. It had been naturalised in Brazil, where it had grown magnificently, in St. Helena, in Java, in Prince of Wales' Island,—but the tea made in these places was very unsatisfactory. Of that made in Prince of Wales' Island (Penang) it was stated that it had "acquired the appalling property of a nauseating and slightly emetic drug." It was, furthermore, very much doubted whether tea grown in India would not be useless in the same way. "Everywhere," said a Calcutta writer in 1834,[2] "it thrives, as far as mere vegetation is concerned, but nowhere except in China has any successful effort yet been made to render it a profitable product of industry. We have a suspicion that this arises from causes which will be found a bar to the profitable cultivation of the plant in India. Admitting that localities for it may exist in our territories, approximating in climate to its native country, we should

[1] **Cf.** Asiatic Journal, Vol. 29 (1839), p. 53, and many other references.

[2] **Calcutta Courier**, 7th February 1834.

fear that, as the value of tea depends upon its aromatic flavour, differences of soil may produce changes as fatal as those which occur in tobacco and in the vine, and that the hyson and pekoe and twankay and souchong of India, will be very little like their high flavoured namesakes of the celestial empire..........."

In spite, however, of a somewhat general feeling at least of doubt as to the likelihood of the success of tea growing in India, there were sufficient believers in its possibility that in January, 1834, the Government of Lord W. Bentinck appointed a committee to consider the question of introducing a supply of plants from China, to decide the most suitable and likely place for growing them, and to make arrangements for bringing the seed, and making the experiment[3]

In some respects this committee acted with more energy than most similar bodies. They issued a circular (March, 1834) asking all opinions which were likely to be of any value as to where tea was most likely to be successful, and they arranged at once that one of their members (Mr. G. J. Gordon)[4] should go to China and bring back plants and seed, and also cultivators from China who knew how the plants should be grown and how the tea should be prepared.

Both these actions of the "Tea Committee" have had results which have continued to this day. The circular was issued and Gordon went to China. The first resulted in the definite decision that the tea plant occurred in Assam; the second brought about the introduction of the first lot of China tea seed,—the curse of the India tea industry.

But the discovery of the tea plant of Assam was only a secondary result of the issue of the circular of March, 1834. Before this, replies were received from people in every corner of India who, on the strength of false analogies of climate and soil, convinced the Tea Committee that

[3] This Tea Committee consisted in the first instance of Mr. James Pattle, Mr. G. J. Gordon and Dr. Lumqua, a Chinese doctor, who had long lived in Calcutta.

[4] At a salary of Rs. 1,000 per month.

the proper places in India for tea cultivation were in order of suitability (1) "On the lower hills and valleys of the Himalaya Range." (2) "On our Eastern Frontier." (3) "On the Neelgherries and other mountains in Central and Southern India." What was meant by the Eastern Frontier I do not know. It seems doubtful whether Assam was referred to. By the Himalayas, however, Darjeeling was certainly not meant, but rather Mussoorie, Dehra Dun and the neighbourhood. The committee, led largely by Dr Wallich, the then Superintendent of the Calcutta Botanical Gardens, maintained the superiority of the Himalayas in this region for several years,—I think, in fact, until the committee was dissolved some years later.

But the circular had been received, among other people, by Captain Jenkins, then in charge of the Assam Valley, and a man of great enthusiasm for the development of that newly-conquered province, and one who knew its possible products better than almost anyone living. He lived at Gauhati, but he knew, as most of those who had had experience of Upper Assam knew, that tea was already existing in the country of the hill tribes (Singphos) at the north-east of the valley, and, not only this, but was used for making tea by the Burmese method.[5] This fact had been known at least since 1815. In that year it was spoken of by Colonel Latter, again in 1818 by Mr. Gardner, again in 1824 by Mr. Bruce who grew it in his garden at Sadiya a year or two later (in 1826). Time and again plants had been sent to Calcutta for identification,—by Mr. David Scott, Commissioner of Assam, by Mr. Bruce, and by others. But there seems to have been an extraordinary reluctance on the part of the botanical authorities in Calcutta to acknowledge the existence of tea in India. The matter could only be settled finally, of course, if flowers and seed were sent,—but it was always apparently the part of the botanists to doubt and deny, rather than to encourage the idea that tea was present in the country.

[5] Letpet Tea.

On the receipt of the Tea Committee's circular, however, Jenkins passed it on to a young officer who was stationed at Sadiya, named Lieutenant Charlton, who had also seen and drunk the so-called tea which was growing in the country of the Singphos and also near the Dibru river. He immediately sent to Calcutta (on 8th November, 1834) not merely the tea but also samples of the *fruit* and leaves of the so-called tea trees, and this enabled the plants to be identified with certainty as tea, identical with that of China.

In informing the Government of this fact the Tea Committee waxed enthusiastic and wrote as follows:— "It is with feelings of the highest possible satisfaction that we are enabled to announce to his Lordship in Council that the tea shrub is beyond all doubt indigenous in Upper Assam, being found there through an extent of country of one month's march within the Honourable Company's territories, from Sadiya and Beesa to the Chinese frontier province of Yunnan, where the shrub is cultivated for the sake of its leaf. We have no hesitation in declaring this discovery......to be by far the most important and valuable that has ever been made in matters connected with the agricultural or commercial resources of this empire. We are perfectly confident that the tea plant, which has been brought to light, will be found capable, under proper management, of being cultivated with complete success for commercial purposes, and that consequently the object of our labours may be before long fully realised."[6]

The effect of this announcement on the policy of the "Tea Committee" and of Government was immediate. Mr. Gordon who had been sent to China to fetch seeds and tea makers was recalled, as his mission was now considered unnecessary, and a scientific expedition was sent to Assam to bring back authentic and full information as to the extent and character of the tea there found.

In accordance with this decision Gordon returned, but not before he had obtained and sent off several lots of tea

[6] Letter from the Tea Committee to the Government of India, 24th December, 1834.

seed from China. As it has often been suggested that he was fooled by the Chinese and put off with inferior seed, it may be well to give a contemporary account, evidently inspired by Gordon[7] himself, of what he did and what seed he got. "The first parcel of the seed was despatched personally by Mr. Gordon, in very good condition, and having been procured from the Bohea hills, is supposed to have been collected from plants bearing only the good sorts of black tea. This seed on its arrival in Calcutta was distributed partly for cultivation in Assam, partly on the Himalaya hills. The second and third batches were both despatched from Canton during Mr. Gordon's absence, and from the channels through which they were procured are supposed to have been only the seeds of inferior kinds of tea. Both these parcels were sown in the Botanic Garden here; the last of them arrived out of season and in such a state as not to vegetate, but from the second batch about a lac of plants were procured, of which about 20,000 were sent up to Assam, as many more to the garden at Mussoorie, and a couple of thousands to Madras." There was evidently more than a reasonable suspicion that part at any rate of these first importations represented not the seed of the best of the Chinese tea plants, but any rubbish which (not even being inspected by the Tea Commissioner in China) could be palmed off on the unsuspecting Indian authorities. This was not the case always with later importations, but some of the first were certainly as doubtful material as could have been obtained.

The recalling of Mr. Gordon from China was a step about the advisability of which much controversy arose later. Wallich, the Superintendent of the Calcutta Botanical Gardens, whose influence was then paramount, held that if tea really occurred in Assam, then there was no need to import seed. He wrote[8]:—"The committee have maturely weighed the subject of the new discovery in Upper Assam

[7] **Calcutta Courier**, September **14, 1835**.

[8] To Mr. Gordon, as Secretary of the Tea Committee on 3rd February 1835.

in all its bearings. The genuine tea grows there, or an indigenous plant which may be cultivated to any extent. There is no ground for supposing that the various sorts of tea seeds imported from China will produce anything but the shrub in its natural state, retaining nothing of the variety whose name the seeds bear: it is therefore useless and unnecessary to import from China at a great expense and great risk what may be had, as it were on the spot, to any extent almost in a state of perfect freshness and strength for vegetating. Your continuance in China, so far as regards supplies of seed, is therefore useless and unnecessary." This policy, as we have since proved by experience, was correct: the reason given for it was as fallacious as could be—and was one of the points which led to bitter controversies a little later between Wallich and Griffith, his colleague on the scientific deputation to Assam.

In the meantime the local progress had been considerable. Tea plants, originally supposed to be only found growing wild in the Singpho hills, had been discovered in the Manipur hills by Major Grant, in the Tippera hills, and in a number of new localities in the Assam Valley. Further Lieutenant Charlton, who had supplied the samples which had finally determined that tea occurred in Assam, had been asked to experiment with the growth of the plant at Sadiya where he was stationed, with Mr. Bruce, who had been in Assam for a number of years on his own business and who had certainly grown the plant since 1826, as his assistant This latter arrangement was not to continue long. In the disturbed state of the country, Charlton had to go out to subdue a rebellion, and in attacking a stockade he was wounded and had to leave the province.[9] Bruce took charge of the experiments, and from this time onward he becomes almost the principal figure in the local development of tea culture for a good many years.

The scientific deputation to Assam to which I have referred was appointed early in 1835 and consisted of Wallich,

[9] **Englishman**, 1st September 1835.

William Griffith,—one of the most distinguished botanists who ever worked in India,—and McClelland, a man of reputation, as a geologist. They left Calcutta on 29th August, 1835, and went straight to Sadiya, arriving in January, 1836. This deputation was not a very happy party. It found the experiments in growing tea at Sadiya in a very crude state. There *had* been tea nurseries at Sadiya but they had been trodden down by cattle, and little could be seen. The country was so disturbed that Wallich got frightened and wished to return without seeing all the country. The others explored the country fairly thoroughly, however, and the reports on what was found by Griffith[10] and McClelland[11] are among the most valuable documents we have as to the condition of indigenous tea in Assam in 1836.

The questions which they set themselves to answer were:—

(1) Is tea indigenous to Assam?

(2) Are the conditions such as to make it probable that a tea industry will succeed?

(3) What are the conditions in Assam under which it is most likely to grow successfully.

(4) Is there any necessity to import Chinese tea seed·

The first of these questions they left doubtful, and doubtful it has remained. They found the tea plants scattered all over the country to the south of the Brahmaputra in Upper Assam, while there were none to the north of the river. They always occurred, however, in the plains in groups, almost as if they had been planted, and only in the Singpho hills did they become apparently more a part of the ordinary vegetation of the country. These groups of tea trees in the jungle, however, were exceedingly common. The "Muttuck" country between the Dibru and Dehing rivers was full of them, and other places like Gabro Purbut

[10] Transaction of Agri-Horticultural Society of India, Vol. V, 1837-38.

[11] Transaction of Agri-Horticultural Society of India, Vol. IV, 1837.

at the foot of the Naga hills where tea had been found, were visited by Griffith and McClelland. But the country had been in a state of war for twenty-five years on and off and completely desolated. The people in the hill round the valley were known to know tea and to drink it. Hence it was quite possible that these were remnants of former tea gardens. In spite of this both Griffith and McClelland considered it probably indigenous.

In discussing the second point, Griffith went at great length into the similarity of Upper Assam to the tea tracts of China. He concluded finally: "(1) that there is a similarity of configuration between the valley of Assam and two of the best known tea provinces of China; (2) that there is a similarity between the climates of the two countries both in regard to temperature and humidity; (3) that there is a precise similarity between the stations of the tea plant in Upper Assam and its stations in those parts of the provinces of Kiangnan and Kiangsee that have been visited by Europeans; (4) that there is a similarity both in the associated and the general vegetation of both Assam and those parts of the Chinese tea provinces situated in or about the same latitude." This conclusion undoubtedly did a good deal to strengthen the confidence in the possibility of Assam as a commercial tea-growing district, though I doubt whether any of these statements are very accurate.

As regards the conditions under which tea would best grow in Assam, McClelland (*loc. cit.*) had nothing to go on except the situation of the indigenous tea which he found. Of this, he said:— "It appears that the tea plant of Assam grows spontaneously under slightly distinct circumstances as follows: (1) in the level plain; (2) on embankments or mounds lightly raised above the plain. Cuju, Noadwar, and Tingrai are examples of the first, Nigroo and Gubrupurbut are examples of the second.[12] The first class of situations are distinguished from the general plain by a porous structure and the peculiar character of maintaining a dry surface

[12] The names of most of these sites will be at once recognised by those who know the Assam tea industry.

under exposure to excessive moisture, the second by a structure less porous than the first. In both the plants are situated at the verge of inundations which prevail during the greater portion of the year on the adjoining lands. The important peculiarity of these sites is that they are less secure from inundation by their elevation than by their structure. Indeed the lower sites are scarcely raised more than a yard above the adjoining flat plains, which are exposed to inundation not merely during falls of rain, but also from the overflowings of the great rivers." It is remarkable how clearly McClelland saw the need for thoroughly efficient drainage if tea is to flourish. It would have been a good thing if everyone since then had seen it equally clearly.

As to whether it was necessary to import Chinese tea seed, there was, as we have already hinted, a violent difference of opinion between Wallich and Griffith. The former held that there was no need: the latter that Chinese seed is required. I have quoted Wallich, I will now quote Griffith. "The most thoroughly philosophical course," said Griffith, "is to cultivate *imprimis,* on the tracts alluded to, the best procurable plant taking at the same time every precaution towards reclaiming the Assam plant...... The first step must be therefore the importation of seeds with a small proportion of the best plant from China: this is still more necessary from the total annihilation of those previously imported,—and the importation must continue to be, for some years, for obvious reasons, an annual one."

Griffith's position was thoroughly logical. A wild plant is not likely to give as good produce as one which has been cultivated for many generations. But the result of its adoption has been disastrous. As a result of it Gordon was sent back to China, for many years China tea seed was brought over regularly, and every thing was done to plant it instead of the "wild" indigenous tea of Assam. Wallich was illogical, but he was right; Griffith was logical, but the result of his recommendation was disastrous. It shows how dangerous it is in such matters to reason by analogy.

The general result of the visit of the scientific deputation to Assam was to commit the Government to go ahead in a definite effort to introduce tea cultivation in Assam. Previously the work had been very half-hearted. A nursery in the compound of the bungalows of Charlton and Bruce at Sadiya or in a small plantation at Chykwa,—the cutting down of the trees in a few of the groups of tea plants in the jungle,—the importation of a few Chinese tea makers, the whole under the general supervision of Bruce,—this was all that had been done, and it had been done very badly. As regards the nursery at Chykwa, Griffith reported out of 20,000 plants put out, in August, 1835, not more than 500 remained alive and those "in the last stage of decline. The ground was literally matted down with low tenacious weeds, and it is a fact that on our arrival at the nursery not a tea plant could be seen owing to the uniform green colour of the surface." As regards the tea colonies in the jungle, he said that Tingri, where operations were commenced, looked unhealthy in 1836. "Great parts exhibited considerable confusion: almost all the tea plants had been cut down: the underwood was cleared away, and all the forest trees either felled or in process of being so, the *debris* being burnt on the spot among the still living bases of the tea plants!"

From this time onward, however, the energy put into the matter was very largely increased. Bruce, as Superintendent of Tea Culture, put a large amount of energy into the work of clearing the tea colonies in the jungle, allowing them to grow, and making tea from them. The following note on his work published in 1839[13] seems to give but a fair account of all that we owe to him.

"Mr. Bruce, a gentleman who by long residence in the province had become habituated to the climate and well acquainted with the country and inhabitants, was appointed Superintendent of Tea Culture. His attention has previously been given to other pursuits, and he does not seem to have possessed any knowledge of botany or horticulture or in-

[13] Asiatic Journal, Vol. 29 (1839), pp. 53-61.

deed any special qualifications for the post, but his intelligence and activity supplied every deficiency, and enabled him to render very valuable service. He discovered that the tea plant, instead of being confined to a few isolated spots, was over a great extent of country[14] and though his researches were at first viewed with great jealousy by the native chiefs, he not only succeeded in removing their prejudices but persuaded them to contribute their hearty assistance to his labours."

I do not pretend that Bruce ever discovered the way to grow and make tea so as to be really profitable. As we shall see, nobody did this really until 1852,—but he was an admirable pioneer, found out the habits of the tea plant, got over many of the initial difficulties, made drinkable tea, and to him almost alone is due the bringing of the cultivation and manufacture to such a point that a commercial company was ready to take it up.

The first tea, good enough to send down to Calcutta, made in Assam, was produced in 1836. Five boxes were made of tea prepared from leaves gathered out of season, dressed according to the process used for black tea, and with a very imperfect apparatus.[15] It was approved in Calcutta. The then Viceroy (Lord Auckland) drank it and pronounced it of good quality, and it was considered by those interested that the question might be regarded as settled that tea could be made in Upper Assam.[16] The following year still better tea was made, and was pronounced to be a mercantile commodity.[17] The difficulty of packing was beginning now to be felt, and remained a serious problem for several years, until tea lead was made on the spot,—a not very easy operation. In 1838 the first tea was sent to England. I will speak of its reception in London a little further on.

[14] He published a map of these in 1838, which shows how widely he must have travelled in what then was almost pathless jungle.

[15] **Calcutta Courier**, 21st November 1836.

[16] **Calcutta Courier**, 21st December 1836.

[17] **Daily News**, Calcutta, 2nd March 1838.

The position of the cultivation and manufacture at the stage we have now reached is well described in a small but very interesting pamphlet published by Bruce in 1838. This[18] gives such an excellent account of what tea culture and manufacture meant to Bruce in those early days that I must quote a few passages.

"The tea plants of Assam have been found to grow, and to thrive best, near small rivers and pools of water, and in those places where after heavy falls of rain, large quantities of water have accumulated, and in their struggle to get free, have cut out themselves numerous small channels. On the top of this land you must fancy a thick wood of all sorts and sizes of trees and amongst these the tea tree, struggling for existence: the ground here and there having a natural ditch cut by the rain water, which forms so many small islands, .... the land being never wholly inundated in the rain, though nearly so. This kind of land is called Coorkah Mutty.[19] I have never met with the tea plants growing in the sun, but invariably under shade, in thick woods, or what we call tree jungle and only there and in no other jungle whatever.... The largest tea tree I ever met with was twentynine cubits high,[20] and four spans round: very few I should say attain that size."

He goes on to say that he had failed always in planting tea when put in the sun: on the other hand, his transplants did very well in the shade. He was astonished at the hardiness of the tea plants and quotes the following experience. In one case the Assamese villagers "took the tea plant to be so much jungle, and therefore nearly cut all of it down close to the ground, and set fire to the whole, and then planted paddy or rice on the spot. The crop of paddy had just been cut and brought in when we saw the plants, the

[18] Entitled "An Account of the Manufacture of the Black Tea as now practised at Suddeya in Upper Assam, by the Chinamen sent thither for that purpose, with some observations on the culture of the plant in China, and its growth in Assam by C. A. Bruce, Superintendent of Tea Culture."

[19] Nowadays still called **Korkani** land.

[20] Say 43 to 44 feet.

shoots were coming up from the roots and old stumps thick and numerous.... I afterwards converted this piece of ground into a tea garden on account of the Government, and now it is one of the finest I have." Bruce says he succeeded in getting tea plants to grow from cuttings, provided they were in the shade. If so, he must have worked very carefully for it is decidedly not easy to do so. In regard to plucking of tea leaf, Bruce does not seem to have attempted to go beyond what was at that time falsely understood to be the Chinese method,—that is to say to pluck the whole of the young shoots as soon as they had four leaves on them, do the same when a second lot of leaves grew, and take a third similar crop,—if it grew after such terrible treatment.

The method of making black tea adopted by Bruce's Chinamen is interesting to those who know the process as carried on at present. Withering of the leaf was always done by preference in the sun and the leaves were taken down and clapped between the hands several times during the process. The preparation for rolling also included a short heating in iron pans over a straw or bamboo fire. The rolling was done, of course, by hand, very much in the manner one sometimes still sees used at the very beginning of the tea season. No definite fermentation process was included and, after rolling, the tea was dried on sieves over charcoal. The drying was done in several stages, and the intermediate times during which the tea got cool gave the chance for some fermentation to go on.

Such were the conditions of production and of manufacture during the succeeding two or three years. New tea colonies were found in the jungle and were opened and extended by local Assamese labour almost entirely in the so-called Muttuck country, and tea was made, in gradually increasing quantity by or under the supervision of a number of Chinese who had been introduced for the purpose. The whole development was assisted by the fact that the British Government took over in the latter part of 1838 the direct administration of the territory of Poorunder Sing, containing the greater part of what is now the Sibsagar district of the Assam valley.

During 1837 nothing really more than samples of tea were made. In 1838, however, enough was produced for a number of boxes to be despatched to England, where their arrival was awaited with great interest. On 6th May 1838, Captain Jenkins,[21] the Commissioner of the Assam valley, announced their despatch. These reached England in the latter part of the year and were brought to auction on 10th January 1839. There were only eight chests and each chest was sold separately. The following contemporary account of the sale will have considerable interest.

"The first importation of tea from the British territories in Assam, consisting of eight chests, containing about 350 lbs., was put up by the East India Company to public sale in the commercial sale rooms, Mincing Lane, on the 10th January, 1839, and excited much curiosity. The lots were eight, three of Assam souchong, and five of Assam pekoe. On offering the first lot (souchong) Mr. Thompson, the sale-broker, announced that each lot would be sold, without the least reservation, to the highest bidder. The first bid was 5*s*. per lb., a second bid was made of 10*s*. per lb. After much competition it was knocked down for 21*s*. per lb., the purchaser being Captain Pidding. The second lot of souchong was bought for the same person for 20*s*. per lb. The third and last lot of souchong sold for 16*s*. per lb., Captain Pidding being the buyer. The first lot of Assam pekoe sold after much competition for 24*s*. per lb., every broker appearing to bid for it: it was bought for Captain Pidding. The second, third, and fourth lots of Assam pekoe fetched the respective prices of 25*s*., 27*s*. 6*d*. and 28*s*. 6*d*. per lb., and were also purchased for Captain Pidding. For the last lot (pekoe) a most exciting competition took place,—there were nearly sixty bids made for it. It was at last knocked down at the extraordinary price of 34*s*. per lb., Captain Pidding was also the purchaser of this lot and has therefore become the sole proprietor of the first importation of Assam tea. This

[21] Letter to Lord Bentinck from Gauhati.

gentleman, we understand, has been induced to give this enormous price for an article that may be produced at 1*s*. per lb., by the public-spirited motive of securing a fair trial to this valuable product of British Assam.[22]

As suggested in the above extract the prices given were purely for the sake of advertisement. The tea was not good but it was a curiosity, and its arrival was followed in the latter part of 1839 by another lot, this time of ninety-five packages eighty-five of which were sold on 17th March, 1840, by auction as before. A very complete account of this consignment was given by the East India Company to the Indian authorities with careful criticism by nearly all the leading London tea brokers.

The tea was evidently much better than the last, and was valued from 2*s*. 11*d*. to 3*s*. 3*d*. per lb. It still fetched, however, a fancy price nearly all going between 8*s*. and 11*s*. per lb. except what was called *toychong*, evidently a very coarse material, which fetched between 4*s*. and 5*s*. per lb. With regard to them Messrs. Twinings and Co. of London[23] well summarised the general opinion by saying, "Upon the whole we think that the recent specimens are very favourable to the hope and expectation that Assam is capable of producing an article well suited to this market, and although at present the indications are chiefly in reference to teas adapted by their strong and useful flavour to general purposes, there seems no reason to doubt but that increased experience in the culture and manufacture of tea in Assam may eventually approximate a portion of its produce to the finer descriptions which China has hitherto furnished."

Thus six years after the Tea Committee was originally formed and experiments commenced, we have really for the first time a reasonable quantity of Indian tea put on the market. So far the Government had borne the whole cost of the experiment, and had every reason to congratulate itself on the progress made. It had been proved that tea existed in

[22] Asiatic Journal, 1839.
[23] Letter dated 12th February 1840.

Assam, that it would grow, that the leaf could be manufactured and that the manufactured tea was a marketable commodity comparable with that obtained from China. It now remained to convert a Government experiment into a real commercial venture,—to take it out of the hands of the experimenters and place it in those of businessmen, who would have to make it pay. Between the present stage and that final one when money could be made from tea culture there was still a long way to go. Many disappointments had to be faced and many losses made, and the preliminary steps only were soon found to have been completed. Twelve years more, in fact, had to pass before tea culture could be considered a commercial success. The story of those twelve years will form the subject of a second article.

## II

In my last article I traced the history of the tea industry in North-east India to the time when tea from the plantations in Assam was really on the market. This point was reached by the end of 1838 or the beginning of 1839, though the public were hardly satisfied of the soundness of the undertaking till a year or so later. At that time it must be remembered the whole of the so-called plantations in the Assam Valley, chiefly consisting of groups of indigenous tea plants in the jungle which had been cleared of other growth and weeds and had been cut down so as to form leaf-bearing bushes, were in the hands of Government under a Superintendent of tea culture. This Superintendent, Mr. C. A. Bruce, the real founder of tea cultivation in Assam, had opened out such areas in many places. Many of his gardens were near Dibrugarh, more near the Tingri and other smaller rivers in Upper Assam, others were at the foot of the Naga hills as far to the south-east as the well-known garden of Gabro Purbut.

All that had been proved, however, by 1839 was that tea would grow, and that commercial tea could be made for which a market existed in London. But the matter was getting beyond the stage at which the Government wished to control it. Their idea was only to prove its success and then hand it over to private enterprise. Early in 1839, hence, both in Calcutta and London, a number of capitalists apparently approached Government for the transfer of the existing plantations to themselves and for the creation of a monopoly of tea cultivation in the Assam Valley in their favour.

The first move was made in Calcutta, where a company termed the Bengal Tea Association was formed in February 1839, with the approbation of the Government.[1] Almost immediately after another company of London merchants came forward for the same purpose. The *Times*, in April 1839, wrote as follows:— "A joint stock company is forming in city for the purpose of cultivating the newly discovered tea plant in Assam. Their intention is, in the first instance, to open a treaty with the Supreme Government in India for the purchase of the East India Company's plantations and establishments in Assam, and afterwards to carry on the cultivation of tea there, for the purpose of importing it into this country. The project has been taken up with so much avidity, principally by the mercantile houses trading with India and the leading firms in the tea trade that all the shares were appropriated in a few days and before any public notice of it had appeared. The capital to be raised is £500,000 and it is stated that a communication has already been opened with the Board of Trade and the East India Company, preparatory to a negotiation for the purchase of the Assam territory."

The two—that is to say the Calcutta and the London companies—combined their forces almost immediately. It was obvious that at the stage things had reached there was no room for two such ventures and by the middle of 1839 they had agreed to join interests. This was suggested, as was stated in a meeting of the Calcutta branch,[2] in order

[1] **Englishman**, June 29th, 1839.
[2] On May 30th, 1839.

that "the junction of such interests as were now combined would induce His Honour in Council to consider that no better guarantee could be given to the Government of Bengal for the early establishment of this important trade upon a bold and energetic scale." At this meeting a resolution was passed "that the Bengal Tea Association do form a junction with the London company on condition that the local management be conducted by a committee of directors to be elected exclusively in this country." Thus was originated the peculiar constitution of the pioneer tea company—the Assam Company—in its early days whereby it had two controlling bodies—one in London and another in Calcutta,—an arrangement which seems almost to have invited disaster.

In the meantime, the formation of the Assam Company in London, though it received the approval of the heads of the East India Company, did not do so without opposition. This was apparently partly due to a fear that the Company would be given a monopoly, and partly to a belief that it had been engineered for reasons not given out to the world. At a meeting of the proprietors of the East India Company (June 19th, 1839) the opposition was led by Sir Charles Forbes, and he got an assurance that no exclusive privilege in Assam would be granted to the Company. This did not satisfy him, however, and he stated that "he feared, although they were told of the immense advantage which must result from this plan, although it was said that the people of this country, as well as the people of India, Mahomedans and Hindoos, would profit to an infinite extent by this scheme,—that it, notwithstanding, would all turn to be a humbug."[3]

It was recognised that apart from actual technical difficulties in the cultivation and manufacture which were not, as we shall see later, sufficiently considered at the time, the chief obstacles to the success of a truly commercial enterprise were the lack of labour and capital. Captain Jenkins, the administrator of Assam, described the country as a land

[3] **Asiatic Journal.** Meeting held June 19th, 1839.

flowing with milk and honey, with provisions abundant and easily procured, and only lacking these two necessaries. The capital was now provided by the Assam Company, the lack of labour remained, and as we know, has remained almost till to-day one of the chief obstacles to the development of the tea industry.

It was well, however, that the difficulties in the provision of labour and in the technical management of tea gardens and the manufacture of tea were not fully realised by the promoters of the proposed company. As it was, there was much enthusiasm both in London and Calcutta, and as a result of the union of the two sets of interests, the Government agreed to hand over two-thirds of the experimental tea gardens in Assam to the new company. This being the case, a "deed of settlement" was made among the subscribers to the Company to remain in force until a charter, or an act of Parliament, was passed constituting them a company as was the usual custom in those days.

The organisation of the Company was peculiar. As already stated it had a double board of directors whose powers were divided as follows. The duties of the Calcutta local directors were "the local management of affairs in India in the purchasing, improving, and clearing lands in Assam and elsewhere in India and of buying, renting, or building necessary warehouses, offices, and other buildings in India and in obtaining, employing and removing officers, managers, clerks, servants, labourers and generally in superintending and conducting all the business and affairs of the Company there, and fulfilling contracts for that purpose." "Provided always," as the deed goes on to say, "that they shall in all respects conform to these presents and any rules and regulations made by a general meeting.... and any directions for their guidance given by the General Directory of the Company."[4]

The Company having been formed, two-thirds of the experimental plantations in Assam were handed over to the

[4] Report Assam Co., for 1840 (London), dated May 7th, 1841.

Company on March 1840, and Mr. Bruce joined them as Superintendent of the Northern Division with headquarters at Jaipur. The other division of the Company's plantations had its headquarters at "Nazeerah"[5] which has remained to this day the headquarters of the Company. A gentleman named Masters was appointed as Superintendent of this division. The arrangement with Government was that the lands were to be occupied for the first ten years rent free, and at the end of this time the assessments were not to be higher than for rice lands generally. The cultivation of the poppy for opium was entirely prohibited.

Labour difficulties began from the first day. Bruce had used local labour, aided by a few Chinese. But in the first report from Masters it was stated that there was little local labour, but that the Assamese were beginning to work, "and for the important art of tea manufacture, they seem particularly adapted, and likely to supply eventually all the labour that will be required."[6] This was obviously, however, not enough and great efforts were made to get labourers from outside. It must never be forgotten that Assam had been almost depopulated before it came under British protection by civil war and by an invasion from Burma. Any large enterprise had therefore in a very large measure to provide its own labour.

The first attempt to fill this need was by the import of Chinese coolies. A large number of Chinese coolies were brought round from Singapore, but "they were selected without discretion. Every man with a tail was supposed to be qualified to cultivàte, manipulate, and prepare tea. They were sent up without adequate control. At Pabna they quarrelled with the natives, or the natives with them: some sixty were captured by the magistrate, and consigned to jail, and the rest refused to proceed without their brethren. Their agreements were therefore cancelled and they returned to Calcutta committing depredations in their progress. On their arrival in the City of Palaces, they seemed

[5] Now generally written Nazira.

[6] Letter quoted in report dated May 7th, 1841 (London).

to revenge themselves on society, for the papers were daily filled with police reports of the outrages they committed. They were at length caught and sent off to the Isle of France, the planters of which will doubtless consider that it is an ill wind, indeed, which blows no one any good."[7] The London report of the Assam Company put it more shortly when it said that the Calcutta Board imported "several hundreds of Chinese." "These men turned out to be of a very bad character; they were turbulent, obstinate, and rapacious. Indeed they committed excesses which on occasions endangered the lives of the people among whom we had sent them, and it was found almost impossible to govern them. So injurious did they seem likely to prove that their contracts were cancelled and the whole gang with the exception of the most expert tea makers dismissed." Thus ended the first attempt to bring Chinese labour to the Indian tea plantations.

But labour had to be obtained if development was to go on, and hence a large number of "Dhangar Coles" were recruited. But misfortune dogged the footsteps of the pioneers. Cholera broke out among six hundred and fifty-two of them who were proceeding to Assam, and the survivors disappeared in one night and no trace of them was ever found. Labourers from Chittagong were also useless. And among such coolies as were on the plantations in Assam, the mortality was very high indeed. Deaths occurred with appalling frequency also among the European and other planters. In the first year the Company lost the services of Dr. Lumqua, a Chinese doctor long established in Calcutta who had consented to assist the Company in its early stages in Assam and of four Europeans from its small staff. The Assam Company, indeed, began very early to feel the difficulties of climate and of labour supply which have been among the greatest which the industry has had to fight.

The absolutely unoccupied character of the country, at any rate in the area worked from Nazira is illustrated by

[7] **Friend of India**, September 9th, 1841.

two letters from Masters. In the first of these he says, "I have now been in this district eighteen months, and know comparatively little about it, owing to the dense tree forest and coarse high grass jungle with which the land is all overrun, so that when travelling one can see nothing but what lies in his immediate route and I am continually finding fresh patches of ground occupied by the sites of former villages or gardens or temples or tanks of beautiful water or small patches of tea plants and immense tracts of waste land." A second letter illustrates another aspect of condition. "It was with great difficulty that I could procure elephants when I first came here: I could not purchase one at any rate.... A herd of elephants, however, having gone off from Jorehaut in that direction" (towards Gabro) "they were followed and thirteen of them secured."

Nevertheless in spite of the labour and health difficulties the Company had a considerable area of tea in cultivation by the end of 1840, and at the annual meeting in Calcutta (August 12th, 1841) there was stated to be 2,638 acres in actual production. The production was, however, by no means intense, for the average number of plants per acre was only 457! As has already been indicated, most of the area consisted of groups of tea plants found in the jungle, cleared and cut down for leaf yielding. The total amount of tea made this year was 10,712 lbs. The cost had, however, been enormous up to the end of 1840. £65,457 had been sent to India from London. Naturally a good deal of this had, however, been absorbed in capital expenditure. A steam boat had been built and purchased in Calcutta of which we shall hear later. A saw mill had been sent to Assam, to be set up at Jaipur, and no less than Rs. 1,23,275 is put down in the Calcutta Board's report for "Labour, lost and unproductive."

At this stage the Company was still sanguine in spite of difficulties, and they ventured to estimate production in future years, as rising to 40,000 lbs. in 1841 and to 320,000 lbs.

in 1845![8] We shall see how this estimate was falsified in every particular.

The condition of the whole enterprise at this time, the way in which the management was in the hands of their Chinese tea makers, and the unsatisfactory character of the European assistants sent to Assam are well shown in the following quotations from letters from Mr. Masters. On February 12th, 1842, he writes to the Directors: — "You will please to observe that these tea makers (Chinese) are very great gentlemen; even those who receive but Rs. 3 per month consider themselves so, and object to do anything else but make tea. When spoken to, they threaten to leave the service if they are insulted by being asked to work. Gradually this will wear away as we shall soon have them under our control, and if they continue saucy, we may take a convenient opportunity of making a strike for two or three months, and when they lose their pay, they will probably become sensible that they are dependent on the Assam Company for their livelihood." Mr. Masters hardly gives one the idea of a tactful manager!

With regard to the European assistants who has been sent he wrote in another letter. "Hitherto I have been overwhelmed with assistants many of whom have been unaccustomed to agricultural employment, but the greatest inconvenience attending the assistant establishment is the unhealthiness of the climate; it so often happens that after much difficulty has been experienced, and the assistant is becoming acquainted with his duty, and he and the natives are becoming a little reconciled, the assistant falls sick, and is obliged to leave his post: if another is sent, the same difficulties and inconveniences are repeated.... It must be evident to the Directors that a passionate European entirely ignorant of the language and entirely ignorant of every part of his duty can but be worse than useless." I can quite understand Mr. Masters' annoyance, but my sympathy goes

[8] Calcutta Board Report, Assam Co., published in **Friend of India**, 9th September 1841. The Shareholders' meeting was held on 11th August 1841.

out to the young Englishman, landed in a very unhealthy country, absolutely in the jungle, with nothing to relieve the tedium of continually driving coolies to work at a job which neither he nor they understand. When we remember that the amount allowed for an assistant's house was but Rs. 300, that there was no sanitation, and that the unacclimatised European was planted down, and got fever, most probably, before he had been there more than a few days, and was never afterwards really free from it,—we could hardly expect anything but despair, irritability, illness and often a speedy death.

In the second London report,[9] though things are still stated to look promising, there begins to be a doubt. Nothing more is said about the labour question and so we may consider that this is temporarily solved. The kind of gardens at this time is well illustrated by figures given both by Masters and Bruce. I quote some at any rate of the names of the gardens, as they may interest those in Assam at the present time. Gabro Purbut consists of 44 *poorahs*,[10] of which 10 *poorahs* were large plants, 20 *poorahs* middling plants, and 14 *poorahs* small plants and seedlings. Satseia had 213 *poorahs*. Cherideo had 23 *poorahs*. Rokanhabbi had 350 *poorahs* nearly all just planted. Deopani had 20 *poorahs*. All these names will be recognised as being still included in the Assam Company's property. Masters states that he planted his seedlings five feet apart, and he considers that the cost of clearing and planting a *poorah* of tea will be Rs. 100, while the annual cost of upkeep would be Rs. 50 per *poorah*. Taking a *poorah* as 1.21 acres, these will be equal to Rs. 83.3 as capital cost and Rs. 41.7 as annual cost of upkeep, per acre.

In the other division, in the control of Mr. Bruce, the sites of several of the gardens will be recognised as being now in the Tingri Tea Company's estates, and also in the

[9] Report dated 9th May 1842.

[10] The figure given for the area of a **poorah** varies. It is sometimes spoken of as 3½ acres. In the present report it is given as 1.21 acres which I think is the figure which should be taken in these reports.

company's working near Jaipur. Kahung had 31 *poorahs* of tea, 11 *poorahs* newly sown. Tingri (including Ballyjan and Tipling) had 34½ *poorahs*. Hoogrijan had 31½ *poorahs*, with an area of newly planted tea. The famous tea seed garden, "Bazaloni" appears in this group in 1841. Near Jaipur we find other gardens whose names still exist. In this section we hear first of the definite planting of China seed. An interesting estimate by Mr. Bruce is that it required one man coolie on Rs. 4 per month to keep one *poorah* of tea in cultivation.

The presage of coming disaster seems to pervade the atmosphere during 1842 and 1843 both in the reports of the Assam Company and in the remarks on the subject in the Calcutta newspapers. There were evidences of mismanagement everywhere. The steamer built for the Company as their means of transit to Assam proved a failure. "The Assam Tea Company," says the *Friend of India*,[11] "after having sent their new steamer on one trip up the Berhampooter, have, on her return, offered her for sale. The cause is not made known,—probably her inability to steam the current of the Berhampooter." The amount of tea made in 1842 was far less than might have been anticipated, and only amounted to 30,000 lbs., while the net cost of the undertaking had been £160,000. Mr. Masters from Assam, evidently feels, from his letters, that there is something unsatisfactory in the methods of tea growing and plucking adopted.

By the latter part of 1843 it was certain that something was amiss. The Calcutta directors sent a commissioner to Assam to see what was wrong.[12] Both Mr. Bruce and Mr. Masters were summarily dismissed, and the report presented for the year 1843 is doleful indeed. "Since we last met," says the report,[13] "your directors have seen much to diminish the confidence which they expressed at the last meeting in the ultimate success of the Company; that confidence was

[11] 19th May 1842.
[12] Mr. J. M. Mackie. He reached Assam in October 1843.
[13] Presented (London) 23rd April 1844.

necessarily founded on statements and calculations prepared in the Province where our operations are carried on. These data have since been altered by the parties who supplied them in many material respects, and the produce of the year has fallen short of the estimate in respect to quantity by one-third; at the same time, the current expenses of the Company appeared not to be diminished." They went even further than this, and wrote:— "We have positively forbidden the local board in Calcutta to pass any more bills upon us, and have enjoined them to reduce their expenditure to the level of the means at their immediate command. We can, therefore, safely pledge ourselves that no further call shall be made upon the shareholders until your directors have shown sufficient grounds for recommending you to prosecute the enterprise in which we have embarked with renewed vigour."

The position was truly perilous for the shareholders. But, to all appearances, a change for the better occurred. The Company had so far *not* been under limited liability. But a special Act of Parliament was passed in 1845[14] which settled their position. It was only to last till April 30th, 1854, declared a capital of fifty lakhs of rupees in shares of five hundred rupees. The cultivation of opium, sugar and coffee was prohibited.

In the meantime expenses at least had been reduced, and this was something. The relationship between expenses and yield was as follows:—

| | *Yield of Tea.* | *Expenditure.* |
|---|---|---|
| 1841. | 10,505 lbs. | £12,984 |
| 1842. | 31,398 „ | £16,560 |
| 1843. | 87,705 „ | £13,146 |
| 1844. | 120,422 „ | £7,284 |

This so pleased the directors that they very foolishly, as is now clear, declared a divident of 10*s*. per share in January

[14] Act XIX of 1845.

1846, though no profit whatever had been made. This was the only dividend the shareholders saw till 1852.

There seems, however, to have been a renewed burst of confidence as a result of the full incorporation of the Company, and of the reduction of expenditure. This spread to the East India Company. Their experiments now having reached what they considered to be a complete success, they now decided to withdraw entirely from their connection with the industry, and the portion of their experimental gardens which they retained were ordered to be sold. The occasion is of such general interest that we may quote their orders on the subject. "The information contained in these proceedings is very satisfactory and gratifying to us. The sales of the tea, both in Calcutta and London, judging from the statements of the cost per pound[15]. . . .confirms the opinion . . . . expressed that the article may under proper management be cultivated at a real remunerative price, and we accede to your[16] proposal that the Government should withdraw from any further connection with the cultivation or manufacture of tea in Assam."

Though the Government, by these orders, indicated that they considered that the industry was established, yet I do not think that this was by any means the case. Certainly after the first extravagance and mismanagement, the prospects appeared a little more hopeful. But though a dividend had been paid, no real profits had been made. The estimates of yield had been considerably falsified, and the same or a greater area showed signs of giving less yield than in previous years. There seems to have still been hopeless mismanagement, but, even more than this, it became increasingly evident that nobody knew how to grow tea so as to maintain the yield of the bushes, let alone increase the amount of tea which could be made for them. The concern had now in fact reached the stage when the method of planting and plucking tea which had been learnt from

[15] The cost of a crop of 96,000 lbs. in 1845 is given as 14 annas per pound with all expenses, including freight and insurance.

[16] The Government of Bengal.

the Chinese who had taught the pioneers, had definitely broken down, and it was evident that unless new methods could be found which would yield more tea and maintain the yield of the bushes better, the industry must close.

The London Directors were the first to see this. Concentration on a smaller area till success was obtained in this matter was their policy, and in 1846 they, hence, closed down altogether the so-called northern and eastern divisions of the company (the Tingri group and the Jaipur group of gardens). But the position was first really faced in the report for 1847, published in 1848. In this the Directors definitely confessed failure, threw the blame on the Calcutta Board, and they go so far as to confess that they are doubtful whether it is worth while to continue, as even with a policy of great economy and very great care over expenditure, it was only just possible to keep the concern from showing a loss. There seemed no confidence as to its future capacity for profit. It is curious to find this only two years after Government had, with a great flourish of trumpets, declared the industry established.

The position is well shown by the following extracts from the Report of the Assam Company for 1847. "The General Directory .... think it proper to mention to you that they find among the proprietors, and even among their own body, a difference of opinion prevails upon the vital question whether it is desirable or not to continue the operations of the Company. On the one hand it is contended that under the present system of management there is at all events no loss, and that the last year was the first in which the expenses in the province were kept within the estimate or nearly so, and the anticipated outturns of produce was not only realised but exceeded, while at the same time there is every reason to expect an annual increase in produce from seedlings, and the vacant spaces in our present cultivation being filled up .... and therefore it would be unwise to throw away all that has been spent on the enterprise at a moment when there appears so little chance of further loss and much reason to hope that some part of the money spent may be redeemed. On the other hand, it appears to be thought by many that there are too small hopes of success and too limited an amount of profit to be anticipated to render it advisable to continue our operations."

The London directors actually in the sequel asked the Calcutta Board to make them an offer for the whole company, and stated that they "would feel inclined to recommend to their shareholders the acceptance of any proposition that would give them a moderate sum per share, rather than depend on the distant prospect of a larger benefit." No offer was, however, made, and both the London and Calcutta authorities determined to risk another year (1848) of work.

We have now reached the lowest point in the fortunes of tea cultivation in Assam. The great hopes and prospects of a successful tea industry seemed to have almost disappeared. The recovery from that position was primarily due in the first instance to two men,—one in Calcutta and one in Assam,—whose confidence in the undertaking, whose business capacity, and whose integrity of character drew the Assam Company from the brink of despair and made a future tea industry in Assam immediately possible. These were Mr. Henry Burkinyoung in Calcutta and Mr. Stephen Mornay who took charge in Assam in 1847. In five years these men made a bankrupt concern into one which it was recognised could at least pay its way. There then followed the improved technical skill and methods introduced and carried out by Mr. George Williamson on the gardens in Assam, which made it into a very profitable industry.

The state of things into which affairs had drifted in 1847 was well described in a Calcutta paper, a year or two later, when the worst was over, as follows[17]:—

"The mismanagement of Joint Stock Companies in India has been so general, and its effects so disastrous to all concerned with, or interested in them, that we regret we cannot afford space at present to detail the measures by which the rapid downward progress of this Company has been

[17] **Friend of India**, 9th May 1850.

so timely arrested, and its rescue from destruction on the very brink of ruin so promptly effected. We presume that all the old hands, when they perceived the inevitable fate awaiting their reckless mismanagement, with the instinct of rats, left the concern, for we find none of their names in the present board or in the management.

"If we are rightly informed, when the present authorities of the company took charge of its affairs, they found that upwards of 21 lakhs of rupees had been expended upon buildings and cultivation, which it was found, on sending a new superintendent to Assam ought not, under judicious and careful management, to have cost one-tenth of that sum; buildings which ought not at that stage of their operations to have been erected, had been so slightly constructed that they were already tumbling down, and but little was to be found of the extensive clearing and planting which had been reported from Assam, and paid for, and even those in existence were in such a neglected state, that another rainy season would have obliterated every trace of them. The credit and resources of the company were exhausted: they were £7,000 in debt in London, Rs. 40,000 in Calcutta, while the indispensable outlay required in Assam to save the miserable wrecks there, almost drove the then local directors to despair, and the more so, because the London Board urged upon them the closing or even total abandonment of the concern. They, however, possessed discernment enough to perceive the capabilities of the enterprise under better management and with a spirit, firmness, and confidence that does them infinite credit, raised funds on their own individual credit and responsibility to make one more effort to retrieve the affairs of the company."

That this was not too dark a picture can be seen from the official documents of the company. Mr. Burkinyoung, the Chairman of the Calcutta Board of Directors, wrote in 1848: "You as well as ourselves, have of course long been aware that whilst the paid up capital of the company had been entirely sunk by the close of the year 1844 or nearly so, its expenditure had not been devoted to the true interests of the undertaking, and the extended properties which such a sum should have opened out so far from having been raised, a most limited and insufficient area of tea cultivation was in possession of the company, the chief portion of the capital having been devoted to extraneous and useless purposes, and, in effect, so far hopelessly squandered." It does seem remarkable, in fact, to find that the area really under cultivation in 1848 was only 400 to 500 *poorahs* (say 300 to 600 acres).

With business management, however, the concern showed a profit of £3,000 in 1848, and the report for that year[18] shows new hope, and new confidence. Out of the debt of £7,000, £2,000 were paid. And the prospects were sufficiently promising to propose a new call of £1 per share (£10,000) to extend the real cultivated area.

On the technical side the production of tea, as will be recognised by all who know tea in Assam in these later days, the authorities were still only feeling their way. The maximum yield per acre on the company in 1848 was 275 lbs. of tea. The largest yield in the year was obtained in April and the season finished in September. The actual yield month by month was as follows:—

| | | |
|---|---|---|
| March | ... ... ... ... | 18,269 lbs. |
| April | ... ... ... ... | 41,125 " |
| May | ... ... ... ... | 36,391 " |
| June | ... ... ... ... | 37,523 " |
| July | ... ... ... ... | 31,920 " |
| August | ... ... ... ... | 26,079 " |
| September | ... ... ... ... | 19,345 " |

To us nowadays this would appear, even with China plant, to show that the bushes were being overplucked in the early part of the season, and were never allowed to grow

[18] Dated London, May 4th, 1849.

properly before the leaf was taken. This state of affairs continued, however, for some years longer.

Progress was very gradual. The Calcutta directors wished to go ahead: the London Board, having had their fingers burnt so many times, held them back. In 1849 the northern and eastern divisions (Tingri, Jaipur, etc.) were re-opened: on this the London Board expressed "their fear as well as displeasure".[19] But the area was slowly extended, and what was more, in spite of the expenditure on this, small profits were made. The crop in 1849 was 216,000 lbs. The debt was reduced in this year to £2,500, and in the next season, with a net profit of £5,025, the whole disappeared. At last the first genuine dividend out of profits was paid in 1852 (for the 1851 season).[20] It only amounted to 2½ percent., but it proclaimed to the world that the company, having made consistent though small profits from 1848 onward, was no longer the bankrupt concern it had been supposed to be, and had, at least, possibilities of success.

This was followed by a dividend of 3 per cent. in the following season[21] and then the two men who had brought the Company from despair to a moderate amount of success —Stephen Mornay in Assam and Henry Burkinyoung in Calcutta—retired. One cannot exaggerate the debt which the tea industry owes to them. Their successors improved their results,—but they it was who made a tea industry appear possible in north-east India.

The new manager in Assam was Mr. George Williamson, perhaps the greatest figure in the development of the Assam tea industry, and afterwards the founder of the Calcutta firm of Williamson Magor & Co.; the managing director in Calcutta was Mr. W. Roberts, afterwards well known for his connection with the Jorehaut and other very successful tea companies. Williamson's report in 1853, after taking charge, was very interesting. He had been there under Mornay and had studied tea planting as nobody had done

[19] Report dated May 3rd, 1850.
[20] Report dated May 7th, 1852.
[21] Report dated May 6th, 1853.

up to that time. He found a yield over the whole of the gardens of 196 lbs. of tea per acre[22] only. The local cost of tea was between five and six annas a pound. He recognised the evil of China plant which had been used in putting out many extensions. Speaking of one garden (Kachari Pookri) he says, "it also possesses an advantage....in having no China plant, the inferior yielding of which in respect to quantity, is now a well-established fact." He notes the great lack of labour, and the unhealthiness of the places, and speaks of serious attacks of cholera "which continued with unremitted virulence for three months."

But so far as I can judge, Williamson's success was primarily due to his recognising that if tea leaf is to be plucked, the tea bushes must first be allowed to grow. The season thus tends to become later. Little tea is obtained in March and April, and when the Directors get alarmed, he re-assures them that all is right. "Injudicious and ignorant plucking may seriously injure the plant and even cause its death by rendering it more liable to be attacked by white ants and worms." The result of his policy was a singular increase of yield per acre. Apart from bad business methods, the non-recognition of that on which Williamson now insisted was, I feel, the biggest cause of the early failures. The lack of technical skill and knowledge had made large success impossible until 1852.

But now with business management, and a man, who had studied the tea bush and its yielding, in charge, things went ahead. The area, crop profit and dividend for the years following are shown below:—

| Year | | Crop | Profit | Dividend |
|---|---|---|---|---|
| 1852 | ... | ... | ... | 3 per cent. |
| 1853 | ... | 366,687 lbs. | £13,262 | 5 " |
| 1854 | ... | 478,258 " | £20,641 | 6 " |
| 1855 | ... | 558,628 " | £11,480 | 7 " |
| 1856 | ... | ... | ... | 8 " |

[22] The figures are given per **poorah.** I have converted these into yields per acre.

With 1856 we reach a point when the pioneering days were over, and at this stage we may leave our study of the early days of the tea industry. It had proved itself so profitable that other companies were being formed, that prospectors for tea were all over the province, and that a regular industry was in the full course of development. It had, however, taken twenty-two years to reach this stage from the time when Government appointed its tea committee in 1834, as I have shown, the establishment of the industry had had by no means a plain course. The Government undoubtedly gave up their experimental work too soon, before the best method of growing tea had been discovered, or any way was really known of maintaining the crop from tea bushes. The Assam Company entered the field too early, before the knowledge of the subject was far enough advanced to make success really possible. It was mismanaged no doubt. The extravagance from 1840 to 1847 was colossal, and deserved the failure it got. But the technical knowledge required for success was hardly there, and even when business methods became perfectly satisfactory there was still only very moderate success until the technical advance had been made which was required for large and continued yields of tea.

The pioneers of the tea industry are nevertheless men of whom we may well be proud. Jenkins who got the experiments established; Bruce who showed that tea making in Assam was possible; Mornay and Burkinyoung who proved that tea would at least pay; and Williamson who showed how to cultivate tea in a really profitable manner,—all these names deserve remembrance and recognition. Building on their foundations progress was rapid. The next ten years showed an almost inconceivable development, and such profits as led to speculation and almost to ruin in 1866 and the years following. That is, however, another story. The foundations of one of the greatest of Indian agricultural industries had been well laid by 1856, and tea cultivation and manufacture had been placed on the track which had led, through many vicissitudes, to the position which it holds to-day.

*Chapter 34*

# THE INDIAN TEA INDUSTRY IN ITS SCIENTIFIC ASPECTS*

I make no apology for bringing before a London audience some account of the achievements of experimental science in connection with the growing and manufacture of tea in India. For I feel that without the careful experiment, which has been made both by professional scientists and by practical planters, the industry, instead of being one of the largest and most successful forms of British enterprise in the tropics and sub-tropics, would probably long ago have been relegated to the limbo of failures. The history of tea in India is, in fact, the story of the introduction of an industry to that country from China, the discovery that almost all the information then existing about how to carry it on was unsuitable and unsatisfactory under the new conditions, the consequent all-but total failure of the industry in its early stages, and, finally, the gradual building up, on a basis of new experiment, of the whole fabric of tea cultivation and manufacture as it is known at present. This present organisation has gradually become more and more efficient; the tea produced has become, on the whole, better and certainly stronger; the yield per acre which can be obtained has steadily increased (though, owing to schemes of restriction, the highest possible yield is not always obtained); and the industry has become one of the triumphs of Western organisation in tropical regions.

* Reprinted with permission from the **Journal of the Royal Society of Arts**, London, Vol. 79 (1931), pp. 469-479.

Considered as an agricultural industry, tea cultivation is in many respects unique. I do not know any other cultivation where the object of the grower is to produce a continual succession of young shoots on a perennial bush or tree, and where, as soon as these young shoots reach a definite stage of growth, they are removed from the plant and form the crop. The only remotely analogous case is that of the mulberry tree grown for feeding silkworms, but the analogy even in this case is very far-fetched. Owing, however, to this uniqueness, many of the agricultural methods adopted in other forms of cultivation are inapplicable without a good deal of modification in tea culture, and it is clear that before the best results can be obtained, special methods of planting, of treating the bush after planting, of pruning, and of the removal of the shoots (plucking), have to be worked out, while the processes of cultivation and manuring have to be adapted to the special case.

The special applications were worked out, for their own conditions, in China many centuries ago where the actual cultivation of tea first became really an industry. For the China conditions of small peasant holdings and indefinite quantities of animal, largely human, manure, I am not sure that these methods have ever been materially improved. And they formed the basis of work when commercial tea began to be grown in India in the forties of the last century. They proved, however, totally inadequate and largely unsuitable, under Indian conditions, and it became necessary to modify largely every one of the methods introduced from China in order to make and maintain tea growing as a paying industry.

The history of the attempts to introduce tea cultivation into India is of extreme interest, and we may spend a few minutes in briefly reviewing the matter.

From its original introduction into use in Europe the supply of tea had been a Chinese monopoly, and the trade in it to England had been a monopoly of the East India Company. In the early part of the nineteenth century, on the renewal of its charter, the East India Company lost its trading monopoly, and as the trade in tea was one of the most valuable parts of its activities, it became anxious to obtain a rival supply entirely within its own control. As a result, great anxiety arose for the production of tea in India, if such production were by any means possible. It was already known that the tea plant would thrive under widely varying conditions. It had been naturalised in Brazil, where it had grown magnificently, in St. Helena, in Java, in Prince of Wales Island, etc.—but the tea produced in all these places was very unsatisfactory. Of that made in Penang, it was stated that it had "acquired the appalling property of a nauseating and slightly emetic drug." "Everywhere," says a Calcutta writer in 1834, "it thrives as far as mere vegetation is concerned, but nowhere, except in China, has any successful effort been made to render it a profitable product of industry. We have a suspicion that this arises from causes which will be found a bar to the profitable cultivation of the plant in India. Admitting that localities for it may exist in our territories approximating in climate to its native country, we should fear that, as the value of tea depends on its aromatic flavour, differences in soil may produce changes as fatal as those which occur in tobacco and in the vine, and that the hyson and pekoe and twankay and souchong of India will be very little like their high-flavoured namesakes of the celestial empire." The writer of this extract was right in that the tea of India is different from that of China, but he had hardly thought of the possibility that Indian tea might, in certain cases at any rate, have a stronger and better flavour than that of China, but such has proved to be the case.

In spite, however, of a somewhat general feeling of doubt as to the likelihood of the success of tea growing in India, there were sufficient believers in its possibility that, in January, 1834, the Government of Lord W. Bentinck appointed a committee to consider the question of introducing a supply of plants from China, to decide the most suitable and likely place for growing them, and to make arrangements for bringing the seed and making the experiment. In some respects this committee acted with more energy than most similar bodies. Events, at any rate, followed very rapidly. They sent one of their members to China to bring plants and seeds, and so brought about the first introduction of China tea seed to India, which has been one of the curses of the tea industry ever since. They recommended that tea cultivation should be started on the lower hills and valleys of the Himalaya range, by which they meant Dehra Dun and Mussoorie, on the Nilgiris and other mountains of Central and Southern India, and on the Eastern Frontier. Whether the last suggestion referred to Assam I do not know. But the discovery of indigenous tea on the borders and even in what was then the new province of Assam, drew attention almost immediately to the special suitability of that part of India to the proposed new cultivation. A scientific commission was sent to Assam to investigate, experimental gardens were established, which were placed in charge of Mr. C. A. Bruce, whose name will always be honoured as perhaps the most effective pioneer of tea cultivation in India. Chinese experts were imported to show how cultivation and manufacture should be conducted, and tea was actually made and sent to Calcutta in 1836, only two years after the tea Committee had been appointed. This original tea must have been very nasty, though the then Viceroy drank it and pronounced it good, but in 1837 merchantable tea was made, and in 1838 it appeared on the London market. In this year there were eight chests in the first importation, which obtained fancy prices, as well as those of the following year.

In 1839 the question of converting the experimental plantations of the Government into a commercial project came to the front, and in 1839 the Assam Company was actually floated to take over two-thirds of the Government plantations in Assam and extend on this basis. In March, 1840, they took charge, and the history of tea cultivation for at least twelve years is the history of the struggles of the Assam Company. They were entirely dependent for technical knowledge of tea cultivation and manufacture on

Chinese imported to India, and I may say at once that the attempts to apply their methods in Assam, without modification, nearly ruined the whole project. The tea was made, and I think it was fairly good tea, but the quantity obtained from the plants was so small, and the methods of working were so expensive that by 1847 the Company was faced with the question of shutting down the whole scheme and acknowledging failure. The London section of the Assam Company wished, at any price, to get out of the whole business, and offered to the Calcutta section to buy the whole concern. This was not taken up, and ultimately it was decided to risk another year, that is to say, the 1848 season.

This was a very sad sequel to the very high hopes of a few years before. No profits whatever had been made. The estimates of yield of tea had always been far in excess of what had been obtained and it had become increasingly evident that nobody knew how to grow tea so as to maintain the yield of the bushes, let alone increase the amount of tea which could be made from them. The concern had now, in fact, reached the stage when the method of planting and plucking tea, which had been learned from the Chinese who had taught the pioneers, had definitely broken down, and it was evident that unless new methods could be found which would yield more tea per acre and maintain the yield of the bushes better, the industry must close.

For the first time, in fact, since the industry was started, experiments in the management of the tea bushes by methods which were unknown to the Chinese growers were undertaken, and gradually results were obtained. I must express my admiration for a number of the planters of those days. Though they had no special scientific training, some of the changes they introduced into tea planting were inspired by as true an instinct as any changes which have been suggested since that time. Perhaps the man whose experimental work inspires me with most admiration, at this time, was George Williamson, who was Superintendent of the Assam Company from 1852, after being assistant on its gardens for several years before. He saw that the failure to get anything like the yield of tea from the bushes in the tea gardens of that epoch was due to the fact that the leaves were taken as soon as they grew and thus there was no yield in the later part of the season, while the bushes quickly deteriorated. After many experiments he devised a system of plucking, which in its essence is in vogue to-day, though there have been many improvements, especially during the last thirty years. In the first year of his superintendency his Directors expressed great anxiety because the crop in the early part of the tea season was smaller than usual: he replied that it was all right and that the crop would come on later. So it did, and now in all the tea districts of North-East India it would seem absurd to get the maximum crop of the season in the month of April, as was done on the gardens of the Assam Company in 1848.

Early in the fifties of the last century, tea growing became a paying proposition, largely because of the experiments of the pioneers in all directions, of which I have just given one example. The yield increased: the yield of the bushes was better maintained: and though deterioration of gardens and bushes was still rapid, yet there was plenty of new land to put under the crop, and the profits were such as to allow tea to be abandoned and to be replaced by new tea without preventing tea plantations from being profitable concerns. To this period, chiefly in the seventies and eighties of the last century, belongs a great extension of the tea industry, but at the same time the lack of anything like satisfactory knowledge of the proper management and maintenance of the tea bushes became more and more evident. For some time attention was chiefly directed to the application of engineering to tea manufacture, that is to say to the replacement of hand rolling, charcoal firing, and other processes of tea manufacture by mechanical methods. This, of course, caused economies in the cost of producing tea. But little by little tea began to suffer from diseases and pests which had not been noticed in the earlier days or which had not done much damage. There was found more and more abandoned tea in North-East India. Many gardens.

perhaps more in Cachar than anywhere else, were found to be so much deteriorated that large areas of old tea gradually relapsed into jungle. And, as a result of all these things, there was great anxiety among progressive people connected with tea growing in India, and it was again and again suggested that scientific advice should be obtained.

The first fruit of this was the engagement of my old friend, Mr. Kelway Bamber, a chemist who was engaged by the Indian Tea Association early in the nineties, and who, after examination of the situation on the spot and examination of many tea soils and of tea itself, wrote a book on the Chemistry and Agriculture of Tea, which may still be consulted with a good deal of advantage. This was issued in 1893. Bamber specially studied the nature of tea soils, and for the first time suggested the lines of manuring the tea bushes for leaf. Mr. Bamber left the country but returned several years later to Ceylon, where he became perhaps the greatest authority on tea manuring for many years.

A year or two later the tea authorities in India became alarmed at the rapid increase in the amount and virulence of the pests and blights which attack the tea plant, and obtained the services of Dr. Watt, later Sir George Watt, to survey the whole ground, at least in Assam, in order to suggest what should be done. He made a most productive tour in Assam and Kangra and collected together all the knowledge then existing in the tea districts regarding the pests and diseases which affect the cultivated tea bush. This, together with many very valuable observations of his own, formed the basis of one of the most valuable publications that has ever been made on tea. He attempted the first real study of the scientific basis of tea growing, specially insisting on the relationship between incorrect methods and the increase of pests and diseases. I want to take this opportunity to say a word or two with regard to the work done by Dr. Watt. When I first went to India in 1900, he was a man near the age of retirement, but was still full of energy and always ready to talk about his tea investigations. He very largely initiated me into the scientific needs of the tea industry, and visited with me two or three of the principal tea growing areas of North-East India. My tours with him were an experience I shall never forget. He was certainly the very best observer I have ever seen, and I always felt that a morning in the field with Dr. Watt was an experience during which I should see probably three or four times as much as I could see when by myself. He has died during the last few months, and though he never did serious work on tea after 1900 his influence still remains in greater measure than is often thought.

After Dr. Watt's work the question of the need for having regular scientific help in connection with the many problems which were rising in connection with tea culture came more and more to the front. And I think that the depression in the tea industry at the end of the last century had something also to do with the resolution to meet this need—at any rate in a tentative fashion. The fact was that in spite of the abandonment of the older and more unproductive tea, the yield per acre hardly rose between 1890 and 1900, while there was a very great fall in the price. Certain diseases became more and more rampant. How to maintain tea gardens at full bearing capacity was a question of great importance. The production of quality in tea was little understood, and in this matter often there were very great and unexpected disappointments. Under these conditions, I was selected in the early part of the year 1900 to go to India as the first scientific officer of the Indian Tea Association. When I went to India, Mr. Bamber, to whom I have already referred, had recently been taken as adviser in Ceylon, but I think I may say that my appointment to India was the creation of the first scientific department wholly devoted to the study of the problems of tea cultivation and manufacture. The department as then constituted was a very small affair. The greater part of the cost was paid by contributions from the local Governments in India and from the local branches of the Indian Tea Association. I was allowed to work in the laboratory attached to the

Indian Museum, Calcutta. And, for a time, I was alone, without even a laboratory assistant, in the work. But gradually the value of the scientific study of tea cultivation was realised and then the workers engaged on the study began to increase in number and the amount spent on the work became larger and larger. At present the Scientific Department of the Indian Tea Association is one of the largest maintained by any similar industry in the world. The central experimental station of this Department at Tocklai in Assam is visited by workers from all over the world, and is one of which the tea industry can well be proud. Unlike similar departments in many other industries, it is not largely maintained by means of Government subsidies, but is almost wholly supported by the tea industry itself. It now consists, according to my latest information, of a cadre of fifteen scientific workers, both Indian and English, and I think I may say that its influence on the progress of the tea industry has been almost incalculable.

So far I have spoken of the organised scientific work in the tea districts of North-East India. Similar work has been organised in South India, and has achieved a good deal of success there. But I prefer not to speak in detail of the scientific organisation there, as I have no direct personal acquaintance with it.

What have been the results achieved by the application of scientific research and experiment under these organised conditions during the last thirty years? I am far from suggesting that all the improvement which has taken place in tea cultivation and manufacture during that time has been due to the results of experiments carried out under the auspices of these scientific departments, as well as those in other tea growing areas like Ceylon and Java, for some of the greatest improvements to which I shall refer later have been developed by highly skilled practical planters without special scientific assistance. But the effectiveness of the advice given has been great, and the experiments undertaken have been the means of introducing a large number of the most useful methods or modifications of methods commonly in use to-day.

It is curious too that the very marked increase in the crop of tea obtained per acre, which has been characteristic of the years since 1905, commenced very shortly after the establishment of the scientific department of the Indian Tea Association. This is shown by figures published, in part by Mr. Carpenter, the present scientific officer of the Indian Tea Association, in 1929. They refer to Assam and Bengal for each five-year period since 1885, at any rate to 1924, with the addition of data for 1926 and 1929.

| | ASSAM | BENGAL |
|---|---|---|
| **Period.** | **lb. per acre.** | **lb. per acre.** |
| 1885 to 1889 .. .. | 316 | 265 |
| 1890 to 1894 .. .. | 358 | 292 |
| 1895 to 1899 .. .. | 361 | 330 |
| 1900 to 1904 .. .. | 416 | 361 |
| 1905 to 1909 .. .. | 477 | 425 |
| 1910 to 1914 .. .. | 531 | 468 |
| 1915 to 1919 .. .. | 614 | 554 |
| 1920 to 1924 .. .. | 527 | 423 |
| 1924 .. .. | 576 | 479 |
| 1926 .. .. | 575 | 503 |
| 1929 .. .. | 603 | 563 |

This certainly in all cases shows a very great increase. Even the lowest figures are far above the yield in the early days of tea, for the *maximum* amount of tea obtained in the Assam Company's gardens in 1848 was 275 lb. per acre. But it will be noticed that the great increase occurs in the years after 1900 and particularly in the years after 1904. Comparing the yield per acre in Assam for 1895 to 1899 with that for 1929, there is an increase of 242 lb. per acre or over 67 per cent. In Bengal, where the figures do not quite mean the same thing, owing to the larger proportion of tea grown in the plains in later years, there has been in the same period an increase of 233 lb. per acre or 71 per cent. of the earlier yield.

The importance of this increase of yield to the tea industry cannot be over-estimated, for at even one shilling a pound it means an increase in gross return per acre of over £12 per acre in the case of Assam, and of nearly £11/10 per acre in the case of Bengal. Undoubtedly, a considerable portion of this increase has been due to the larger proportion of young tea existing in the later years. But except in the war years, it is not due to coarser plucking, that is to say to the production of lower quality of tea.

The value of the scientific work has been shown in a number of different directions, of which I shall be only able to speak of a few. In the first place, we have learnt better than before the characters which are required in good tea soils, so that it would be now, I think, inexcusable if tea were planted in unsuitable positions. This is important, for in another part of the world recently I have found it taken for granted, as it was in India in the earlier days, that where tea will grow, there it is likely to be a commercially paying crop. Even thirty years ago, I saw tea planted where nobody would put it nowadays, at any rate if they took competent advice. And there are few crops whose soil requirements are so specific as in the case of tea if a full crop is to be obtained. First, the soil must be acid—a most unusual requirement in agricultural crops. Again, a tea soil must be very deficient in lime—again a most unusual requirement. I have very often seen the addition of lime recommended on soils for tea, because this would be required by most agricultural crops. In the case of tea soils, I do not say that the application of lime is never useful. Far otherwise. But it is much more rarely useful than has usually been supposed, and than would be the case with most other crops. The absence of this knowledge has often led to useless or even injurious recommendations for the manuring of tea, but it was one of the earliest things I recognised when I began the systematic examination of tea soils thirty years ago.

There is another result of the study of tea soils which is of interest. I found a good many years ago that the quality, and especially the flavour, of tea seemed to be closely connected with the amount of useful phosphoric acid in the soil. In other words, the tea districts which gave high quality tea all had soils which contained a rather large proportion of this constituent. On the other hand, there were some districts at that time which, on this basis, should have produced much higher quality of tea than was the case, and I ventured to suggest that while the chance of producing a much higher quality of tea than was customary was small in some areas, in others, such as the Dooars, there was every reason to suppose that a much higher grade of tea was obtainable. Results have, in this case, justified the prophecy, and the special attention which was consequently paid to this matter in the districts in question.

But the assistance which has been rendered by scientific work has been specially great in the matter of tea manuring. When I went to India in 1900 any systematic manuring of tea was in its infancy. Much use had certainly been made of peat and other forms of "bheel soil," and there was a certain amount of cattle manure and of oilcake applied. But practically no artificial manures were used, and there was little recognition of the necessity of planning regular manuring of tea if the best results are to be obtained and the bushes saved from rapid deterioration. The recognition of this fact came only slowly, but all progressive planters understand it now, and there is at present a larger consumption of artificial manures in the Indian planting districts than, I think, in the whole of the remainder of the country.

There are certain features of this development of manuring, largely as a result of experiments done by the scientific departments to which I have referred and also by the scientific advisers who are now employed by a number of the large tea companies, which are particularly interesting. The first of these is the more complete use of green manuring than in any other agricultural industry. This method of manuring consists, as is well known, in growing a crop of some kind among the main crop, and when grown, burying this, in whole or in part, in the land, for the sole purpose that it may, by its decomposition, serve as a manure to the main crop. With tea, such manuring is done, not only with annual crops, but with bushes which remain growing among the tea for several years and are periodically pruned and the prunings buried, and also with trees which remain permanently growing among the tea but which manure the ground with their leaves and also with the nitrogen they are able to fix from the air. Taking the last case first, I may say that the possibilities in this direction were first realised by the late Sir James Buckingham, whom all those who have been connected with tea for many years remember with respect and reverence. The trees which he used are still used, but the list has been largely increased, and the growing of such trees among tea, after many years of controversy, is now well recognised in most districts as desirable, and as leading to a larger yield of leaf. Other forms of green manuring are now recognised as an essential part of any scheme of maintenance of tea at its highest pitch of efficiency.

Another feature of the development of tea manuring is that of a regular rotation of manures, differing, of course, on different classes of land, which have proved themselves of the greatest advantage in tea culture. In the Handbook of Information issued by the Scientific Department of the Indian Tea Association, special prominence is given to this matter and very interesting details are given as to the stage when each constituent of manures is likely to be of advantage. These are important, for the production of a crop which consists of green leaf demands a very different set of principles from those we are accustomed to apply with more ordinary crops.

May I pause here to congratulate the present Scientific Officer of the Indian Tea Association, Mr. P. H. Carpenter, on the very great advances in the systematisation of tea manuring which he has made in the last few years? The result has been that I feel that tea manuring is on perhaps a sounder basis than that with any other tropical crop.

I would like to speak a few words with regard to the influence of scientific work on the process of tea pruning, though a full treatment of it would demand a lecture all to itself. It will be recognised that the process of pruning a plant for leaf is altogether a different matter from that of pruning it for fruit. Up to 1900, however, there had been little study of any practical problem where pruning for leaf was demanded, and the greatest confusion of thought and practice existed in this matter. As a consequence, there were very extraordinary results obtained, and variations of practice were extreme. I attempted, for I think the first time in 1903, along with Dr. Watt, to lay down the principles on which correct tea pruning should be based, and these have formed the basis of very great changes in the processes in vogue when I joined the tea industry. My successors as Scientific Officers to the Indian Tea Association carried matters much further in their publication in 1914, and the principles involved and the results of varied practices, in giving healthy tea bushes and a high yield of tea, are now pretty well known. This is not to say that the last word has been spoken on the subject. Tea pruning is still a matter of active controversy and of no less active experiment—but we have reached a point far in advance of that which existed thirty years ago.

The success of tea culture depends, however, primarily on the success attained in producing the continual succession of young tea shoots for plucking. It is well known that in order to make tea, the young growing tea shoot is plucked from the bush, the amount taken including the unopened leaf bud and the youngest two or three leaves on the shoot. If three such leaves are taken, the plucking is said to be coarse: if only two such leaves are plucked, it represents fine plucking. So far everybody knows. But endless variations occur and have occurred, all of which caused variations in the quality and in the amount of tea which could be obtained from a bush. I have already referred to the early methods applied in Assam which almost led to the collapse of the industry. At first, in fact, as far as I can learn, it was customary to take the shoots as soon as they had grown, right from the beginning of the season. The result was a very early crop, reaching its maximum in April or May in each year, after which the bushes refused to give much more leaf, and, furthermore, the bushes themselves quickly deteriorated. The success of tea growing in India, in fact, dates from the time when it was recognised that considerable growth must be allowed to occur and remain on the bushes before any could be taken—which secures much more later growth and preserves the bushes themselves. But this leaving of growth on the bush may easily be overdone, and then growth tends to cease, and no more shoots are produced. I have seen this many times in the olden days in India, and have come upon it also recently in studying tea in Russia. Between these two extremes lies the correct method of plucking tea, but the variations in practice have been very great. Mr. Carpenter still considers the absolute best method of plucking very much a matter of controversy. But I would say that in few things has more progress been made than in plucking to obtain the largest quantity of fine tea, during the last thirty years. Many experiments have been done by the scientific officers in India, but I think that the largest part of this progress has been the result of the work of practical planters, especially in Upper Assam. It is now possible to obtain yields of tea from the bushes, plucking fine, which are far larger than could be obtained in the olden days while plucking coarse.

*Chapter* 35

# THE WOBURN EXPERIMENTAL STATION (1876-1956)*

In the early 'seventies of last century British agriculturists were agitated by changes in the relation between landlord and tenant of farms, first by the Irish Land Act of 1870 and then by the Agricultural Holdings Act of 1875. The second Act made it compulsory to compensate an outgoing tenant for the unexhausted value of foods bought during his tenancy. After discussions on this subject, Sir John Lawes of Rothamsted, whose authority at that time was very great indeed, published a paper on the "Valuation of Unexhausted Manures" in the *Journal of the Royal Agricultural Society* (I) that contained a table purporting to give the estimated money value of the manure obtained from the consumption of 1 ton of each of the commoner farm foods. Farmers, however, at once asked whether the values assigned by Lawes were really true. Does the outgoing tenant get his due, or is the landlord or the incoming tenant over-charged? These questions were raised at a meeting of the Royal Agricultural Society in November 1875, when it was urged that the subject needed settling by direct experiment on different soils and under different conditions.

The then Duke of Bedford (Hastings Russell) offered the use of land on his estate at Woburn and funds to pay the cost of such experiments, provided that the Royal Agricultural Society would direct and manage them. The offer was gratefully accepted, and the Society asked Sir John

* Reprinted with permission from the **Report of the Rothamsted Experimental Station** for 1958 (Harpenden), 1959, pp. 223-234.

HAROLD H. MANN in 1955-56
At the age of 84.

Lawes and Dr. A. Voelcker (the consulting chemist to the Society) to draw up a scheme for the proposed experiments, which, it was recognized, must continue for many years. The land generously placed at the disposal of the Society was known as Crawley Mill Farm. It consisted of 90 acres, of which 67 were arable and 23 grass. To this a specially suitable field (Stackyard Field) was added in the parish of Woburn, which became the main experimental field of the Station. By the autumn of 1876 the experiments started. The venture had the cordial interest of Lawes and Gilbert of Rothamsted, and it was directed by Dr. Voelcker under the supervision of a special committee of the Royal Agricultural Society. From the start it was decided not to confine the experiments to finding the residual values of foodstuffs, but to duplicate, on the very different soil at Woburn, many of the tests already made at Rothamsted. The central problem, however, was the valuation of unexhausted manures, to which no less than 16 acres of the main experimental field were devoted.

The site chosen for the first experiments has special interests. It lies chiefly on the lower greensand, and so consists of a light sandy loam. This is important when results are compared with those obtained at Rothamsted and other centres. This soil was carefully studied many years later by E. M. Crowther (2), who described it as follows: "The mechanical composition is uniform down to 24 inches, below which there is a sudden fall in the silt content and an increase in the coarse sand and stones .... The rapid decrease in the organic matter with depth and the poverty of the deep soil were shown by analysis." The chief chemical point to note is the small amount of calcium in any form, and the slight acidity, though this is rarely enough to prevent the healthy growth of most ordinary farm crops. The experimental field was in good heart, and drainage was excellent.

Though this type of soil is typical of most of the Station, there is some heavy silt based on Oxford clay, but this was, and still is, little used for experiments.

The experiments during the first twenty years studied three questions (3). The first, already referred to, was the relative manurial value of animal feeding with rich and poor foodstuffs. The second to repeat, with modifications, experiments with continuous wheat and barley given various fertilizers and manures of the type originally done by Lawes and Gilbert at Rothamsted. The third, introduced a little later, concerned so-called green manuring, that is to say the attempt to use one crop to manure the ground for a more important one to follow. Except for animal-feeding experiments, these represented the staple work at Woburn till 1898.

In 1897 two important things happened. First, the Royal Agricultural Society received a bequest from Mr. E. H. Hills of £10,000 to study the manurial effect of the "tertiary constituents of plant ash" better known in later years as trace elements. It was decided to do the desired experiments at Woburn, so a small laboratory and a well-equipped pot-culture station were built and opened in January 1898, when I took charge as resident chemical assistant. Up to that date all the necessary chemical and other analytical work had been done at the laboratory of the Society in London; the establishment of the laboratory made possible a much closer scientific control of the work done than previously.

The establishment of the laboratory, and the development of pot culture on a considerable scale, opened a new phase, which permitted the terms of the Hills bequest to be fulfilled and to study specific problems as they arose in the course of the work at Woburn. With the appearance of new and interesting phenomena in the original field experiments, as well as in others afterwards established, there was a period of great activity and 1898-1914 was perhaps the busiest period in the history of the Station till 1928. I left in March 1900 to go to India; my successor as resident chemist, Mr. H. M. Freear, obtained results of great interest, some of which were published in the *Journal of the Royal Agricultural Society,* but others even now remain in manuscript. Of these I may mention a detailed study of the development and meaning of the acidity acquired in soil from the application of ammonium salts, the development of clover sickness from the repeated growing of clover on the same land, the relative effect of magnesium and calcium salts on the growth of crops, the work with trace elements and that with the field experiments originally started in 1876-77. The outbreak of war and the death of Mr. Freear in 1914 inevitably decreased the work and new developments were few in the next years. Also in 1912 the Duke of Bedford, who had generously paid all the costs of the experiments at Woburn, decided to withdraw this financial support, although he allowed the farm and buildings to remain with the Royal Agricultural Society for experimental work. From 1912, the cost of the Station was borne by the Society until 1921, when it withdrew from the control and management. Dr. J. A. Voelcker, the consulting chemist to the Society, then maintained the Station at his expense, except for a small subsidy from the Ministry of Agriculture, given on the understanding that the work would be supervised from Rothamsted. Since 1921 the annual reports of the work at Woburn have been issued as part of the Rothamsted reports.

The general results of the work of the Station, excluding the results of pot experiments, to the year 1926 were published by Russell and Voelcker in 1936 under the title *Fifty Years of Field Experiments at the Woburn Experimental Station*. This book is now difficult to get, as the whole stock at the publishers was destroyed by enemy action in the Second World War, yet the fact that it still exists, with a statistical portion by W. G. Cochran, makes it unnecessary for me to describe the work of the Station in detail up to that time. But another period of great activity started in 1928, and I shall summarize part of what has been done since then.

Before doing so, however, I wish to acknowledge the great debt owed to the late Dr. J. A. Voelcker, who from the death of his father in 1884 was Director of the Station till 1936, first under the Royal Agricultural Society and then under Rothamsted. His zeal and his interest in the experimental work never flagged; to the last year of his life he visited and inspected the work month by month, and there is no doubt that but for him the Station would have ceased to exist.

After I retired from my post in India, Sir John Russell invited me to return to Woburn and take charge of the experimental work there. I did this in 1928, when the farm was managed from Rothamsted, but in 1936 I also became responsible for running the whole farm, an arrangement which continued till 1946. Since then Mr. J. R. Moffatt has directed all the farming operations, both of the field experiments and the non-experimental crops. For the next 10 years, I was concerned only with the experimental work in the field and the laboratory, and at the end of 1956, when I officially retired, Mr. C. A. Thorold took over these duties.

Since 1926 the major changes and developments in the field experiments have been the adoption of the modern experimental designs developed by R. A. Fisher and his successors at Rothamsted. The new designs were applied to many problems that widened the interests of the Station. Among new experiments have been the study of what happens under a well-planned rotation of field crops, well manured but without adding extra organic manures for a long period; the first direct tests in this country of the effects of so-called ley farming on the fertility of the land; the relative value of coarse organic manures, including sewage sludge and several composts, in increasing the productivity of poor agricultural land; and the suitability of a number of new or lesser-known crops, such as soybeans, maize grown for grain and several fodder crops, for slightly acid land. Irrigation trials have included tests on different amounts and times of watering on potatoes, sugar beet, barley, wheat and beans, as well as on both a mixed grass and clover sward and one composed of grass only.

In the laboratory and pot-culture station the most important questions studied have been: (*a*) the comparative value of green manuring by various types of crops and mineral fertilizers as sources of nitrogen; (*b*) the effect of soil acidity on the composition of barley and on crop failures; (*c*) the reasons for the failure of clover when grown frequently on the same land, and (*d*) the competition between weeds and grain crops.

## Chief Work of the Station

This outline history of what, except for Rothamsted, is the oldest agricultural experimental station in Britain, now requires an account of the more important results obtained during the eighty years of its existence. For many of the experiments it will suffice to do little more than refer to the published records of what was done and the conclusions reached. At first nearly all these were published in the *Journal of the Royal Agricultural Society,* and were summarized in 1936 in the book *Fifty Years of Field Experiments,* by Russell and Voelcker. Since then most of the work, apart from the sketchy accounts in the *Rothamsted Annual Reports,* has been issued in a series of papers, mostly in the *Journal of Agricultural Science.* Some of the conclusions will now be summarized and references made to the sources where further details can be seen. Some notes will also be given of work which has not so far been published.

### *Valuation of the unexhausted value of manures*

What are the conclusions from the Woburn experiments on the primary problem for which the station was established, namely the relationship between the chemical composition of feeds given to animals and the value of the manure they produce? These experiments, continued with various modifications from 1876 to 1932, always gave similar results. The tests made consisted of crops grown in a four-course rotation, usually either the Norfolk four-course rotation or some modification of it, in which the roots were fed off to sheep, and in the earlier forms of the experiment the wheat crop was also manured. Contrasting plots manured with dung made with high-nitrogen oil-cake (decorticated cotton-cake was used at first) and similar plots with dung made with low nitrogen foods like maize meal or grain of any sort, should show for each following year whether the high-nitrogen foods gave a more powerful or more lasting manure.

After fifty years, Voelcker summarised the results as follows:

> "A review of the results forces one to the conclusion that the experiments have entirely failed to show any marked superiority of cake feeding over corn feeding on the soil. The most that can be said is that the barley immediately following the sheep-folding showed a small advantage for cake as against corn. But no subsequent crops benefited. The highest gain in barley from cake feeding did not exceed 6 bushels per acre and even when there was as much difference as the equivalent of 2 cwt of nitrate of soda between the cake and corn feeding, the crop difference only came to 1 to 2 bushels per acre of barley.
>
> "No explanation can as yet be given for the failure to recover the added fertilizer constituents. It was at first supposed that the feeding of the land was too liberal and the crops too large to indicate any difference between the higher and the lower manuring. But when the feeding was reduced and only given once in the rotation the differences were not more strongly marked, not even when growing barley without manure for a number of years. The substitution of green crops like mustard for clover in the rotation did not help though the wheat crop following clover was larger than that following mustard. All said and done, there remains the fact that much more nitrogen has gone into the land by the use of cake than by that of corn and yet for some reason or other it has not become available for use." (4)

This quotation summarized the position when the work had continued for fifty years. The question has since been approached by Mann and Barnes by means of pot experiments; it seems now clear that when organic manures containing more than so much nitrogen (usually about 2 per cent nitrogen of the dry matter) are applied to land the nitrogen is soon lost either as elementary nitrogen (5) or possibly as nitrous oxide (6), and that within a few months the land is in the same condition as if a less-rich material had been added. In fact, it seems likely that unless the nitrogen in a rich manure is captured by growing plants shortly after it is mixed with the soil it is lost for ever (7). It seems likely that this is what happened in the long-continued experiments noted above and what made the original tables of Lawes quite wrong in the manurial value they gave to the richer forms of organic manure.

*Continuous growing of wheat and barley*

The results obtained for the first 50 years of growing wheat and barley continuously on the light, slightly acid and lime-free soil at Woburn were thoroughly analysed by Russell (8). The chief effect sought in the original plan of the experiments was that of nitrogen, and phosphatic and potash manures were always applied together. The nitrogen was, however, added in three different forms, and in four different quantities, as nitrate of soda, as sulphate of ammonia and as farmyard manure. The first two supplied nitrogen at the rate of 41 and 82 lb. nitrogen/acre each year, and the farmyard manure at rates not exactly determined, but probably about 53 and 105 lb./acre.

During the first 15 years the yields were high, and those of winter wheat on this light land, usually not considered as suitable for this crop, were almost as great as of barley. The three manures differed in effectiveness among themselves. Sulphate of ammonia was somewhat inferior to nitrate of soda and farmyard manure considerably inferior, yet each of them acted similarly on both crops. The different manuring made very little difference to the proportion of grain to total produce.

After 15 years something started to go wrong with the plots given ammonium salts. The crops failed in patches and resowing did not give a crop. This was attributed to exhaustion of calcium, and in 1897, after 20 years of cropping, lime was added to one-half of each of the plots. The yield recovered to that in the early years of the experiments. This was 1891, a time when the question of soil acidity and its influence on yield was first considered and experiments by Wheeler (9) in Rhode Island, U.S.A., showed how injurious acidity could be. I remember going to Stackyard Field in 1898 with a book of litmus papers and finding that the plots treated with ammonium salts had soil which reddened litmus paper. The further addition of ammonium salts reproduced the trouble, and within a few years the yields had again become very small; finally, the crop failed completely, particularly the barley, and the plots carried simply a mass of spurrey (*Spergula*). The need to apply lime in British agriculture was greatly emphasized by these results, for though medieval agriculture made the marling of land very important, the reason was hardly understood. Nowadays, the importance of dressings of lime on many, if not on most, of the soils of Britain is fully realized.

It has long been accepted that cereals should not be grown frequently on the same land, particularly on light land. The results of the Woburn and Rothamsted experiments allowed Russell (8 & 10) to discuss the subject. He concluded that they can be grown for several years provided that weeds and serious pests and diseases can be controlled. This, though, is difficult, and yields at Woburn have undoubtedly been greatly affected by weeds, and by take-all (*Ophiobolus graminis*), which has been very serious. Whether acidity, weeds and diseases are the only causes of declining yields, I doubt.

De Candolle suggested a century ago that plants excrete from their roots something injurious to their own species but not to plants of a different kind. This certainly seems to happen with some crops, notably with clover (see below), and recent experience with both wheat and barley on the light land at Woburn makes me favour this as a possible explanation with these two crops. It is of interest that only one treatment has maintained yields reasonably well, and that is farmyard manure applied each year at the rate of 8 tons/acre.

*Green manuring experiments*

The practice of growing one crop to prepare the ground for a second and more important one is common, and has in many places entered into general use. It has, though, often failed to increase the yield of the succeeding crop or to increase the fertility of the soil.

A striking example of green manuring failing to give benefit occurred in the long-continued experiments at Woburn conducted by Voelcker, in which vetches and mustard were used to prepare the land for winter wheat. Whether the green crop was ploughed into the land or fed to sheep, and whether or not phosphates and potash were given, the following crop of wheat was small and became poorer and poorer as the experiment was continued. Further, wheat after the vetches, particularly early in the experiments, consistently yielded less than after mustard, even though the nitrogen buried with the leguminous plants was double or treble that with mustard. The history of these experiments up to the year 1932 was considered by Crowther and Mann in 1935 (11), who concluded that a green manure crop must be large to be effective, that when the green manure has a low carbon-nitrogen ratio (like vetches) its manurial value is short-lived and soon lost, and that its value will show only in the presence of a crop that uses the nitrogen as the green manure rots and it becomes available.

Further work with field crops and in pot cultures since 1933 confirmed the conclusions of Crowther and Mann and showed that beneficial results come from green manures containing a high percentage of nitrogen only when the following crop is taken very soon after the green manure is buried. The crop-producing power of nitrogenous manures depends, in the main, on two factors, namely: (1) the length of time they have been in the soil before the next crop is sown, and (2) their content of nitrogen. With materials containing much nitrogen, the maximum benefit is obtained after they are buried for only a short time, and the benefit decreases with increasing time before a crop is present, even when there is no loss by drainage. The time after ploughing in that gives the maximum benefit to the succeeding crop increases as the percentage of nitrogen in the green manure decreases. All manures containing much nitrogen, whether organic or inorganic, show this decreased effectiveness as the interval between their application and the sowing of the crop increases. However, when the nitrogen in plant or animal residues is less than about 1.8 per cent of the dry matter, their immediate use as manures decreases the crop-producing power of soil. From this amount to between 4 and 6 per cent of nitrogen the crop-producing capacity increases and does so more rapidly than the actual nitrogen increase. When the percentage of nitrogen goes above this, the crop yield does not increase, though the recovery of nitrogen may be greater. Different plant materials with the same percentage of nitrogen did not behave exactly alike, though their behaviour was similar. There is no evidence that any of the materials used, including farmyard manure, greatly increased the amount of *residual* nitrogen in the soil. After the first crop was taken, whether the green manures contained much or little nitrogen, the addition had no further effect.

*The importance of organic manures in land fertility*

Whether organic manures are needed to maintain the fertility of land is often discussed and an experiment was started in 1930 in which crops in a six-course rotation were supplied with inorganic fertilizers but no organic manures. The experiment was originally designed to test the effect of weather on crops, but its chief interest changed to the function of organic manures and the extent to which yields can be maintained without them. The results after 25 years are being examined by others, but I can say that there is as yet little sign that the land is deteriorating. Certainly the yields of four of the six crops grown, namely barley, rye, potatoes, and sugar beet, are being maintained. Conclusions are less definite with winter wheat and clover, because the wheat has been consistently poor, for some unknown reason, and the clover is so seriously attacked by clover rot (*Sclerotinia trifolium*) each year that no valid conclusions can be drawn about soil fertility.

Another experiment on the importance of organic manures in maintaining the fertility of light soil has given different results. An area that had become very poor, and from which only meagre crops were obtained for some years, was put down in 1942 to an experiment with market-garden crops. Various coarse organic manures were annually applied in large quantities to see whether their use increased the fertility of the land. The experiment, now 17 years old, was designed to answer the following questions: (1) whether the presence of organic matter is essential to maintain the productivity of such a soil; (2) whether different organic manures are of equal value in increasing and maintaining fertility; (3) whether fertility increases immediately or steadily over a long period of applying organic manures; (4) whether the organic manure has a specific effect not determined by its content of plant nutrients.

Most of these questions have been answered in part, and some answers are unequivocal. The answers vary with different crops; of the four grown, namely globe beet, green peas (seeded on the plots), cabbages and transplanted leeks, cabbages and peas do not respond specifically to the organic manures, but globe beet and leeks do. With cabbages the same yields were obtained by extra mineral fertilizers as with organic manures, and peas were little affected by manuring. With globe beet and leeks, however, organic manures gave large increases in yield that were unobtainable with inorganic fertilizers. Different organic manures are difficult to compare accurately, but on the basis of the amount of organic matter added, farmyard manure was similar to a compost made with straw and weeds and activated with dung, and much superior to sewage sludge and a compost activated with sewage sludge. Fertility seems only slightly increased by annually repeated applications of the manures, and the increase from manuring, except with globe beet, is now almost the same as it was in the first 3 years. Doubling the dose of manure does not double the response, but increases it by about 60 per cent. The question of a value in the organic manures other than their recognized content of nutrients remains unanswered, but the experiment is being continued, and a complete answer may be obtained. In the meantime Mann and Barnes (13) studied the permanence of the organic matter in the soil. The soils after each treatment were analysed after 9 years, when: (1) Different materials were found to divide into two classes, different in activity. Farmyard manure is much more active than sewage sludge or either of two forms of compost, which do not differ much from each other, though the least active is the compost made with sewage sludge. (2) A top dressing of sulphate of ammonia decreases the rate at which the last three disappear from the soil. (3) Doubling the amount of manure added led to their more rapid disappearance.

*Ley farming*

I now turn to one of the most interesting field experiments of the last few years, namely that dealing with the value of ley farming. This was started in 1937, and a paper by Boyd and Mann (14) gives the full results. The leys used were a 3-year grazed area and a 3-year crop of lucerne cut for hay; the arable cropping with which to compare the results consisted of potatoes, a winter cereal crop and either a 1-year ley or another tillage crop. The relative effect of these crop sequences was measured by test crops of potatoes followed by barley, uniformly treated except that 15 tons of farmyard manure were applied for the potato crop on half of each plot. Without farmyard manure, potatoes after the grazed ley yielded consistently more than after the three tillage crops by about 3 tons/acre (23 per cent); after lucerne it was about 2 tons/acre more than after the tillage crops, and after the 1-year ley it was less than 1 ton more. With farmyard manure the benefit from leys was less, about 2 tons/acre for both lucerne and the grazed ley. The average effect of the farmyard manure (at 15 tons/acre) was 2.8 tons of potatoes, except after the grazed ley where it was only 1.6 tons/acre. Effects of the previous cropping on the yield of barley were small in the early years of the experiment, but later the yield after leys and lucerne exceeded that after tillage crops by about 15 per cent.

*Irrigation experiments*

In 1951 a new and important development was the start of experiments to measure the value of irrigating field crops and grass leys at Woburn, and this has since continued with some modifications. I shall not discuss the results in detail, as the work continues and will be described by H. L. Penman, under whose general direction it is done. However, the results with different crops can be stated; in spite of the fact that 6 years out of 8 have had dry periods in March and April, irrigation has had little effect on cereals; its value has varied from year to year with sugar beet and grass, but in 6 out of 8 years it has been highly profitable with potatoes, whether early or main crop. One point of peculiar interest is that, in two seasons when a mixed grass and clover sward was grown, the proportion of clover in the cut herbage was considerably greater on the irrigated plots. T. W. Barnes uses the experiment to study the uptake of nitrogen by grass and how it is affected by watering.

*New and unusual crops*

In recent years several unusual crops have been grown to see how far they have any use on a slightly acid light soil such as that at Woburn. Almost all available varieties of *soybeans* have been tried, but none proved promising enough to suggest their cultivation on a large scale. On the other hand, *maize grown for grain* gave good results when the hybrid varieties now early enough for the English climate were used. Using the best of these varieties, developed from American types in Holland, we have got an average yield of nearly 35cwt./acre of dry grain in good summers, falling to about 15 cwt./acre in cold, wet years. The crop suits smallholders, as the amount of seed required is small, it can be grown in almost any odd corner and the maize ears can be used for chickens or even for pigs without threshing. Another crop which seems to have possibilities is the *Jerusalem artichoke* in the form of a specially luxuriant variety developed in Germany. This has given an average yield of leafy tops of over 3 tons of dry matter/acre, which are readily eaten by pigs and make good silage, and over 12 tons of tubers, which from all tests elsewhere are as good as potatoes for pig feeding. It is clear, however, that this crop is only suitable where the soil is light and well drained.

Special interest attaches to efforts to extend the range of fodder crops suited to semi-acid land, and several of these have possibilities in this country. One is the *sweet lupin,* which has given us good results in many years. Though the fodder produced is not at first relished, animals soon get used to it and then eat it readily. In our tests we have been able to get an average of 12.5 tons of green fodder/acre containing just under 2 tons/acre of dry matter. This crop, which is now used fairly extensively in New Zealand, seems to have considerable possibilities on the more acid lands, especially in the eastern counties of England. *Serradella* is similarly suitable to acid soils, and we have grown it for a number of years as a fodder crop for use in September when grass is beginning to fail. The seed used at Woburn was originally obtained from the U.S.A., but it has been maintained with home-grown seed obtained in the drier years. The crop contrasts with lupins in that it is eaten greedily by all classes of stock. In wet years, such as 1954 and 1956, it gave large yields of green stuff in September; in 1956 30 tons of green forage were grown containing 11 per cent dry matter. No seed was obtained in this wet year, but in 1955 it was abundant and of good quality. Of other successful fodder crops for acid soils, I might mention *Birdsfoot Trefoil,* which has grown well, though slowly, but which gave in October 1956 4.7 tons of green stuff/acre, containing 16.6 per cent of dry matter. The seed was obtained from the U.S.A., where much work has been done to improve the crop, and the improved types seem to have prospects as fodder crops in Britain.

*Studies of acid soils and the barley crops grown on them*

Barley grows normally on Woburn soil until the acidity falls below pH 5.4. This is so whether normality is judged by the vegetative growth, by the character of the earing, by the composition of the plants during growth or by that of the straw and grain at harvest. When acidity reaches a pH value between 5.4 and 4.7 abnormalities occur. The plants are dwarfed, and the grain-bearing capacity is greatly affected. The number of leaves, shoots and heads per plant are not much decreased, but the plants are dwarfed and sickly, with the leaves yellow at the tips and purple in the veins. These effects all increase with increasing acidity beyond what may be called the "critical point". In the more acid soils the amount of oxide of iron and alumina in plants may be three times the normal percentage, but the total ash content is decreased, for silica, calcium and phosphorus are all decreased; so also is the percentage of nitrogen (15).

When an acid soil is treated with calcium carbonate the pH value rises to a point at which good growth of barley is again possible, but adding more calcium carbonate does not further increase the yield. Extreme acidity of the soil, even when equivalent to a pH value of 3.5, is no bar to nitrification. The amount of soda removed by drainage from these acid soils is little affected by the pH value of the soil. On the other hand, the leaching of potash is much greater than has been generally thought. The more acid the soil, the higher the proportion of magnesium to calcium in the drainage water. The amount of alumina in the drainage waters remains almost constant till the pH value is below 4.6, but at 4.1 it increases to nearly four times the previous amount, though even then it is very small (16).

*Competition of weeds with arable crops*

A unique series of studies has been made by Mann and Barnes on the competition of barley with various plants, weeds such as spurrey (*Spergula arvensis*), mayweed (*Metricaria inodora*), *Holcus mollis*, *Agrostis gigantea*, chickweed (*Stellaria media*) and clover. No evidence was found except perhaps with *Holcus mollis* that any plant acted specifically on the barley. The roots usually intertwine with no sign of mutual attraction or repulsion. With abundant nitrogenous manure, clover behaves like the other plants (17).

*Soil sickness from frequent growing of clover*

In 1860 Lawes and Gilbert drew attention to the failure of red clover when frequently grown on the same land. In the early days of the Woburn Experimental Station this problem received attention, and its study was taken up seriously by Mann about 20 years ago. A first result was published in 1938 (18), but the work has continued ever since, and a further note was published in 1950 (19). The general conclusions that have been reached can be summarized as follows: though the experimental results do not conclusively establish the cause of the failure, they do indicate clearly the direction in which a solution to the problem of this and similar soil sicknesses will be found. The evidence is strong that the trouble is not pathogenic but is induced directly by the continued or frequent growth of clover on the same soil. The soil is not rendered unsuitable for gramineous plants like barley or ryegrass, but radishes, lettuce and spinach beet are affected almost to the same extent as clover. The advent of the sickness is slightly accelerated by increasing the temperature at which clover is grown, but is still more accelerated by growing many plants in a little soil.

Three methods have been found by which soil seriously affected by this clover sickness can be, at least partially, restored to a healthy condition. The first is by heating the soil, preferably moist, to 70° C. for some hours. The second is by treating the soil with a heavy dressing of farmyard manure or of a compost activated with dung. This result is not a matter of increasing plant nutrients, as it cannot be achieved by even large dressings of mineral fertilizers. A third, less-effective method, is to add calcium carbonate and decrease acidity.

All attempts to leach out any causative substance from the soil have failed. It is not easily oxidized. Treatment with toluene followed by the removal of the antiseptic agent has no effect, but treatment with formaldehyde gave some improvement. I suggest no pathogenic organism is concerned but that the cause is an organic product of the growing clover which is heat-sensitive, insoluble in water, not affected by antiseptics, tending to be more active in slightly acid soils and whose activity is somewhat reduced by certain colloid organic materials like farmyard manure or charcoal, but not by all.

## CONCLUSION

I have dealt with the principal work done at the Woburn Experimental Station during its 80 years of existence. Beyond this, many studies have been made, both in the field and in the laboratory by members of the Rothamsted staff, and others which have had a temporary, though often important, connotation. The record of these can be found in the annual reports of the Station issued up to 1920 in the *Journal of the Royal Agricultural Society*, and, since then, in the annual reports of Rothamsted. As for the former class of experiments, they have been published in various journals under the names of those responsible.

The future is likely to show the continuing importance of the work, and I hope to see a gradually increasing amount of scientific activity at Woburn, which will lead, in ever-greater measure, to results important to the agricultural industry in Britain.

## REFERENCES


1. LAWES, J. B. (1875). On the valuation of unexhausted manures. **J. R. agric. Soc. 2,** 1.
2. CROWTHER, E. M. (1936). The soils of the Woburn plots. (In: **Fifty Years of Field Experiments at the Woburn Experimental Station.**) London: Methuen, pp. 315-345.
3. RUSSELL, E. J. (1936). **Ibid.,** pp. viii-ix.
4. VOELCKER, J. A. (1936). **Ibid.,** pp. 23-25.
5. MANN, H. H. and BARNES, T. W. (1951). The Behaviour of nitrogeneous manures in the soil. I. **J. agric. Sci. 41,** 309
6. ARNOLD, P. W. (1954). Losses of nitrous oxide from soil. **J. Soil Sci. 5,** 116.
7. MANN, H. H. & BARNES, T. W. The Behaviour of nitrogenous manures in the soil. III. (not yet published).
8. RUSSELL, E. J. (1936). (In: **Fifty Years of Field Experiments at the Woburn Experimental Station.**) London: Longmans, pp. 213-260.
9. WHEELER, H. J. (1891— ). **Rep. R. I. agric. Exp. Sta.**
10. MANN, H. H. (1934). Rothams. Conf. **17**, 30-34.
11. CROWTHER, E. M. & MANN, H. H. (1936). **J. R. agric. Soc. 94,** 94—128.
12. MANN, H. H. (1958). Field studies in green manuring. I. **Emp. J. exp. Agric. 26,** 274.
13. MANN, H. H. & BARNES, T. W. (1956). The Permanence of organic matter in the soil. **J. agric. Sci. 48,** 160.
14. MANN, H. H. & BOYD, D. A. (1958). Some results of an experiment to compare ley and arable rotations at Woburn. **J. agric. Sci. 50,** 297-306.
15. MANN, H. H. (1937). The character of barley grown on soil made acid with sulphate of ammonia. **J. agric. Sci. 27,** 108.
16. MANN, H. H. & BARNES, T. W. (1940). Studies of soil after fifty years of wheat and barley cropping. **J. agric. Sci. 30,** 345.
17. MANN, H. H. & BARNES, T. W. (1945-53). The competition between barley and certain weeds under controlled conditions. **Ann. appl. Biol. 32,** 15; **34,** 252; **36,** 273; **37,** 139; **39,** 111; **40,** 566.
18. MANN, H. H. (1938). Investigations on clover sickness. **J. agric. Sci. 28,** 457.
19. MANN, H. H. (1950). Notes for paper on clover sickness. **Ann. appl. Biol. 37,** 327.

# BIBLIOGRAPHY OF THE WORKS OF HAROLD H. MANN

**Books**


1903.— With Watt, Sir G. **The Pests and Blights of the Tea Plant** 2nd ed. Calcutta; Office of the Superintendent Government Printing, pp. 1-429 + i-xv.

1907.— **The Tea Soils of North-East India and their treatment** Calcutta, Indian Tea Association, pp. 1-284 (Scientific Department Publication No. 22).

1907.— Edited by Dr. H. H. Mann: Mukerji, N. G., **Handbook of Indian Agriculture;** 2nd ed. Calcutta pp. xii + 706.

1915.— Edited by Dr. H. H. Mann: Mukerji, N. G., **Handbook of Indian Agriculture;** 3rd ed. Calcutta pp. xiv + 620.

1917.— With Sahasrabuddhe, D. L., Kanitkar, N. V., Tamhane, V. A., et al. **Land and Labour in a Deccan Village** (Pimpla Saudagar) (Study No. 1). Bombay, University of Bombay, Economic Series No. 1 pp. 184 + i-iv.

1921.— With Kanitkar, N. V., **Land and Labour in a Deccan Village** (Jategaon Budruk), (Study No. 2), Bombay, University of Bombay, Economic Series No. III, pp. vii + 182.

1925.— **Statistical Atlas of the Bombay Presidency.** 3rd ed., Bombay; Government Central Press; pp. 194.

1955— **Rainfall and Famine.** A Study of Rainfall in the Bombay Deccan, 1865-1938. Bombay; Indian Society of Agricultural Economics, pp. 1-47.


**Articles, Reports, Memoranda**

I. METHOD AND PHILOSOPHY


1911.— "The philosophy of a man of science" **Indian Interpreter,** (Madras), Vol. VI, No. 1, pp. 18-28.

1911.— "The social study of a people" **Indian Interpreter,** (Madras), Vol. VI, No. 2, pp. 79-83.

1961.— "The investigation of economic conditions in under-developed countries" **Economic Development and Cultural Change** (Chicago), Vol. X, No. 1, pp. 102-104.

## II. SOCIAL STUDIES AND HISTORY

### A. Village Studies and Rural Economy

#### 1. England

1905.— "Life in an agricultural village in England." **Sociological Papers,** etc., 1904, vol. I, pp. 161-193 (London, Sociological Society).

#### 2. India

1907.— "Note on the diet of tea garden coolies in Upper Assam, and its nutritive value." **Journal and Proceedings,** Asiatic Society of Bengal, (Calcutta) Vol. III, pp. 103-108.

1916.— "The 'Mahars' of a Deccan village." **Social Service Quarterly,** (Bombay), Vol. II, No. 1, pp. 1-8.

1916.— "The Economics of a Deccan village." **Indian Journal of Economics** (Allahabad), Vol. I, Part 4, December 1916, pp. 409-433.

1919.— "Economic conditions in some Deccan areas." **Agricultural Journal of India** (Calcutta), Vol. 14, pp. 804-810. (Reproduced from Poona Agricultural College Magazine, April 1919).

1920.— "Efficiency of agricultural labour." **Indian Journal of Economics** (Allahabad), Vol. II, Part 4, pp. 456-461.

1920.— "The effect of rise of prices on rural prosperity." With Kanitkar, N. V., **Proceedings of the third Annual Economic Conference held in Madras in 1919-1920,** pp. 30-40; also published as Chapter VIII of **Land and Labour in a Deccan Village,** Study No. 2, 1921 (q.v. above).

1923.— "A Deccan village under the Peshwas." **Indian Journal of Economics.** (Allahabad), Vol. IV, Part 1, October 1923, pp. 1-17.

1923.— "Interest payable by Deccan Cultivators." **Bombay Cooperative Quarterly,** Vol. VII, pp. 1-6.

1924.— "The economic progress of the rural areas of the Bombay Presidency 1911-22". **Bombay; Government Central Press,** pp. 62.

1942.— **"Annual Administration Report of the Department of Agriculture." H.E.H. the Nizam's Government, for the year 1348 Fasli (1938-39)** (Review-article), **The Asiatic Review,** London, Vol. XXXVIII, No. 133, January 1942, pp. 106-108.

1948.— "Village betterment in the New India and Pakistan." Proceedings of the East India Association (London), as published in the **Asiatic Review**, London, Vol. XLIV, No. 158, pp. 154-170.

1957.— "The General report of the Committee of Direction: All-India Rural Credit Survey." **Indian Economic Review,** (Delhi), Vol. III, No. 4, pp. 76-80.

1958.— "The economic results and possibilities of irrigation.' **Indian Journal of Agricultural Economics**, (Bombay) Vol. 13 (2), pp. 1-6.

1960.— "The central mechanized farm at Suratgarh, Rajasthan. 1956-1958." **Economic Weekly,** (Bombay), Vol. XII, pp. 1827-1828.

3. Middle East and U.S.S.R.

1938.— "The collective farm system in Russia." **Indian Cooperative Review,** (Madras), Vol. IV, No. 2, pp. 149-155.

1946.— "The new agricultural land law in Turkey." **Bombay Cooperative Quarterly,** Vol. XXX, No. 2; pp. 77-81.

1946.— "Wheat in the Middle East." **Empire Journal of Experimental Agriculture** (Oxford), Vol. XIV, pp. 31-42.

1946.— "Millets in the Middle East." **Empire Journal of Experimental Agriculture** (Oxford), Vol. XIV, pp. 208-216.

1947.— "Agricultural improvement in Turkey," (originally, perhaps a radio broadcast delivered earlier—**Editor**).

1947.— "Pulse grain crops in the Middle East." **Empire Journal of Experimental Agriculture** (Oxford), Vol. XV, pp. 249-259.

B. Town studies in India

1903.-- With Goodwin, A. E., **"The drink problem in Calcutta" (No. 1),**

"General outlines" by H. H. Mann, pp. 1-3;

"The quality of country spirit consumed", by H. H. Mann, pp. 7-9. (Reprinted from **The Englishman,** Calcutta, 1903)

**The Drink problem in Calcutta, No. II, A Year's Work of the Drink Problem Investigation Committee.** "Investigation Results. (Grogshops of) No. 13 Ward" by H. H. Mann, pp. 7-9 (The Englishman Press, Calcutta).

1904.— **The Drink problem in Calcutta, No. III**

"The Grogshops of Ward No. 6, pp. 1-12;

"Special report on Calcutta rum-shops and their customers", pp. 13-16 (The Drink Problem Investigation Committee, Edinburgh Press, Calcutta).

1904.— **Memorandum relative to the preparation of pure distillery spirit.** Calcutta, B. S. Press.

1908.— (?) **"The Intoxicating drinks of India",** Poona, Israelite Press, pp. 12.

1912.— "The untouchable classes of an Indian city". **The Sociological Review,** (London), Vol. V, No. 1, pp. 42-55.

1912.— "The untouchable classes of an Indian city". **The Hindustan Review,** (Allahabad), Vol. XXV, No. 151 & 152, pp. 233-240 & 317-324.

1914.— "The supply of milk to Indian cities." **Agricultural Journal of India,** (Calcutta), Vol. IX. Part 2, pp. 160-177. (A paper read at the provincial Agricultural Conference, Poona, September 1913).

1916.— "The housing of the untouchable classes in an Indian city." **Social Service Quarterly,** (Bombay), Vol. I, pp. 1-10.

1919.— "The sources of the milk supply of Poona city." **Agricultural Journal of India,** Vol. XIV, pp. 628-638.

1919.— "The provision of milk cattle stables in Bombay." Report sent to the Government of Bombay, letter No.

13,855, to the Secretary, Department of Agriculture, November 5th, 1919, pp. 19.

1919.— **Report on investigations with regard to social welfare work at Jamshedpur**, Jamshedpur Social Welfare Series (Bombay), pp. 120.

1919.— "Note on the increase in the cost of living in the Deccan". **Indian Journal of Economics**, (Allahabad), Vol. II, No. 1, pp. 1-5.

C. History and Antiquities of India

1904.— With Edwards, W. N. "An ancient Assamese fortification and the legends relating thereto". **Journal of the Asiatic Society of Bengal**, (Calcutta), Vol. 73, Part 1, (3) pp. 254-261.

1906.— With Hooper, D. "Earth-eating and the earth-eating habit in India." **Memoir, Asiatic Society of Bengal**, Vol. I, No. 12, pp. 249-270.

1918.— "The early history of the tea industry in North-East India." **Bengal Economic Journal**, (Calcutta), Vol. II, No. 1, pp. 44-59.

1920.— "India in the sixteenth century". **The Indian Review**, (Madras), Vol. XXI, pp. 481-482.

1938.— "Report on the Administration of H.E.H. the Nizam's Dominions for the year 1933-34." (Review-Article) **The Asiatic Review**, (London), Vol. XXXIV, No. 118, April, 1938, pp. 404-406.

III. AGRICULTURE AND NATURAL SCIENCE.

D. Chemistry and Botany

1894.— "Action de certaines substances antiseptiques sur la levure." **Annales de l'Institut Pasteur**, (Paris), Vol. VIII, pp. 785-795.

1895.— With Ingle, H. "Stereoisomeric Osazones". Part I. **Journal of the Chemical Society**, (London), Transactions Vol. 67, pp. 606-616.

1896.— With Ingle. H. "A new modification of Benzilosazone." **Proceedings of the Chemical Society**, (London), Vol. XI, pp. 111-112.

1913.— With Sahasrabuddhe, D. L., and Patwardhan, V. G. "Studies in the Chemistry and Physiology of the Leaves of the Betel-vine **(Piper Betle)** and of the Commercial Bleaching of Betel-vine Leaves." **Memoirs of the Department of Agriculture in India,** (Pusa) Chemical series, Vol. III, No. 2, pp. 15-63.

1916.— With Patwardhan, V. G. "Studies in the Chemistry and Physiology of the Leaves of the Betel-vine **(Piper Betle)** and of the Commercial Bleaching of the Betel-vine Leaves". Part II. **Memoirs of the Department of Agriculture in India** (Pusa) Chemical series, Vol. IV, No. 7, pp. 281-322.

1914.— With Kanitkar, N. V. "Notes on the Fat of **Garcinia indica,** the so-called **kokam** butter." **Journal and Proceedings Asiatic Society of Bengal** (New Series) Vol. X, No. 8, pp. 293-294.

1919.— With Kanitkar, N. V. "Safflower oil as a drying oil." **Journal of the Society of Chemical Industry,** (London), Vol. XXXVIII, pp. 36-38.

1920.— "Variation in the Flower of **Jasminum malabaricum,** Wight." **Journal of the Linnean Society.** (Botany) London, Vol. 45, pp. 155-158.

1928.— With Patel, M. L. "Studies in Gujarat Cottons, Part V. Variability in certain economic characters, particularly in seed weight and weight of lint per seed in pure strains of Broach-Deshi Cotton." **Memoirs of the Department of Agriculture in India** (Pusa) Botanical series, Vol. XV, No. 7, pp. 161-217.

E. Tea

1. Tea in India

(a) Culture and General Studies

1903.— With Watt, Sir G. "The principles of tea pruning." **Agricultural Ledger,** (Calcutta), Vol. X (Vegetable Products Series No. 73), pp. 1-36.

1906.— "The treatment of deteriorated tea." **Indian Tea Association,** (Calcutta) Scientific Department Publication, No. 4, pp. 1-24.

1906.— "The renovation of deteriorated tea." **Agricultural Journal of India,** (Calcutta), Vol. I, Part 2, pp. 83-96.

1907.— "Experiments in heavy pruning in Assam." **Indian Tea Association,** (Calcutta), Scientific Department Publication, No. 3, pp. 1-20.

1908.— "Cultivation of Tea", pp. 218-239 of **The Commercial Products of India,** (by Sir George Watt), London, John Murray, pp. 1189 + viii.

1905.— With Hutchinson, C. M. "The objects and plan of the Heeleaka Experimental Station." **Indian Tea Association,** (Calcutta), Scientific Department Publication, No. 2, pp. 1-11.

1907.— With Hutchinson, C. M. "The Heeleaka Experimental Station: Investigations during the seasons of 1905 and 1906." **Indian Tea Association,** (Calcutta) Scientific Department Publication, No. 2, pp. 1-35.

1918.— "The early history of the tea industry in North-East India." **Bengal Economic Journal,** Vol. II, pp. 44-59.

1931.— "The Indian Tea Industry in its Scientific Aspects." **Journal of the Royal Society of Arts,** Vol. 79, pp. 469-483.

1961.— "Classification of the Tea Plant." **Nature,** (London) Vol. 191, p. 409.

1927.— "Report by Dr. H. H. Mann on his visit to Tocklai in April, 1927." **Indian Tea Association,** (Calcutta), pp. 1-27.

(b) Chemistry and Manufacture.

1901.— "Studies in the Chemistry and Physiology of the Tea Leaf." Part I. "The Enzymes of the Tea Leaf." **Journal Asiatic Society of Bengal,** Vol. 70, Part II, No. 2, pp. 154-166.

1903.— "Het Ferment in het Theeblad en zijne verhouding tot de Hoedanigheid der Thee", pp. 70-79 in **Verslag eener reis de thee-districhten van Britisch-Indie en Ceylon** (by Netscher, H. J. Th., and Holle, A. A.), Batavia, pp. 94.

1901.— "The Ferment of the Tea Leaf, and its relation to quality in Tea." **Indian Tea Association,** (Calcutta) Scientific Department Publication, No. 2, pp. 1-12.

1903.— "The Ferment of the Tea Leaf", Part II. **Indian Tea Association,** (Calcutta) Scientific Department Publication, No. 5, pp. 1-15.

1904.— "The Ferment of the Tea Leaf" Part III. **Indian Tea Association,** (Calcutta) Scientific Department Publication, No. 6, pp. 1-12.

1906.— "The Fermentation of Tea" Part I. **Indian Tea Association,** (Calcutta) Scientific Department Publication, No. 12, pp. 1-22 (Reprinted July 1913).

1907.— "The Fermentation of Tea" Part II. **Indian Tea Association,** (Calcutta) Scientific Department Publication, No. 1, pp. 1-17 (Reprinted August 1913).

1907.— "The factors which determine the quality of tea." **Indian Tea Association,** (Calcutta) Scientific Department Publication, No. 4, pp. 1-29 (Reprinted September 1913).

(c) Soils and Fertilisers

1901.— "The Tea Soils of Assam and Tea Manuring." **Indian Tea Association,** (Calcutta) Scientific Department Publication, No. 1, pp. 137.

1903.— "The Tea Soils of Cachar and Sylhet." **Indian Tea Association,** (Calcutta) Scientific Department Publication, No. 4, pp. iv + 89.

1907.— "The Tea Soils of North-East India and their treatment." **Indian Tea Association,** (Calcutta) Scientific Department Publication, York, Delittle, Fenwick & Co., pp. 1-284.

1960.— With Gokhale, N. G. "Soils of Tea Growing Tracts of India." **Journal of the Indian Society of Soil Science,** Vol. VIII, pp. 191-200.

1906.— With Hutchinson, C. M. "Green manuring in tea culture in India." **Indian Tea Association,** (Calcutta) Scientific Department Publication, No. 13, pp. 1-43.

1918.— "Memorandum on the use of, and trade in, artificial manures in Bengal, Eastern Bengal and Assam, pp. 22-42 in **Memorandum on the use of artificial manures on the tea estates of Assam and Bengal** — Decade 1907-1917 (by Hope, G. D.) **Indian Tea Association,** (Calcutta) Scientific Department Publication, Pamphlet No. 2 (published in Proceedings of the 3rd Annual Meeting of the Board of Agriculture held in Cawnpore on 18 February, 1907 and succeeding days).

(d) Pests and Diseases

1903.— With Watt, Sir G. **The Pests and Blights of the Tea Plant** 2nd ed. Calcutta; Office of the Superintendent, Government Printing, pp. i-xv + 429.

1900.— "A suggested remedy for thread blight on tea" one page in **Thread Blight** (The subjoined memo by Mr. H. H. Mann, B.Sc. F.I.C., is published for the information of members (signed) H. M. Haywood). **Indian Tea Association,** (Calcutta) Circular No. 75.

1901.— "**Red Rust:** A serious Blight of the Tea Plant." **Indian Tea Association,** (Calcutta), pp. 1-16.

1904.— With Hutchinson, C. M. "Red Rust: A serious Blight of the Tea Plant" 2nd ed. **Indian Tea Association,** (Calcutta) Scientific Department Publication, No. 4, pp. 1-26.

1907.— With Hutchinson, C. M. "Cephaleuros virescens, Kunze: The 'Red Rust' of Tea". **Memoirs of the Department of Agriculture in India** (Pusa) Botanical series, Vol. I, No. 6, pp. 1-35.

1906.— "The Blister Blight of Tea." **Indian Tea Association,** (Calcutta) Scientific Department Publication, No. 3, pp. 1-13.

1902.— "The 'Mosquito-Blight' of Tea: Investigations during the season of 1902." **Indian Tea Association,** (Calcutta), pp. 1-14.

1904.— "The 'Mosquito-Blight' of Tea: Investigations during the season of 1903." Part II. **Indian Tea Association,** (Calcutta), pp. 1-20.

1905.— "The 'Mosquito-Blight' of Tea: Part III. Investigations during the season of 1904 and 1905." **Indian Tea Association,** (Calcutta), Scientific Department Publication, No. 1, pp. 1-18 + appendix.

1902.— "Note on a disputed point in the Life-History of **Helopeltis theivora.**" **Journal Asiatic Society of Bengal,** Vol. 71, Part II, No. 3, pp. 133-134.

1903.— "Notes on **Helopeltis theivora**, the 'Mosquito-Blight' of Tea." **Indian Museum Notes,** Vol. 6, No. 1, pp. 5-13.

1906.— "The Proportion between the Sexes in **Helopeltis theivora**, Waterhouse." **Journal and Proceedings, Asiatic Society of Bengal** (New Series), Vol. II, No. 5, pp. 177-181.

1907.— "Individual and seasonal variations in **Helopeltis theivora**, Waterhouse, with description of a new species of **Helopeltis." Memoirs of the Department of Agriculture in India,** (Pusa) Entomology series. Vol. I, No. 4, pp. 275-337.

1910.— "Note on Mr. Antram's Report on work in connection with the Mosquito-Blight of Tea", pp. ii-v, in "The Mosquito-Blight of Tea **Helopeltis theivora,** Waterhouse. Investigations during 1908-1909 (By C. B. Antram)." **Indian Tea Association,** (Calcutta) Scientific Department Publication, No. 30.

1906.— With Antram, C. B. "The 'Red Slug' Caterpillar; a serious pest of the Tea Plant." **Indian Tea Association,** (Calcutta) Scientific Department Publication, No. 5, pp. ii + 12.

1907.— With Green, E. E. "The Coccidae attacking the Tea plant in India and Ceylon". **Memoirs of the Department of Agriculture in India,** (Pusa) Entomology series, Vol. I, No. 5, pp. 337-355.

2. Tea in U.S.S.R., and Asia (excluding India)

1904.— "An account of Formosa Tea Soils", pp. 24-30 in **Report on the cultivation and manufacture of Formosa Colong Tea** (by James Hutchison), Calcutta, Newman & Co., 30 pp.

1932.— "The Soviet Tea Industry". **British Russian Gazette and Trade Outlook,** Vol. VIII, (New Series), No. 6, pp. 143-146 and 149.

1932.— "The Recent Tea Developments in Georgia". **Quarterly Journal Scientific Department Indian Tea Association,** Part 1, pp. 55-69.

1932.— "Notes on Soviet Russian Tea Industry". **The Tea and Coffee Trade Journal,** Vol. 62, pp. 378-381.

1932.— "The recent tea developments in Russia" (containing extracts from reports by Dr. H. H. Mann). **The Spice Mill,** Vol. 55, pp. 1042-1051, and 1166-1170.

1933.— "The climatic and soil requirements of tea". **Empire Journal of Experimental Agriculture,** (Oxford), Vol. I, No. 3, pp. 245-252.

1935.— "Tea Soils". **Imperial Bureau of Soil Science,** (Harpenden), Technical Communication No. 32, pp. 1-64.

3. Tea in Africa

1932.— "Tea Growing in Nyasaland" (1), pp. 1-17; "Note on Tea Possibilities in the Area near Nkata Bay" (2), pp. 18-21. **Nyasaland Protectorate Department of Agriculture Bulletin** (New Series), No. 4.

1933.— **Report on Tea Cultivation and its development in Nyasaland,** Westminster; Crown Agents for the Colonies, for Government of Nyasaland, pp. 1-41.

1933.— **Report on Tea Cultivation in the Tanganyika Territory and its development**, Westminster; Crown Agents for the Colonies. for Government of Tanganyika Territory, pp. 1-54.

F. Agriculture in India (excluding Tea)

(a) General

1909.— With other Members of Committee of Board of Agriculture. **Report on the Introduction of Improvements into Indian Agriculture by the Work of the Agricultural Departments.** Calcutta; Superintendent Government Printing, 23 pp.

1910.— **Second Report** (as above), 9 pp.

1912.— **Third Report** (as above), 15 pp.

1914.— **Fourth Report** (as above). 11 pp.
Calcutta. Superintendent Government Printing.

1910.— "The introduction of improvements into Indian agriculture". **Agricultural Journal of India,** Vol. V, Part I, pp. 6-18.

1912.— "Local Bodies as Agents in agricultural improvement" **Agricultural Journal of India,** Vol. VII, Part 2, pp. 280-285.

1915.— "Lines of development of Indian agriculture" Address delivered to the **Indian Science Congress** (Section of Agriculture and Applied Science) on January 14th, 1915, and specially revised by Dr. H. H. Mann for the **Mysore Economic Journal,** Vol. 1, pp. 11-18.

1908.— **Report on the Proceedings of the Agricultural Conference** held at Ahmedabad on the 26th and 27th November 1907 (statements by Dr. Mann on pp. 10-11 & 17-18.) Bombay; Government Central Press, 37 pp.

1908.— "The Agricultural Conference at Ahmedabad in November 1907". **Agricultural Journal** of India, Vol. III, Part I, pp. 41-52.

1914.— **Investigations of the first Importance**

**Investigations of smaller importance at present**

Proceedings of the Board of Agriculture in India held at Coimbatore on the 8th December 1913, and following days.

Appendix B Subject III. Programmes of the Provincial Agricultural Veterinary Departments and of Native States Departments of Agriculture for 1914-15. VII Agricultural Chemistry. p. 85.

Calcutta; Superintendent Government Printing.

1927.— "Evidence Taken in the Bombay Presidency" (oral evidence by Dr. Harold Mann, pp. 95-102). **Royal Commission on Agriculture in India**, London; H.M.S.O., Vol. II, Part II (Bombay), pp. xiii + 541 & i-cxxviii.

1929.— "The Agriculture of India". **Annals of the American Academy of Political and Social Science**, (Philadelphia), Vol. 145, pp. 72-81.

1957.— "Reminiscences". **Poona Agricultural College Magazine**. Vol. 48, pp. i-ii.

1907.— (Edited by Dr. H. H. Mann) **Handbook of Indian Agriculture** (by N. G. Mukerji) 2nd ed. Calcutta, pp. xii + 706.

1915.— (Edited by Dr. H. H. Mann) as above, 3rd ed., pp xiv + 620.

(b) Crops: 1. Potatoes

1919.— With Nagpurkar, S. D. "Notes on the 'Ring Disease' of Potato". **Agricultural Journal of India**, Vol. XIV, pp. 388-394.

1920.— With Nagpurkar, S. D., and Kulkarni, G. S. "The 'Tambera' Disease of Potato". **Agricultural Journal of India,** Vol. XV, pp. 282-288.

1921.— With Nagpurkar, S. D., et al. "Investigations on potato cultivation in Western India". **Bombay Department of Agriculture Bulletin,** No. 102, pp. 145.

1925.— With Joshi, W. V. "Further investigations on potato cultivation in Western India". **Bombay Department of Agriculture Bulletin**, No. 121, pp 37.

1922.— With Nagpurkar, S. D. "Further investigations on the **Fusarium** blights of potatoes in Western India". **Agricultural Journal of India**, Vol. XVII, pp. 564-576.

2. Rubber

1906.— "Assam Rubber and its commercial prospects". **Agricultural Journal of India**, Vol. I, Part IV, pp. 390-398.

1907.— "The tapping of Assam rubber (**Ficus elastica**)" **Agricultural Journal of India**, Vol. II, Part III, pp. 277-279.

1907.— "The cultivation of Para Rubber in North-East India" **Agricultural Journal of India**, Vol. II, Part III, pp. 273-276.

3. Sisal

1904.— With Hunter, J. "Sisal-Hemp Culture in the Indian Tea Districts". **Indian Tea Association**, Publication No. 3, pp. 1-41.

1907.— "The development of sisal hemp cultivation in India" **Agricultural Journal of India**, Vol. II, pp. 323-333.

4. Sugar Cane

1927.— Preface to "Furnaces for making **Gul** or crude sugar in the Bombay Presidency" (by P. C. Patil), one page. **Bombay Department of Agriculture Bulletin**, No. 144.

1928.— "Sugar-Cane cultivation in the Nizam's Dominions" **Hyderabad Agricultural Department Bulletin** No. 2.

5. Tobacco

1926.— With Patel, M. L., and Majumdar, V. M. "The treatment of patches of inferior tobacco in North Gujarat". **Bombay Department of Agriculture Bulletin** No. 129. pp. 13.

1926.— With Patel, M. L., and Majumdar, V. M. "The Improvement of tobacco in Northern Gujarat". **Bombay Department of Agriculture Bulletin** No. 132, pp. 22.

6. Fodder crops

1916.— "Fodder crops of Western India". **Bombay Department of Agriculture Bulletin** No. 77, pp. 142.

1921.— "Fodder crops of Western India". **Bombay Department of Agriculture Bulletin** No. 100, pp. 230.

1926.— "Elephant Grass. A new and useful fodder crop in Western India". **Bombay Department of Agriculture Bulletin** No. 127, pp. 1-7.

1927.— With Burns, W. "The Locust attack of 1926-27 in Sind, Kathiawar and Gujarat". **Agricultural Journal of India**, Vol. XXII, pp. 325-332.

(c) Livestock

1911.— With Meggitt, A. A. "The composition of the milk of some breeds of Indian cows and buffaloes and its variations" Part I. The milk of some breeds of Indian cows. **Memoirs of the Department of Agriculture in India**, (Pusa) Chemistry series, Vol. II, pp. 1-61.

1912.— With Meggitt, A. A. "The composition of the milk of some breeds of Indian cows and buffaloes and its variations" Part II. The milk of some breeds of Indian Buffaloes. **Memoirs of the Department of Agriculture in India** (Pusa) Chemistry series, Vol. II, pp. 195-258.

1927.— "Development of the Poultry Industry in India" **Agricultural Journal of India,** Vol. XII, pp. 109-113.

(d) Soils, Irrigation and Erosion Problems

1910.— With Tamhane, V. A. "The salt lands of the Nira Valley". **Bombay Department of Agriculture Bulletin** No. 39, pp. 35.

1912.— With Joshi, N. V., and Kanitkar, N. V. "The **'Rab'** System of Rice Cultivation in Western India". **Memoirs of the Department of Agriculture in India** (Pusa) Chemistry series, Vol. II, pp. 137-193.

1915.— With Paranjpe, S. R. "Artificial Manures: Experiments on their value for crops in Western India". **Bombay Department of Agriculture Bulletin** No. 76, pp. 55.

1918.— With Paranjpe, S. R. "Artificial Manures: Experiments on their value for crops in Western India" No. II. **Bombay Department of Agriculture Bulletin** No. 89, pp. 33.

1917.— "The present position of artificial manures in India" **Mysore Economic Journal,** Vol. III, pp. 1-8.

1920.— "The improvement of the manure supply in the intensively cultivated tracts of the Bombay Presidency, including the development of the organisation for advice as to manures. The question of freight of manures" pp. 9-13 in: **Proceedings of the Bombay Provincial Board of Agriculture** held at Poona on the 9th and 10th June 1920.

Bombay; Government Central Press, pp. 46.

1915.— "Experiments with the Automatic Water Finder in the Trap Region of Western India". **Bombay Department of Agriculture Bulletin** No. 72, pp. 17.

1925.— With Sahasrabuddhe, D. L. **Experiments with the Automatic Water Finder in the Trap Region of Western India** Poona; Yeravda Prison Press, pp. 39.

1915.— "Well waters from the Trap Region of Western India" **Bombay Department of Agriculture Bulletin** No. 74, pp. 66.

1917.— With Paranjpe, S. R. "Intermittent Springs at Rajapur in the Bombay Presidency". **Journal of the Bombay Branch of the Royal Asiatic Society,** Vol. XXIV, pp. 14-32.

1917.— With Paranjpe, S. R. "The Hot Springs of the Ratnagiri District". **Journal of the Bombay Branch of the Royal Asiatic Society,** Vol. XXIV, pp. 185-212.

1929.— "A preliminary note on soil questions in certain irrigated areas in Hyderabad." Secunderabad, pp. 18.

1930.— "Note on the erosion problem in the Karnatak". **Hyderabad Agricultural Department Bulletin** No. 5, pp. 6.

1955.— "Rainfall and Famine. A study of rainfall in the Bombay Deccan, 1865-1938". **Indian Society of Agricultural Economics** (Bombay),, pp. 47.

F. Agriculture in India (excluding Tea) (Contd.)

(e) Statistics and Reports


| | | | | | | | | | | | |
|---|---|---|---|---|---|---|---|---|---|---|---|
| 1909. | "Report of Poona Agricultural College to June 30th, 1909" pp 43-49 | **Annual** | **Report** | **Department** | **of** | **Agriculture**, | **Bombay** | **Presidency** | for | Year | 1908-1909 |
| | "Report of Agricultural Chemist for year 1908-1909" pp 50-54 | " | " | " | " | " | " | " | " | " | " " |
| 1910. | "Report of Poona Agricultural College for year ending June 30th, 1910" pp 13-21 | " | " | " | " | " | " | " | " | " | 1909-1910 |
| | "Report of Agricultural Chemist for year 1909-1910" pp 46-49 | " | " | " | " | " | " | " | " | " | " " |
| 1911. | "Report of Poona Agricultural College for year ending June 30th, 1911" pp 17-21 | " | " | " | " | " | " | " | " | " | 1910-1911 |
| | "Report of the Vernacular Agricultural School, Poona" pp 22-24 | " | " | " | " | " | " | " | " | " | " " |

| | | | | | | | | | | | |
|---|---|---|---|---|---|---|---|---|---|---|---|
| | "Report of Agricultural Chemist for year 1910-11" pp 60-64 | **Annual** | **Report** | **Department** | **of** | **Agriculture**, | **Bombay** | **Presidency** | for | Year | 1910-1911 |
| 1912. | "Report of Poona Agricultural College for year ending June 30th, 1912" pp 17-21 | ,, | ,, | ,, | ,, | ,, | ,, | ,, | ,, | ,, | 1911-1912 |
| | "Account of the Vernacular Agricultural School, Poona, 1911-12" pp 22-24 | ,, | ,, | ,, | ,, | ,, | ,, | ,, | ,, | ,, | ,, ,, |
| | "Report of Agricultural Chemist for year 1911-1912" pp 67-70 | ,, | ,, | ,, | ,, | ,, | ,, | ,, | ,, | ,, | ,, ,, |
| 1913. | "Report of Poona Agricultural College for year ending June 30th, 1913" pp 17-22 | ,, | ,, | ,, | ,, | ,, | ,, | ,, | ,, | ,, | 1912-1913 |
| | "Account of Vernacular Agricultural School for year ending June, 30th, 1913" pp 23-25 | ,, | ,, | ,, | ,, | ,, | ,, | ,, | ,, | ,, | ,, ,, |
| | "Report of Agricultural Chemist for year 1912-13" pp 61-64 | ,, | ,, | ,, | ,, | ,, | ,, | ,, | ,, | ,, | ,, ,, |

1915. "Report of Poona Agricultural College for year ending 30th, June 1914" pp 17-23 **Annual Report Department of Agriculture, Bombay Presidency** for Year 1913-1914

"Account of Vernacular Agricultural School, Loni, Poona for 1913-1914" pp 24-26 ,, ,, ,, ,, ,, ,, ,, ,, ,, ,, ,,

"Report of Agricultural Chemist for year 1913-1914" pp 67-69 ,, ,, ,, ,, ,, ,, ,, ,, ,, ,, ,,

1916. "Report of the Agricultural College for year 1914-1915" pp 13-18 ,, ,, ,, ,, ,, ,, ,, ,, ,, 1914-1915

"Report of Marathi Agricultural School, Loni, Poona for 1914-15" ,, ,, ,, ,, ,, ,, ,, ,, ,, ,, ,,

"Report of Agricultural Chemist for year 1914-15" pp 65-67 ,, ,, ,, ,, ,, ,, ,, ,, ,, ,, ,,

1916. "Report of Agricultural College for year 1915-16" pp 13-19 ,, ,, ,, ,, ,, ,, ,, ,, ,, 1915-1916

| | | |
|---|---|---|
| | "Report of Marathi Agricultural School, Loni, Poona for 1915-16" pp 20-22 | **Annual Report Department of Agriculture, Bombay Presidency** for Year 1915-1916 |
| | "Report of Agricultural Chemist for year 1915-16" pp 64-66 | ,, ,, ,, ,, ,, ,, ,, ,, ,, ,, ,, |
| 1918. | "Report of Poona Agricultural College for year 1916-17" pp 9-13 | ,, ,, ,, ,, ,, ,, ,, ,, ,, 1916-1917 |
| | "Report of Marathi Agricultural School. Loni, Poona for year 1916-17" pp 14-16 | ,, ,, ,, ,, ,, ,, ,, ,, ,, ,, ,, |
| | "Report of Agricultural Chemist for year 1916-17" pp 51-53 | ,, ,, ,, ,, ,, ,, ,, ,, ,, ,, ,, |
| 1919. | "Report of Poona Agricultural College for year 1917-18" pp 13-17 | ,, ,, ,, ,, ,, ,, ,, ,, ,, 1917-1918 |
| | "Report of Marathi Agricultural School, Loni. Poona. for 1917-18" pp 18-20 | ,, ,, ,, ,, ,, ,, ,, ,, ,, ,, ,, |

| | | | |
|---|---|---|---|
| | "Report of Agricultural Chemist for year 1917-18" pp 72-74 | **Annual Report Department of Agriculture, Bombay Presidency** for Year | 1917-1918 |
| 1920. | "Annual Report of Agricultural Department for year 1918-19" pp 1-16 | ,, ,, ,, ,, ,, ,, ,, ,, ,, | 1918-1919 |
| | "Report of Marathi Agricultural School, Lonikalbhor, Poona District for year 1918-19" pp 23-26 | ,, ,, ,, ,, ,, " " " " " | " |
| 1922. | "Annual Report of Department of Agriculture, Bombay Presidency, 1920-21" pp 1-11 | ,, ,, ,, ,, ,, ,, ,, ,, ,, | 1920-1921 |
| 1923. | "Annual Report of Department of Agriculture, Bombay Presidency, 1921-22" pp 1-11 | ,, ,, ,, ,, ,, ,, ,, ,, ,, ,, | 1921-1922 |
| 1924. | "Annual Report of Department of Agriculture, Bombay Presidency, 1922-23" pp 1-30 | ,, ,, ,, ,, ,, in the ,, ,, ,, ,, | 1922-1923 |

1926. "Annual Report of Department of Agriculture, Bombay Presidency, 1924-25" pp 1-26 — **Annual Report Department of Agriculture, Bombay Presidency** for Year 1924-1925

1927. "Annual Report of Department of Agriculture, Bombay Presidency, 1925-26" pp 1-30 — ,, ,, ,, ,, ,, ,, ,, ,, ,, ,, ,, 1925-1926

1928. "Annual Report of Department of Agriculture, Bombay Presidency, 1926-27" pp 1-32 — ,, ,, ,, ,, ,, ,, ,, ,, ,, ,, ,, 1926-1927

1925. Season and Crop Report of the Bombay Presidency for the year 1924-25. — Bombay; Government Central Press, pp iii + 104

1926. Season and Crop Report of the Bombay Presidency for the year 1925-26. — ,, ,, ,, ,, pp ii + 89

1928. Season and Crop Report of the Bombay Presidency for the year 1926-27. — ,, ,, ,, ,, pp 89

(e) Statistics and Reports (contd.)

1925.— **Statistical Atlas of the Bombay Presidency.** 3rd ed. Bombay; Government Central Press, pp. 194.

1927.— Preface to: "The crops of Bombay Presidency; Their geography and statistics II" (by G. R. Ambekar) 2 pp.; **Bombay Department of Agriculture Bulletin** No. 146.

1928.— Foreword to: "Studies in the cost of production of crops in the Deccan" (by P. C. Patil, in collaboration with T. G. Shirname and T. B. Pawan) 2 pp.; **Bombay Department of Agriculture Bulletin** No. 149.

1928.— Preface to: "The crops of Sind; Their geography and Statistics" (By G. R. Ambekar) one page; **Bombay Department of Agriculture Bulletin** No. 150.

1927.— "The origin, plan and progress of the Sakrand Agricultural Research Station, Sind". **Bombay Department of Agriculture Bulletin** No. 145, pp. 30.

1931.— **General Note on Provisional Figures for Standard Outturns of Crops in the Dominions.** Khurdad; Statistics Department, H. E. H. the Nizam's Government, pp. 1-64.

1928.— "The origin, plan and progress of the Sakrand Agricultural Research Station, Sind". **Annual Report of Department of Agriculture in the Bombay Presidency** for year 1926-27, pp. 240-263.

G. Agriculture in the Middle East

1946.— "Wheat in the Middle East". **Empire Journal of Experimental Agriculture** (Oxford), Vol. XIV, pp. 31-42.

1946.— "Millets in the Middle East". **Empire Journal of Experimental Agriculture** (Oxford), Vol. XIV, pp. 208-216.

1947.— "Pulse grain crops in the Middle East". **Empire Journal of Experimental Agriculture** (Oxford), Vol. XV, pp. 249-259.

1950.— "World Cereals Today: The Millets". **World Crops,** Vol. II, pp. 97-101 (Dr. H. H. Mann replied to Dr. L. Nicholls, **World Crops,** Vol. II, p. 224).

1938.— "The Collective Farm System in Russia". **Indian Co-operative Review,** Vol. IV, pp. 149-155.

1946.— "The New Agricultural Land Law in Turkey". **Bombay Co-operative Quarterly**, Vol. XXX. pp. 77-81.

H. Agriculture in England

1. Crops (a) Temperate crops

1942.— With Russell, E. W., and Keen, B. A. "Studies in Soil Cultivation XI." The effect of inter-tillage on the sugar-beet crop. **Journal of Agricultural Science**, Vol. 32, pp. 330-337.

1945.— With Barnes, T. W. "Manuring for the production of sugar-beet seed". **Agriculture** (London), Vol. 52, pp. 400-404.

1951.— "The effect of manures on the bolting of the beet plant" **Annals of Applied Biology** (London), Vol. 38, pp. 435-443.

1939.— With Martin, J. T., and Tattersfield, F. "The manurial requirements of Pyrethrum (**Chrysanthemum cincrariae-folium** Trev.)". **Annals of Applied Biology** (London), Vol. 26, pp. 14-24.

1944.— "History of a lucerne crop". **Agriculture** (London), Vol. 51, pp. 32-35.

(b) Exotic crops

1941.— "Soya bean culture in Great Britain". **Nature** (London), Vol. 147, pp. 660-662.

1944.— "Latest about Soya Beans". **Countryman** (Idbury), Vol. 30, p. 196.

1949.— "Soya Beans in England". **Countryman** (Idbury), Vol. 40, pp. 287-288.

1950.— "Growing Maize for Grain in England". **Countryman** (Idbury), Vol. 41, pp. 319-321.

(c) Plant competition

1945.— With Barnes, T. W. "The competition between barley and certain weeds under controlled conditions". **Annals of Applied Biology** (London), Vol. 32, pp. 15-22.

1947.— With Barnes, T. W. "The competition between barley and certain weeds under controlled conditions" II. Competition with **Holcus mollis. Annals of Applied Biology** (London), Vol. 34, pp. 252-266.

1949.— With Barnes, T. W. "The competition between barley and certain weeds under controlled conditions" III. Competition with **Agrostis gigantea. Annals of Applied Biology** (London), Vol. 36, pp. 273-281.

1950.— With Barnes, T. W. "The competition between barley and certain weeds under controlled conditions" IV. Competition with **Stellaria media. Annals of Applied Biology** (London), Vol. 37, pp. 139-148.

1952.— With Barnes, T. W. "The competition between barley and certain weeds under controlled conditions" V. Competition with clover considered as a weed. **Annals of Applied Biology** (London), Vol. 39, pp. 111-119.

1953.— With Barnes, T. W. "The mutual effect of ryegrass and clover when grown together". **Annals of Applied Biology** (London), Vol. 40, pp. 566-572.

2. Soils (a) Acidity

1937.— "The character of barley grown on soil made acid with sulphate of ammonia". **Journal of Agricultural Science,** Vol. 27, pp. 108-122.

1940.— With Barnes, T. W. "Studies of soil after fifty years of wheat or barley cropping, especially of soil made acid with sulphate of ammonia". **Journal of Agricultural Science**, Vol. 30, pp. 345-386.

1939.— "The weed herbage of a slightly acid arable soil". **Journal of Ecology**, Vol. 27, pp. 89-113.

1957.— "Weed herbage of slightly acid arable soils as affected by manuring". **Journal of Ecology**, Vol. 45, pp. 149-156.

1955.— "Variation of sulphur in soils under a continuous grain crop". **Journal of Soil Science** (Oxford). Vol. VI, pp. 241-247.

(b) Cropping systems

1934.— "Dangers of deterioration under continuous cropping". **Rothamsted Conference Reports** 17, **Modern Changes in the Treatment of Light Soils;** Harpenden, pp. 30-34.

1943.— "The influence of fallowing on the yield of wheat or barley on very exhausted land". **Journal of Agricultural Science**, Vol. 33, pp. 207-212.

1934.— With Crowther, E. M. "Green Manuring and Sheep Folding on Light Land". **Journal of the Royal Agricultural Society of England**, Vol. 94, pp. 128-151.

1958.— "Field studies in green manuring" I. **Empire Journal of Experimental Agriculture** (Oxford), Vol. XXVI, pp. 274-282.

1959.— "Field studies in green manuring" II. **Empire Journal of Experimental Agriculture** (Oxford), Vol. XXVII, pp. 243-251.

1958.— With Boyd, D. A. "Some results of an experiment to compare ley and arable rotations at Woburn". **Journal of Agricultural Science,** Vol. 50, pp. 297-306.

1951.— With Barnes, T. W. "The behaviour of nitrogenous manures in the soil" Part I. The loss of manurial nitrogen. **Journal of Agricultural Science**, Vol. 41, pp. 309-314.

1956.— With Barnes, T. W. "The permanence of organic matter added to soil". **Journal of Agricultural Science**, Vol. 48, pp. 160-163.

### (c) Diseases

1948.— With Garrett, S. D. "Soil conditions and the take-all disease of wheat" X. Control of the disease under continuous cultivation of a spring-sown cereal. **Annals of Applied Biology** (London), Vol. 35, pp. 435-442.

1938.— "Investigations on Clover Sickness". **Journal of Agricultural Science,** Vol. 28, pp. 437-455.

1950.— "Notes for paper on clover sickness". **Annals of Applied Biology** (London), Vol. 37, pp. 327-328.

## H. Agriculture in England (Contd.)

### 3. Reports

1934. With Moffatt, J. R. "Woburn Farm" **Report of the Rothamsted Experimental Station** for 1933 pp 91- 96

1938. "Woburn Experimental Farm Report for 1936-7" „ „ „ „ „ „ „ 1937 pp 108-116

1939. "Woburn Experimental Station Report for 1937-38" „ „ „ „ „ „ „ 1938 pp 92-100

1946. "Woburn Experimental Station" „ „ „ „ „ „ „ 1939-1945 pp 253-261

1948. "Woburn Experimental Farm" „ „ „ „ „ „ „ 1946 pp 88- 92

1948. "Woburn Experimental Station" „ „ „ „ „ „ „ 1947 pp 91- 94

1949. "Woburn Experimental Station" „ „ „ „ „ „ „ 1948 pp 112-114

1950. "Woburn Experimental Station" „ „ „ „ „ „ „ 1949 pp 114-118

1951. "Woburn Experimental Station" „ „ „ „ „ „ „ 1950 pp 128-132

1952. "Woburn Experimental Station" „ „ „ „ „ „ „ 1951 pp 149-153

1953. "Woburn Experimental Station" „ „ „ „ „ „ „ 1952 pp 151-155

1954. "Woburn Experimental Station" „ „ „ „ „ „ „ 1953 pp 157-161

1955. "Woburn Experimental Station" „ „ „ „ „ „ „ 1954 pp 167-171

1956. "Woburn Experimental Station" „ „ „ „ „ „ „ 1955 pp 175-181

1957. "Woburn Experimental Station" „ „ „ „ „ „ „ 1956 pp 205-211

1959. "Woburn Experimental Station" „ „ „ „ „ „ „ 1958 pp 223-234

**Obituaries**

1935.— Dr. J. Walter Leather, **Nature** (London), Vol. 135, p. 58.

1938.— John Augustus Voelcker, **Proceedings of the Linnean Society** (London), Session 150, 1937-38, pp. 344-345.

1942.— Herbrand Arthur Russell (1858-1940) 11th Duke of Bedford **Proceedings of the Linnean Society** (London), Session 153, 1940-41, pp 296-299.

1952.— James Insch (1877-1951) **Proceedings of the Linnean Society** (London), Session 163, 1950-51, p. 253.

# INDEX